Agricultural Science: A Global Overview

Editor: Oliver Adams

CALLISTO
REFERENCE

www.callistoreference.com

Callisto Reference,
118-35 Queens Blvd., Suite 400,
Forest Hills, NY 11375, USA

Visit us on the World Wide Web at:
www.callistoreference.com

ISBN: 978-1-64116-305-7 (Hardback)

Cataloging-in-publication Data

Agricultural science : a global overview / edited by Oliver Adams.
    p. cm.
Includes bibliographical references and index.
ISBN 978-1-64116-305-7
1. Agriculture. 2. Agriculture--Research. I. Adams, Oliver.
S439 .A37 2020
630--dc23

Table of Contents

Preface VII

Chapter 1   **Sugarcane expansion and farmland prices in São Paulo State**..1
              Alex Wilhans Antonio Palludeto, Tiago Santos Telles, Roney Fraga Souza and
              Fábio Rodrigues de Moura

Chapter 2   **Lessons for promotion of new agricultural technology: a case of Vijay wheat
              variety**..13
              Krishna P. Timsina, Yuga N. Ghimire, Devendra Gauchan, Sanjiv Subedi and
              Surya P. Adhikari

Chapter 3   **Millets: a solution to agrarian and nutritional challenges** 23
              Ashwani Kumar, Vidisha Tomer, Amarjeet Kaur, Vikas Kumar and Kritika Gupta

Chapter 4   **Small farmers' preferences for weather index insurance**..37
              Kenneth W. Sibiko, Prakashan C. Veettil and Matin Qaim

Chapter 5   **Climate change impacts and adaptation among smallholder farmers** 51
              Celia A. Harvey, Milagro Saborio-Rodríguez, M. Ruth Martinez-Rodríguez,
              Barbara Viguera, Adina Chain-Guadarrama, Raffaele Vignola and
              Francisco Alpizar

Chapter 6   **The determinants of crop yields in Uganda: what is the role of climatic and
              non-climatic factors?** 71
              Terence Epule Epule, James D. Ford, Shuaib Lwasa, Benon Nabaasa and
              Ambrose Buyinza

Chapter 7   **Agricultural extension and its effects on farm productivity and income**...........................88
              Gideon Danso-Abbeam, Dennis Sedem Ehiakpor and Robert Aidoo

Chapter 8   **Evidence of rapid spread and establishment of *Tuta absoluta* (Meyrick)
              (Lepidoptera: Gelechiidae) in semi-arid Botswana**..98
              Honest Machekano, Reyard Mutamiswa and Casper Nyamukondiwa

Chapter 9   **Agricultural history nexus food security and policy framework**......................................110
              Msafiri Yusuph Mkonda and Xinhua He

Chapter 10   **Correlations of cap diameter (pileus width), stipe length and biological
              efficiency of *Pleurotus ostreatus* (Ex.Fr.) Kummer cultivated on
              gamma-irradiated and steam-sterilized composted sawdust as an index of
              quality for pricing** 121
              Nii Korley Kortei, George Tawia Odamtten, Mary Obodai,
              Michael Wiafe-Kwagyan and Deborah Louisa Narh Mensah

Chapter 11   **Promoting sustainable agriculture in Africa through ecosystem-based farm management practices**..129
Caesar Agula, Mamudu Abunga Akudugu, Saa Dittoh and Franklin Nantui Mabe

Chapter 12   **Effect of climate-smart agricultural practices on household food security in smallholder production systems**..140
Bright Masakha Wekesa, Oscar Ingasia Ayuya and Job Kibiwot Lagat

Chapter 13   **Barriers to and determinants of the choice of crop management strategies to combat climate change**..154
Zerihun Yohannes Amare, Johnson O. Ayoade, Ibidun O. Adelekan and Menberu Teshome Zeleke

Chapter 14   **Effect of planting time on growth, yield components, seed yield and quality of onion (*Allium cepa* L.)**..165
Maria Tesfaye, Derbew Belew, Yigzaw Dessalegn and Getachew Shumye

Chapter 15   **Factors affecting adoption of upland rice in Tselemti district, northern Ethiopia**...173
Hadush Hagos, Eric Ndemo and Jemal Yosuf

Chapter 16   **Assessment of production potential and post-harvest losses of fruits and vegetables**..182
Hagos Abraha Rahiel, Abraha Kahsay Zenebe, Gebreslassie Woldegiorgis Leake and Beyene Weldegerima Gebremedhin

Chapter 17   **Effect of inter- and intra-row spacing on yield and yield components of mung bean (*Vigna radiata* L.) under rain-fed condition**..195
Asaye Birhanu, Tilahun Tadesse and Daniel Tadesse

Chapter 18   **Effect of split application of different N rates on productivity and nitrogen use efficiency of bread wheat (Triticum aestivum L.)**...202
Fresew Belete, Nigussie Dechassa, Adamu Molla and Tamado Tana

Chapter 19   **Survey of mushroom consumption and the possible use of gamma irradiation for sterilization of compost for its cultivation**...212
Nii Korley Kortei, George Tawia Odamtten, Mary Obodai, Michael Wiafe-Kwagyan and Juanita Prempeh

Chapter 20   **Influence of productive resources on bean production in male- and female-headed households in selected bean corridors**...219
Scolastica Wambua, Eliud Birachi, Ann Gichangi, Justus Kavoi, Jemimah Njuki, Mercy Mutua, Michael Ugen and David Karanja

Permissions

List of Contributors

Index

Preface

The main aim of this book is to educate learners and enhance their research focus by presenting diverse topics covering this vast field. This is an advanced book which compiles significant studies by distinguished experts in the area of analysis. This book addresses successive solutions to the challenges arising in the area of application, along with it; the book provides scope for future developments.

Agriculture is the practice of cultivating plants and livestock. Various foods, fuels, fibers and raw materials are cultivated. Modern agriculture relies on intensive farming practices. Conventional agricultural practices can have a negative impact on the environment. This has led to an increased emphasis on regenerative, organic and sustainable agriculture. Alternative technologies of agriculture comprising of selective breeding, integrated pest management and controlled-environment agriculture are being actively researched and studied. Sustained agricultural productivity is vital to food security. There are numerous risk factors which can affect food security such as droughts, fuel shortages, economic instability, etc. This book contains some path-breaking studies in the field of agricultural science. The topics covered herein deal with the core aspects of agriculture. It is a vital tool for all researching or studying this field as it gives incredible insights into emerging trends and concepts.

It was a great honour to edit this book, though there were challenges, as it involved a lot of communication and networking between me and the editorial team. However, the end result was this all-inclusive book covering diverse themes in the field.

Finally, it is important to acknowledge the efforts of the contributors for their excellent chapters, through which a wide variety of issues have been addressed. I would also like to thank my colleagues for their valuable feedback during the making of this book.

Editor

Sugarcane expansion and farmland prices in São Paulo State, Brazil

Alex Wilhans Antonio Palludeto[1], Tiago Santos Telles[2*], Roney Fraga Souza[3] and Fábio Rodrigues de Moura[4]

Abstract

Background: Brazil is the world's largest sugarcane producer, and its production is concentrated in south-central and northeast regions, particularly in the state of São Paulo. The land use change, principally from the increasing sugarcane production, may reflect in the farmland prices. The aim of this study is to evaluate the extent to which agricultural land prices in São Paulo are determined by variations in cultivation and prices of three products that represent a significant share of agriculture in the state: sugarcane, soy and corn, in a low-inflation environment.

Methods: Analysis is based on data from the Rural Development Offices (EDR) from 1997 to 2013. A simple panel data model is constructed with land price as the dependent variable, subdivided, according to the definition of the São Paulo State Institute of Agricultural Economics, into first- and second-class croplands. Cultivation area, unit price of the products, and lease value are explanatory variables, according to each crop. Inflation and the overall production value of São Paulo's farming production, excluding the production values of corn, soy and sugarcane, also serve as explanatory variables.

Results: The results show that in São Paulo, although part of the land price variation can be explained by the variables associated with their productive use, the impact of inflation indicates that land's function in storing value contributes significantly to land prices.

Conclusions: The most prominent conclusion is that expansion in sugarcane cultivation has led to higher farmland prices in the state of São Paulo.

Keywords: Land use, Land value, Land price, Land profitability, Bioenergy crops, Panel data

Background

Studies of land use changes and farmland prices are very useful to policy makers, who need to be able to identify the determinants of such changes. Changes in land use patterns significantly affect both the environment (biodiversity, water pollution, soil erosion, and climate change) and economic and social welfare [9].

In Brazil, the large fluctuation in rural land prices during the periods of high inflation that extended throughout the 1970s, 1980s, and early 1990s (associated with the modernization of farming production that began in the 1960s) stimulated several studies on the Brazil-ian land market. The analyses by Sayad [45, 46], Oliveira and Costa [34], Pinheiro and Reydon [36], Rezende [43], Egler [15], and Bacha [1] are representative of this period. Although different approaches were taken, these studies are dedicated primarily to the identification of factors beyond those strictly related to farming activity that acted to determine land prices in the country. In particular, the literature began to consider the impact of macroeconomic variables on land prices.

Beginning with the arrival of the Real Plan and subsequent monetary stability in 1994, however, increased interest rates and low exchange rates in the country led to a new dynamic in the agricultural land market [58]. As noted by Reydon and Plata [41], in the years immediately following implementation of the Real Plan, there was a marked drop in agricultural land prices in the country. According to these authors, this devaluation occurred as a result of the reduced inflation rate, which diminished

*Correspondence: telles@iapar.br
[2] Instituto Agronômico do Paraná, C.P. 10.030, Londrina, Paraná CEP 86057-970, Brazil
Full list of author information is available at the end of the article

the demand for assets that serve as a store of value, such as land. They also concluded that the price drop was the result of increased interest rates, which made financial assets more attractive than land because the former began to yield better profit expectations to investors. In this context, the influence of agricultural land in storing value decreased, even if it was not completely eliminated. Hence, the authors suggest that the demand for agricultural land became more closely related to its productive dimension than to its capacity to store value, which is an economic dimension linked to speculation[1] [40].

In fact, with the subsequent exchange devaluation that began in 1999, exports were favored and the prices paid for important crops such as sugarcane, corn and soy increased, followed by a rise in the price of agricultural land in the primary producing regions for these commodities [4, 30]—among them, the state of São Paulo.[2] The trend of increased demand for agricultural land for production purposes gained importance in the state of São Paulo, especially from the expansion of sugarcane farms [13, 20] destined for the production of biofuels and agro-energy [17]. One of the reasons for this expansion is related to the liberalization of sugar exports starting in 1990, following the intense process of trade liberalization of the Brazilian economy at the time. In 2007, the state of São Paulo was responsible for 60% of the sugarcane production in Brazil and 53% of the total area occupied by this crop. The advancement in sugarcane farming activity is occurring mostly in areas formerly used to raise cattle, and primarily through land leasing, which maintains the market for land used for sugarcane production on the rise [19, 32, 50]. Furthermore, São Paulo has one of the most complete logistics infrastructures in the country [49], which reduces the costs of agricultural activities. Associated with the good performance of the farming sector, the demand for land certainly reflects the favorable nature of this environment, which in turn has been reflected in higher land prices since 2000 [54].

Several studies conducted in Brazil after 1995, especially those by Plata [37], Reydon and Plata [41], Gasques and Bastos [20], Zilli et al. [58], and Ferro and Castro [18], indicate that the determinants of land prices in Brazil may be connected primarily to its role as a production factor, suggesting that a similar movement may have occurred in the state of São Paulo.

This study thus aims to contribute to the recent literature on land prices in Brazil by examining the determinants of agricultural land prices in the state of São Paulo using an empirical evaluation of the effects of variables associated with productive activity on land prices. The analysis is based on data from the Rural Development Offices (Escritórios de Desenvolvimento Rural—EDR)[3] from 1997 to 2013. The specificity of the analysis conducted here lies precisely in the adoption of a data disaggregation level based on EDRs, in contrast to the literature on land prices for Brazil in general.

More specifically, this paper analyzes the extent to which land prices in São Paulo are determined by variations in the area of cultivation and in the prices linked to three of the products that represent a significant share of agriculture in the state: sugarcane, soy, and corn, in a low-inflation environment. An econometric model is constructed with land price, the dependent variable, subdivided, according to the definition of the Institute of Agricultural Economics (Instituto de Economia Agrícola—IEA) of the state of São Paulo, into first- and second-class croplands. Explanatory variables include the cultivation area of each of the crops considered, the unit price of their products, the land-lease value according to each crop, and the inflation rate and overall production value of São Paulo's farming production, excluding corn, soy, and sugarcane production values. The overall production value for each unit is incorporated into the model to reflect the income associated with other crops (e.g., coffee, oranges, and cattle), whose importance in the agricultural production of the state cannot be disregarded.

Methods
Conceptual framework
Land value has been a privileged subject of analysis in numerous economic studies ever since the conceptualization of a tripartite division of factors of production into land, labor, and capital [47]. In general, several theoretical approaches to the land market have considered the negotiated asset price to be a direct or indirect result of its potential earning stream [29, 51]. That is, even considering some assumptions with respect to the overall operation of the economy and the role that land plays in it, it is possible to identify a common element with

---

[1] When the word speculative or associated terms are used in this study, one should consider them as a generalization for the several markets of the classic definition given by Kaldor [27], p. 1: "Speculation, for purposes of this article, may be defined as the purchase (or sale) of goods with a view to resale (or repurchase) at a later date, where the motive behind such action is the expectation of a change in the relevant price relative to the ruling price and not a gain obtained through their use, or any type of transformation performed on them or their transfer between different markets."

[2] Because of the modernization of the production process, agriculture in São Paulo is considered among the most developed in the country.

---

[3] The EDR is a group of municipalities defined by the Institute of Agricultural Economics of the state of São Paulo that includes the Houses of Agriculture (Casas de Agricultura), which are present in all municipalities of São Paulo State.

respect to land price determinants in several theoretical approaches. That common element is the fact that land price is determined by the earnings that it generates to those who make use of it. From this perspective, land value is dictated by the production capacity of the land.

The relative consensus that had been established in the literature, however, was shattered in the mid-1950s, when many empirical studies found that land prices in the USA rose well above that which would be justified by the earnings from the land use, contrary to what was suggested by the theories of the time [10, 48]. Because of what was known as the "land price paradox" (the name given to this phenomenon in the specialized literature), various scholars began to consider factors not strictly related to land production capacity as determinants of land value. However, many of these authors restricted the influence of these other determinants to the impact they had on the agricultural sector itself. It was thus found that other factors, in addition to production-related ones, could influence land price. Studies by Scofield [48], Chryst [10] and Traill [52] provide some examples of the way land price determinants began to be considered.

Scofield [48] emphasized that land had a tendency to be more highly valued than the income growth derived from its productive use. To the author, price-sustaining policies, technological advances, and even the use of land as a store of value, e.g., as protection against increased inflation rates, changed the land price and could thus be considered among the elements determining it. Although Chryst [10] argued that land price should reflect the earnings the land is able to generate, including production increases, the author also considers non-agricultural earnings in land price formation. From this same perspective, Traill [52] found that in England, the increase in land prices in the 1960s was much higher than any increase in earnings from agricultural activities. In this context, the relation between the profitability of agricultural activities and land price was not as direct as it had been.

Other authors, such as Tweeten and Martín [53], Reinsel [38], Reinsel and Reinsel [39], Doll et al. [14], Just and Miranowski [26], and Weersink et al. [55] investigated the impact of public policy on land price. For example, Reinsel and Reinsel [39] observed that the present value of the land earnings stream, farm credit, the interest rate, and the inflation rate acted as land price determinants. Thus, the authors highlight factors associated with agricultural activities and those related to public policies in the sector. In the same vein, Doll et al. [14] performed an empirical assessment of the evolution of land price in the USA and developed a model that includes variables directly associated with agricultural activity, the interest rate, and other variables that reflect the existence of different government incentives. In general, they found that public policies directed toward the agricultural sector, especially credits and government subsidies, stimulated the demand for land, which raises its price.

Indeed, several factors are currently considered in the literature to be determinants of land prices: institutional aspects [25], spatial influence [8, 16, 23, 42, 57], international investment [2, 44], and the rental price of land [24, 51], among others factors.

The observation that in Brazil, during periods of high inflation after the modernization of Brazilian agriculture during the 1960s, land prices lost their relationship to the earnings level of agricultural activity led authors such as Sayad [45] and Telles et al. [51], among others, to argue that, in general terms, speculation, in addition to factors associated with agriculture, contributed to explanation of land value changes in the country.

Sayad [45] noted that, while serving as a store of value in periods of accelerated inflation, demand for land can exist regardless of the prevailing conditions in the production sphere. Land income was relatively constant during different economic cycles, making land a high-demand asset in periods of cyclical decline. Egler [15], in turn, draws a parallel between the land market and financial markets, noting the importance of interest rate movements as a land price determinant. To Reydon et al. [42], the determination of land value depends (in addition to the prices of agricultural products and inputs) on the actual interest rate, farming credits, and technological innovations. Reydon and Plata [41], as another example, state that between 1966 and 1975, growth in land price was influenced by technological innovation that changed the way in which agricultural activities were conducted. For Novo et al. [33], buying land and investing in farming is based on several reasons, including capital protection (land value used to increase over time and is considered a safe asset to invest money earned in urban business), social recognition (to pursue a farm is a clear sign of wealth to the urban society), leisure weekends and vacation activity, and nostalgia. Oliveira and Costa [34] also highlight that, in addition to agricultural product prices and inputs, the transportation infrastructure was one of the determining factors in land price.

In summary, land is a production factor and its price therefore reflects the income generated by the production activities it enables. However, land is also a financial asset used as a store of value, most importantly in periods of high inflation rates and uncertainties related to the economic environment. Land is also used as a guarantee for obtaining credit and government subsidies. Because of these additional financial factors, land is often used as the target of speculation [31]. Moreover, land prices reflect governmental policies, such as taxes, subsidies, technical assistance, and other programs that may be directly or indirectly related to agriculture. The costs and benefits of such policies are often capitalized in the land prices affected by them. As a result, land price result from a broad range of factors, which makes the task of measuring its determinants more complex.

Nevertheless, with the economic stability resulting from the Real Plan (1994) in Brazil, studies indicated that land price determinants became increasingly associated with production factors, i.e., the income obtained from agricultural activities [6, 7, 18, 37, 41, 58]. This argument is even more relevant for the state of São Paulo, a region where a strong expansion in corn, soy, and especially sugarcane crops has occurred [13]. Thus, due to the data limitation for the selected unit of analysis, i.e., the EDRs, we opted for the use of variables associated with these cultures of the set of possible factors that may contribute to the determination of land prices.

Data and descriptive statistics
The data used in this study are presented in Table 1. They were acquired from the IEA and from the Getúlio Vargas Foundation (Fundação Getúlio Vargas—FGV). It is important to note that all variables measured in monetary terms were deflated and are shown in constant Reais (2013 BRL). The land price is divided into first-class croplands and second-class croplands according to the specifications of the EDR, and according to the classification established by the IEA as follows: (1) "First-class cropland: First-class cropland is potentially fit for annual crops, perennial crops and other uses, and supports intensive management of crop practices, tillage,

etc. It is medium to high productivity land that can be mechanized, being flat or slightly sloping and with deep and well-drained soil" (IEA); (2) "Second-class cropland: Although potentially fit for annual and perennial crops and other uses, second-class cropland has more serious limitations than first-class cropland. It may have mechanization problems because of steep slopes. However, the soil is deep, well drained, fertile, while sometimes requiring some type of compost" (IEA).

Table 2 Descriptive statistics of the variables used *Source*: **prepared by the authors based on IEA and FGV data**

Variables	Observations	Mean	SD	Minimum	Maximum
P1	679	15,685.45	8887.64	1550.87	63,380
P2	679	12,665.57	7426.64	1436.03	51,670
Area_Sugarcane	670	107,022.92	107,921.65	4.00	488,500
Area_Corn	680	24,506.59	28,170.08	98.00	173,500
Area_Soy	680	14,043.86	30,503.19	0	169,200
Price_Sugarcane	680	62.19	9.20	47.30	80.42
Price_Corn	680	28.09	3.91	21.08	35.10
Price_Soy	680	53.15	7.91	39.07	67.01
Lease_Sugarcane	591	772.51	284.65	248	3468
Lease_Corn	625	445.70	149.96	93.89	1220
Lease_Soy	420	506.65	161.51	248.44	1205
Vp_Total	680	683,100[a]	342,322[a]	60,170[a]	2,236,000[a]

[a] Values divided by 1000

Table 1 Information on the data used *Source*: **prepared by the authors**

Model variables	Data used	Unit	Source
P1	First-class cropland price	2013 BRL per hectare	IEA
P2	Second-class cropland price	2013 BRL per hectare	IEA
Area_Sugarcane	Sugarcane cultivation area	Hectare	IEA
Area_Corn	Corn cultivation area	Hectare	IEA
Area_Soy	Soy cultivation area	Hectare	IEA
Price_Sugarcane	Sugarcane price	2013 BRL per ton	IEA
Price_Corn	Corn price	2013 BRL per bag	IEA
Price_Soy	Soy price	2013 BRL per bag	IEA
Lease_Sugarcane	Sugarcane leasehold	2013 BRL per hectare per year	IEA
Lease_Corn	Corn leasehold	2013 BRL per hectare per year	IEA
Lease_Soy	Soy leasehold	2013 BRL per hectare per year	IEA
Vp_Total	Total agricultural production value (excluding corn, soy, and sugarcane)	2013 BRL	IEA
IGP	IGP-DI (base = 2013)	2013 = 100	FGV
T2011	Dummy for 2011		
T2012	Dummy for 2012		
T2013	Dummy for 2013		

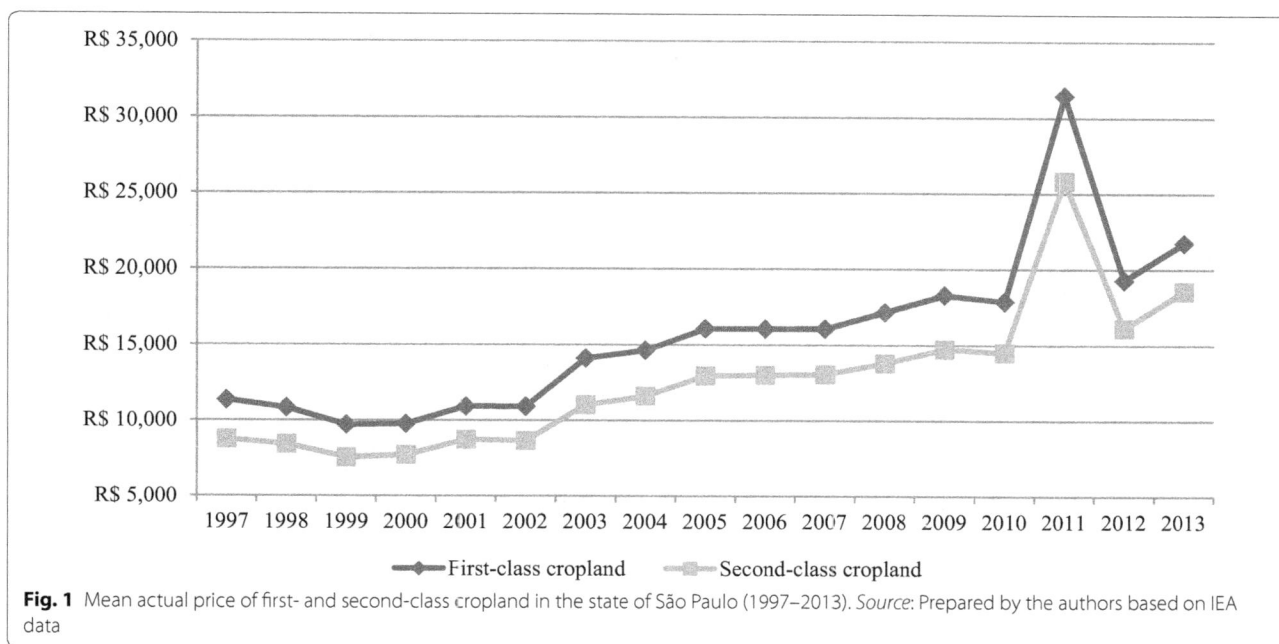

Fig. 1 Mean actual price of first- and second-class cropland in the state of São Paulo (1997–2013). *Source*: Prepared by the authors based on IEA data

The descriptive statistics of the variables used are shown in Table 2. It should be noted that because the data used do not have the same number of observations for all variables (Table 2), the panel described in the subsection below is unbalanced. Sugarcane, corn, and soy crops are predominant in certain regions. Therefore, variables have large ranges. As an example, the soy cultivation area (Area_Soy) has a minimum value of 0 hectares and a maximum value of 169,200 ha.

Figure 1 graphically illustrates the evolution of the mean real price of first- and second-class lands in the state of São Paulo between 1997 and 2013. Similar price behavior for both classes of land is observed over the same time period. Between 1997 and 2013, there was an increase in value of 91.64% for first-class croplands and 112.58% for second-class croplands. The data show that, between 1997 and 1999, land was devalued, a movement similar to the one previously discussed for Brazil as a whole. After 2000, an increase in the value of land began, and between 2005 and 2007 land prices stabilized. In 2008, there was a slight increase in land prices in relation to the previous period, followed by a slight decrease in 2010. In 2011, there is a large increase in land value, most likely caused by the high prices of agricultural commodities, especially sugarcane, corn, and soy, in addition to the speculative effect inherent to land price. In 2012, there was a drop in land prices, possibly resulting from the drop in commodity prices, legal insecurity (such as the acquisition of land by foreigners), and environmental impediments resulting from the new Forest Code. Finally, in 2013, prices

started to rise again, regaining the upward trajectory that began in 2000.[4]

The EDR is chosen as the unit of analysis because it is the smallest unit for which reliable data are available over the period of analysis (1997–2013), which provides robustness to the model results. With respect to the explanatory variables, the crops chosen (corn, soy, and sugarcane) represent the main temporary[5] agricultural activities of the state. As illustrated in Fig. 2, the share of the sum of corn, soy, and sugarcane production values in the total production value of São Paulo's farming activity increases over time, from approximately 30% in 1997 to 55% in 2013. This movement is credited primarily to sugarcane. When considering shares of cultivated area, the importance of these three crops is also evident [13]. As displayed in Fig. 3, the share of the sum of corn, soy, and sugarcane crop areas in the total agricultural area of São Paulo, equivalent to the sum of the temporary crop

---

[4] Naturally, as this is the mean price of the EDRs in the state of São Paulo, important individual differences may remain hidden. For example, with the exception of the years after 2011, São Paulo and Campinas show a continuous drop in land price. The allocation of land in the rural areas of these EDRs may not be fully directed to agriculture and may include activities related to urban life, such as clubs, parks.

[5] The study's choice to analyze temporary crops reflects their importance within the state of São Paulo's farming activity, as well as the assumption that the impact of these crops on the formation of land prices is best captured by the proposed model. If considered permanent crops, it would be necessary to take into account the different temporalities of each crop as a way to get a reasonable approximation of the activity earnings, a fact that would make the model and the interpretation of its results more complex.

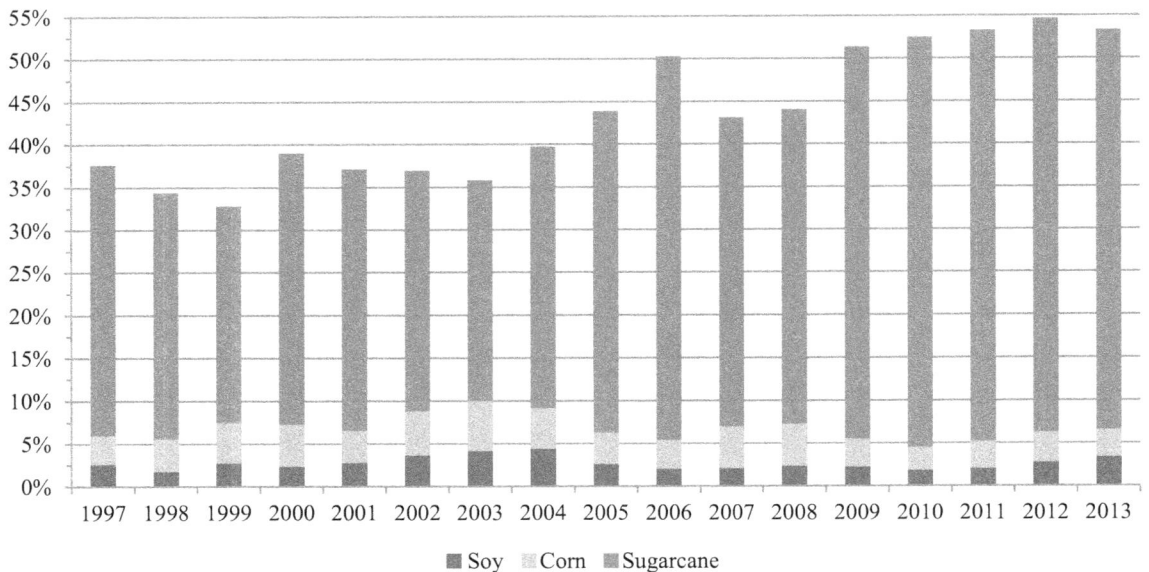

Fig. 2 Share of corn, soy, and sugarcane production values in the agricultural production value for the state of São Paulo (1997–2013). *Source*: Prepared by the authors based on IEA data

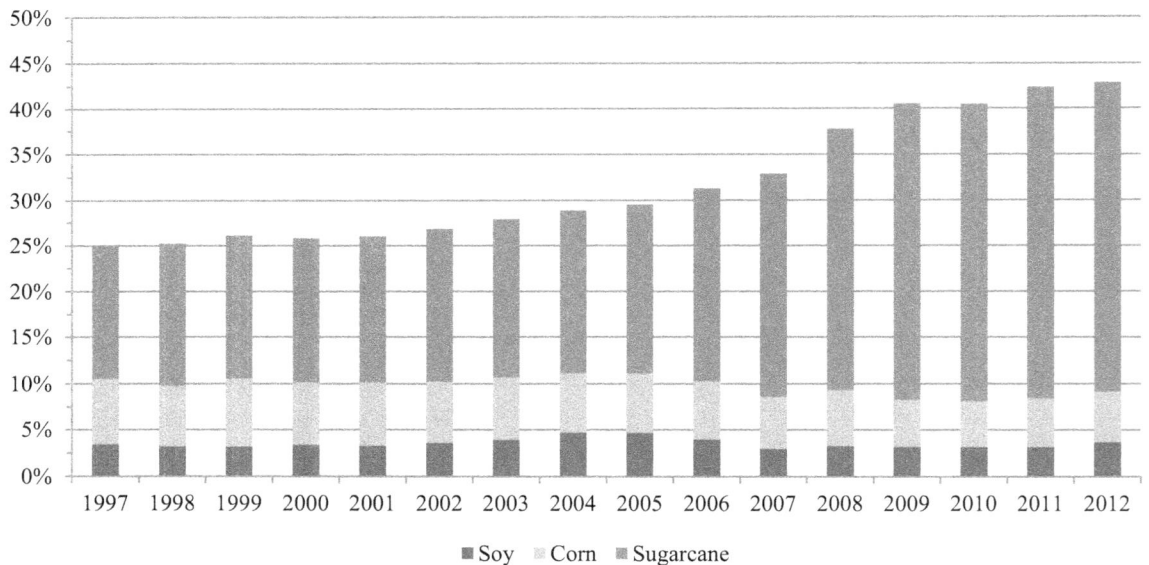

Fig. 3 Share of cultivation area dedicated to corn, soy, and sugarcane in the total agricultural area of the state of São Paulo (1997–2012). *Source*: Prepared by the authors based on Brazilian Institute of Geography and Statistics (Instituto Brasileiro de Geografia e Estatística (IBGE)) data

areas, the permanent crop areas and pastures, increases from less than 25% in 1997 to more than 40% in 2012.

The area, the unit price, and the leasehold price of each of these three crops are variables used to assess the impact of variations in the earnings of these activities on land prices in the state of São Paulo. The inclusion of

the total agricultural production value of the state of São Paulo captures the added effect of other activities.

Finally, the general price index of domestic availability [Índice Geral de Preços-Disponibilidade Interna (IGP-DI)] represents the role of the land as a store of value, since the demand for land reflects the willingness of agents to protect their wealth, depending on the rate of inflation. Moreover, as a way to isolate the dramatic rise in land prices in 2011 (Fig. 1), dummy variables were constructed for the years 2011, 2012, and 2013.

Empirical model

This study applies panel data econometric techniques, in which time series and cross-sectional data are combined, allowing higher degrees of freedom, reducing the collinearity between explanatory variables, and controlling for unobservable heterogeneity present in the units of analysis [21, 22, 56]. In this case, the units of analysis are the EDRs in the state of São Paulo.

The empirical analysis applies the following econometric model:

variance for each EDR) and first-order autocorrelations. In this analysis, tests are conducted to determine the presence of these error structures. If the results indicate the presence of heteroskedasticity between EDRs and/or first-order autocorrelations, appropriate methods are implemented to correct the model.

Model selection tests

To determine the appropriate model for the data, the following procedures were employed: (1) the Breusch–Pagan Lagrangian multiplier test was applied in order to choose between a pooled ordinary least squares (pooled OLS) model, which does not present unobservable individual effects, and the random effects model I. If the null hypothesis is rejected, the random effects model is

$$\begin{aligned} logP_{it} = {} & \beta_0 + \beta_1 log\text{Area\_Sugarcane}_{it} + \beta_2 log\text{Area\_Corn}_{it} + \beta_3 log\text{Area\_Soy}_{it} \\ & + \beta_4 log\text{Price\_Sugarcane}_{it} + \beta_5 log\text{Price\_Corn}_{it} + \beta_6 log\text{Price\_Soy}_{it} \\ & + \beta_7 log\text{Lease\_Sugarcane}_{it} + \beta_8 log\text{Lease\_Corn}_{it} + \beta_9 log\text{Lease\_Soy}_{it} \\ & + \beta_{10} log\text{Vp\_Total}_{it} + \beta_{11} log\text{IGP}_{it} + \varepsilon_{it} \end{aligned} \tag{1}$$

where $i = 1, 2, \ldots, 40$, refers to each of the 40 EDRs selected, $t = 1, 2, \ldots, 17$, indexes years 1997–2013, β_0 is the intercept, β_1 to β_{11} are the coefficients of the covariables defined as potential determinants of land price, β_{12}, β_{13}, and β_{14} are the coefficients of the dummy variables for year 2011, 2012, and 2013, μ_i captures the unobservable individual effects of the EDRs, and ε_{it} is the idiosyncratic error term. The number of observations is not the same for all variables in the analysis, thus leaving the panel unbalanced (Table 2).

The unobserved heterogeneity of the EDRs can be addressed by means of a fixed effect, or may be treated as a random variable. If the specific effect of the units is defined as a random variable, the random effects model is used, in which $\mu_i \sim \text{IID}(0, \sigma_\mu^2)$, $\varepsilon_{it} \sim \text{IID}(0, \sigma_\varepsilon^2)$, with μ_i assumed to be independent from both the idiosyncratic error term and the independent regressors of μ_i and ε_{it}. The fixed effects model assumes μ_i to be a fixed parameter in time, estimable for each unit, with $\varepsilon_{it} \sim \text{IID}(0, \sigma_\varepsilon^2)$ and the independent regressors of ε_{it}. In the fixed effects model, no independence is assumed between covariables and the unobservable heterogeneity of the units. The parameters estimated using the random effects model become inconsistent if there is correlation between individual effects and covariables. In this situation, the fixed effects model should be used in order to generate consistent coefficients [3].

Data in the cross-sectional and time series format are likely to have complex error structures, such as the presence of heteroskedasticity among the units (specific

used; (2) if the random effects model is favored over the pooled OLS model, the Hausman test is then applied in order to choose between the fixed effects model and the random effects model; (3) if the fixed effects model is chosen over the random effects model, an F test is then used to determine whether at least one fixed effect is different from zero; (4) the hypothesis of heteroskedasticity between individuals is tested. If the null hypothesis of heteroskedasticity is rejected, the structure of the error variance–covariance matrix is corrected, incorporating the variance estimate for each EDR; and (5) the data are tested for first-order autocorrelations. If first-order autocorrelations are found to be present, the first-order autocorrelation error coefficients are estimated. Table 3 shows the results of these applied tests.

Based on the Breusch–Pagan LM test and the Hausman test, the unobserved heterogeneity of individuals is incorporated using the fixed effects model. However, the heteroskedasticity and autocorrelation tests demonstrate the existence of two non-spherical disturbances in the model. Because of this result, three different estimation methods are used to correct the error structures in the model as follows: (1) fixed effects (within) [FE AR(1)] to control for first-order autocorrelations; (2) panel-corrected standard errors (PCSE) to correct for heteroskedasticity and the autocorrelations; and (3) feasible generalized least squares (FGLS) to correct for heteroskedasticity and autocorrelations.

The FGLS method that is applied was developed by Parks [35]. It corrects the error variance–covariance

Table 3 Tests used to identify the model Source: **prepared by the authors**

Test	Null hypothesis	Alternative hypothesis	Test value		Conclusion
			P1	P2	
Breusch–Pagan LM	Pooled OLS	Random effects	498.35 $p = 0.000$	391.61 $p = 0.000$	OLS is rejected in favor of random effects
Hausman	Random effects	Fixed effects	86.29 $p = 0.000$	108.48 $p = 0.000$	Random effects is rejected in favor of fixed effects
F test for detection of fixed effects	All fixed effects are equal to zero	At least one fixed effect is different from zero	68.35 $p = 0.000$	59.99 $p = 0.000$	Lack of fixed effects is rejected
Greene for heteroskedasticity	Same variance for each individual	The variances of individuals are not equal	413.41 $p = 0.000$	180.27 $p = 0.000$	Homoskedasticity is rejected in favor of heteroskedasticity among individuals
Wooldridge for autocorrelation	No first-order autocorrelations	First-order autocorrelations	50.386 $p = 0.000$	79.732 $p = 0.000$	Lack of first-order autocorrelations is rejected

matrix structure in the presence of heteroskedasticity and first-order autocorrelations. The FGLS estimator generates asymptotically efficient coefficients and unbiased standard errors. This is considered a feasible method because the process that generates the data is not known a priori. Therefore, the variance–covariance matrix elements must be estimated. The PCSE method was developed by Beck and Katz [5], who questioned the efficiency of Parks' FGLS method in several situations commonly encountered in empirical studies. According to the authors, the FGLS method significantly underestimates the variance of parameters, thus inflating the reliability of estimates. Beck and Katz [5] developed an alternative estimator using OLS that corrects for heteroskedasticity and autocorrelations, thus generating more accurate error estimates with little or no efficiency loss when compared to FGLS. The situations under which Parks' method (1967) produces more efficient estimates, according to the Monte Carlo simulations made by Beck and Katz [5], are those with extreme heteroskedasticity or extreme contemporary correlations. In order to control for fixed effects in the FGLS and PCSE methods in the present analysis, intercept dummies are inserted for the EDRs.

It is important to emphasize that FGLS and PCSE methods, in addition to estimating the specific variance of each unit, also allow [contrary to the FE AR(1) method] the estimation of the autocorrelation parameter of disturbances for each panel unit. The FE AR(1) method estimates a single autocorrelation parameter for all units. The estimate of different parameters for the correction of the error structure generates some loss in degrees of freedom. In a panel with many units and a short temporal window, there is a critical loss of parsimony when estimating the variance and autocorrelation of each unit. Therefore, the loss of degrees of freedom may render the estimation of the entire error structure impossible. In the present study, there are 40 units of analysis and 17

time periods. In order to estimate the autocorrelation coefficient and the specific variance for each EDR, 80 degrees of freedom are lost. Adding the estimated coefficients and fixed effects dummies, there is a total loss of 130 degrees of freedom when using the FGLS and PCSE methods to correct the entire error structure as identified. Thus, given the total available observations, the estimate of all parameters for the model correction is adequately parsimonious.

Results and discussion

Primary regression results are shown in Table 4. The data demonstrate, in general, that both models fit reasonably well. Using the FE AR(1) and PCSE methods, the data show that, for first-class croplands, the percentage variation in the independent variables explains, on average, 97 and 99% of the percentage variation of land prices, respectively.

The coefficients associated with the area of sugarcane cultivation, sugarcane price, soy price, the IGP-DI, and the total production value were significant for all estimates at the 1% level, including the dummies. The coefficient of the soy cultivation area was also significant at the 1% level for both classes of land, except for the FE AR(1) method for first-class croplands, which was not significant. Sugarcane leasehold was significant at 1% only in the FGLS method, significant at 5% in the PCSE model for both classes of land, at 5% in the FE AR(1) model for second-class croplands, and at 10% in this same model for first-class croplands. The coefficient for soy leasehold was significant at the 5% level in the FGLS model and at 10% in the PCSE model for first-class croplands. In the case of second-class croplands, the soy leasehold coefficient was significant only in the FGLS model at the 10% level.

The coefficients of the corn cultivation area, corn price, and corn leasehold were not significant for any of the methods. There is therefore no statistical evidence of the

Table 4 Estimates of the model parameters. *Source*: prepared by the authors

Variables	P1 (First-class cropland price)			P2 (Second-class cropland price)		
	FE AR(1)	PCSE	FGLS	FE AR(1)	PCSE	FGLS
Intercept	0.02416 (0.062)	3.4682*** (1.028)	1.97274 (0.876)	0.03615 (0.050)	3.55010 (1027)	2.0945** (0.915)
Price_Sugarcane	0.26090*** (0.067)	0.23288*** (0.062)	0.24315*** (0.052)	0.26495*** (0.066)	0.24573*** (0.065)	0.25992*** (0.055)
Price_Corn	0.01582 (0.057)	− 0.01448 (0.050)	− 0.00602 (0.042)	0.01666 (0.057)	− 0.02361 (0.052)	− 0.00847 (0.045)
Price_Soy	0.22768*** (0.063)	0.2746*** (0.051)	0.27197*** (0.041)	0.22237*** (0.062)	0.28032*** (0.052)	0.27343*** (0.045)
IGP	0.82923*** (0.060)	0.72557*** (0.034)	0.74679*** (0.030)	0.84081*** (0.061)	0.75657*** (0.036)	0.77879*** (0.032)
Vp_Total	0.25616*** (0.032)	0.10688** (0.043)	0.16672*** (0.036)	0.25569*** (0.032)	0.10488** (0.044)	0.16977*** (0.038)
Area_Sugarcane	0.11624*** (0.028)	0.09807*** (0.023)	0.10064*** (0.019)	0.12252*** (0.028)	0.09338*** (0.023)	0.08439*** (0.021)
Area_Corn	0.06761 (0.042)	− 0.01166 (0.033)	0.00476 (0.027)	0.03236 (0.042)	− 0.03105 (0.033)	− 0.03170 (0.029)
Area_Soy	0.02049 (0.016)	0.04422*** (0.013)	0.03608*** (0.010)	0.02058*** (0.004)	0.04190*** (0.013)	0.042417*** (0.010)
Lease_Sugarcane	0.07078* (0.039)	0.07066** (0.034)	0.08083*** (0.029)	0.08977** (0.039)	0.07850** (0.034)	0.09110*** (0.032)
Lease_Corn	− 0.0001 (0.037)	− 0.02019 (0.032)	− 0.01676 (0.028)	− 0.00159 (0.037)	− 0.01605 (0.033)	− 0.00517 (0.029)
Lease_Soy	0.00732 (0.034)	0.04943* (0.029)	0.05109** (0.024)	0.00134 (0.033)	0.04004 (0.030)	0.04582* (0.026)
T2011	0.36371*** (0.035)	0.40579*** (0.032)	0.38605*** (0.027)	0.36712*** (0.035)	0.40801*** (0.033)	0.038111*** (0.029)
T2012	− 0.21345*** (0.041)	− 0.1937*** (0.034)	− 0.1948*** (0.028)	− 0.1964*** (0.041)	− 0.18658*** (0.036)	− 0.1975*** (0.031)
T2013	− 0.1089*** (0.039)	− 0.1056*** (0.032)	− 0.1005** (0.026)	− 0.0780*** (0.040)	− 0.07820** (0.034)	− 0.0853*** (0.029)
R^2	W. = 0.9702 B. = 0.0577 O. = 0.3478	0.9978	–	W. = 0.9698 B. = 0.0568 O. = 0.3800	0.9972	–

W. = R^2 within. B. = R^2 between; O. = R^2 overall

*** Significant at 1%; ** significant at 5%; * significant at 10%. Standard errors in parentheses

effect of these variables on land price, a fact that indicates that corn cultivation may not impact land prices in the state of São Paulo.

All significant coefficients showed the expected signs. In fact, positive variations in the area of sugarcane and soy cultivation, in production values, in the leasehold value of these crops, and in the value of total agricultural production and inflation, are reflected in positive variations in first- and second-class cropland prices. An increase in the area under cultivation is associated with an increase in the demand for land that, *ceteris paribus*, leads to an increase in land value. An increase in the price of a certain agricultural product, with all other factors constant, causes land prices to increase by raising the production-related income. Similarly, if the landowner obtains a higher income through leasehold, the land thus

provides greater earnings potential and its price therefore increases. Rising inflation results in an increased demand for land as a store of value, raising the real price of land.

Additionally, coefficients related to explanatory variables exhibit relatively similar values for both classes of land in each of the three models, indicating that land prices of first- and second-class cropland move in parallel to changes in the variables considered. For example, using the PCSE, an increase of 1% in the price per ton of sugarcane is expressed in a corresponding increase of approximately 0.23% in the first-class cropland price and 0.25% increase in the second-class cropland price. Although the variation in cultivated area exerts a relatively lesser impact on land price, the direction is similar. That is, assuming an increase of 1% in the sugarcane cultivation area, increases of approximately 0.10% in the first-class cropland price and 0.09% in the second-class cropland price are expected.

Particular attention should be paid to variables associated with the sugarcane industry, which were significant in all models. Coefficient levels show that the price of sugarcane, the cultivation area, and the leasehold associated with this crop are major determinants, under all three methods, of the price of the first- and second-class croplands in the state of São Paulo.[6]

The recent development of the biofuels sector, which is served primarily by the production of sugarcane, has been one of the most dynamic agricultural sectors in the state of São Paulo [11, 28]. Because the ethanol sector represents such an important element of agricultural production in the state of São Paulo (making the state the country's primary producer), it is noteworthy that regression analysis measured a notable effect from this sector on land prices in recent years.

On the one hand, it appears that some of the recent variation in land price has been associated with elements related to the agricultural activity itself, thus indicating that the weight of production factors in determining land price should be considered. The identification here of significant associations between production activity and land prices is supported by numerous studies of the Brazilian market, even though these studies consider different units of analysis and employ different methodologies. Dias et al. [12], in a study on land prices in Brazil between 1966 and 1998, showed that both the index of prices received and paid by producers and land productivity positively affect land price. Similarly, another depiction of the importance of the factors related to the productive use of land is found in the recent study by Ferro and Castro [18]. Considering the price of soy as one of the explanatory variables of the model for the determination of land price in Brazil from 2000 to 2010, the authors note that "land price is strongly related to income that can be obtained with this factor" [18].

However, on the other hand, the data evidence the influence of the inflation rate in determining land prices, a phenomenon primarily related to the use of land as a store of value. Even if the impact of inflation on the land market was greatly reduced in periods of relatively low and stable inflation, the evidence found in the three methods considered here suggests the contrary, i.e., that the inflation rate, among the variables in the model, has the highest influence on land price. As shown in Table 3, a variation of 1% in the IGP-DI is reflected in a variation of, at least, 0.73% in prices for both classes of land. Therefore, one cannot discard the suggestion of several authors, such as [45] and [51], who assigned an important weight to the function of land as a store of value and identified speculation as a key determinant of land prices in Brazil. The rise in inflation, while signaling an environment of greater instability for holders of wealth, nonetheless, increases the demand for assets that serve as stores of value, land being among them. Therefore, the real land price increases through the increased profitability associated with its productive use, and also through the demand of those who see land as an asset with valuation prospects that allow them to protect their wealth from inflation.

Model results indicate that land prices in recent years in the state of São Paulo are determined by a combination of production factors and other variables not inherent to the production process, such as inflation. In fact, as demonstrated, inflation has a greater impact on land price than do the production factors. To more accurately assess the extent to which the demand for land is derived from production factors and from factors associated with speculation would require consideration of a large set of variables. However, for the level of disaggregation considered herein, i.e., EDRs of the state of São Paulo, incorporating variables that express the speculative behavior of agents, in addition to inflation, is a difficult task to perform. This undertaking awaits further analysis.

The empirical evaluation of the determinants of land prices in the state of São Paulo indicates that the variables associated with income from agricultural activity, particularly sugarcane and soy, contribute to the changes in land prices. However, even in the recent period of low and relatively stable inflation, land prices respond to factors indirectly related to its productive use. Strong evidence for the weight of inflation in the determination of land prices and for the importance of land as a store of value is found in the data.

The results suggest that land prices in the state of São Paulo are determined by a combination of production factors and factors linked to the function of land as a store of value, with greater influence of the latter, as demonstrated by the notable impact of inflation. The novelty of the simple model presented here results from the adoption of a panel data approach for EDRs. As evidenced by the significance of results presented here, it would be of interest to see further studies that employ data with this level of disaggregation. However, a better understanding of how these factors influence land use patterns over time and space would help policy makers in evaluating existing practices or in drawing up new sustainable environmentally policies.

---

[6] Even if, for some methods, the coefficient of the sugarcane price is lower than that associated with soy, potentially suggesting that the price of soy has a greater influence on land price, the linear restriction test indicated that, for each type of land, the impact of the sugarcane and soy prices on land price can be considered equal. A similar relation was not found with respect to the other variables.

Authors' contributions
All authors contributed equally to this study. AWAP and TST contributed to study conception. AWAP, TST, RFS, and FRM helped in study design. AWP and TST contributed to acquisition of data. RFS and FRM contributed to analytical methods and performed modeling and data processing. AWP, TST, and RFS analyzed and interpreted the data. AWP and TST drafted the manuscript. AWP, TST, and RFS helped in critical revision. All authors read and approved the final manuscript.

Author details
[1] Instituto de Economia, Universidade Estadual de Campinas, Rua Pitágoras 353, Campinas, São Paulo CEP 13083-857, Brazil. [2] Instituto Agronômico do Paraná, C.P. 10.030, Londrina, Paraná CEP 86057-970, Brazil. [3] Faculdade de Economia, Universidade Federal de Mato Grosso, Avenida Fernando Corrêa 2367, Cuiabá, Mato Grosso CEP 78060-900, Brazil. [4] Departamento de Economia, Universidade Federal da Grande Dourados, C.P. 364, Dourados, Mato Grosso do Sul CEP 79.804-970, Brazil.

Competing interests
The authors declare that they have no competing interests.

Funding
This study was financially supported by the Universidade Estadual de Campinas (UNICAMP).

References
1. Bacha CJC. A determinação do preço de venda e de aluguel da terra na agricultura. Estud Econ. 1989;19:443–56.
2. Baker TG, Boehlje MD, Langemeier MR. Farmland: Is it currently priced as an attractive investment? Am J Agric Econ. 2014;96:1321–33.
3. Baltagi B. Econometric analysis of panel data. 3rd ed. Chichester: Wiley; 2005.
4. Barretto AGOP, Berndes G, Sparovek G, Wirsenius S. Agricultural intensification in Brazil and its effects on land-use patterns: an analysis of the 1975–2006 period. Global Change Biol. 2013;19:1804–15.
5. Beck N, Katz J. What to do (and not to do) with time series cross-section data. Am Polit Sci Rev. 1995;89:634–47.
6. Camargo AMMP, Camargo FP, Siqueira ACN, Camargo Filho WP, Francisco VLFS. Agricultural land valorization according to regional use of soil in the state of São Paulo. Inf Econ. 2004;34:28–40.
7. Camargo FP. Análise do mercado de terras agrícolas nas regiões do estado de São Paulo, 1995 a 2006. Inf Econ. 2007;37:50–63.
8. Cavailhès J, Thomas I. Are agricultural and developable land prices governed by the same spatial rules? The case of Belgium. Can J Agric Econ. 2013;61:439–63.
9. Chakir R, Le Gallo J. Predicting land use allocation in France: a spatial panel data analysis. Ecol Econ. 2013;92:114–25.
10. Chryst WE. Land values and agricultural income: A paradox? J Farm Econ. 1965;47:1265–73.
11. Deuss A. The economic growth impacts of sugar cane expansion in Brazil: an inter-regional analysis. J Agric Econ. 2012;63:528–51.
12. Dias GLS, Vieira CA, Amaral CM. Comportamento do mercado de terras no Brasil. 1st ed. Santiago: CEPAL; 2001.
13. Dias LCP, Pimenta FM, Santos AB, Costa MH, Ladle RJ. Patterns of land use, extensification, and intensification of Brazilian agriculture. Global Change Biol. 2016;22:2887–903.
14. Doll JP, Widdows R, Velde PD. The value of agricultural land in the United States: a report on research. Agric Econ Res. 1983;35:39–44.
15. Egler CAG. Preço da terra, taxa de juro e acumulação financeira no Brasil. Rev Econ Polit. 1985;5:112–35.
16. Eagle AJ, Eagle DE, Stobbe TE, van Kooten GC. Farmland protection and agricultural land values at the urban-rural Fringe: British Columbia's agricultural land reserve. Am J Agric Econ. 2015;97:282–98.
17. Ferreira Filho JBS, Horridge M. Ethanol expansion and indirect land use change in Brazil. Land Use Policy. 2014;36:595–604.
18. Ferro AB, Castro ER. Determinantes dos preços de terras no Brasil: uma análise de região de fronteira agrícola e áreas tradicionais. Rev Econ Soc Rural. 2013;51:591–609.
19. Ficarelli TRA, Ribeiro H. Dinâmica do arrendamento de terras para o setor sucroalcooleiro: Estudo de casos no estado de São Paulo. Inf Econ. 2010;40:44–54.
20. Gasques JG, Bastos ET. Terra: preços no Brasil. Agroanalysis. 2008;28:14–5.
21. Greene WH. Econometric analysis. 6th ed. Prentice Hall: Pearson; 2008.
22. Gujarati DN, Porter D. Basic econometrics. 5th ed. Irwin: McGraw-Hill; 2009.
23. Hausman C. Biofuels and land use change: sugarcane and soybean acreage response in Brazil. Environ Resour Econ. 2012;51:163–87.
24. Hennig S, Breustedt G, Latacz-Lohmann U. The impact of payment entitlements on arable land prices and rental rates in Schleswig-Holstein. Ger J Agric Econ. 2014;63:219–39.
25. Hüttel S, Wildermann L, Croonenbroeck C. How do institutional market players matter in farmland pricing? Land Use Policy. 2016;59:154–67.
26. Just RE, Miranowski JA. Understanding farmland price changes. Am J Agric Econ. 1993;75:156–68.
27. Kaldor N. Speculation and economic stability. Rev Econ Stud. 1939;7:1–27.
28. Kohlhepp G. Análise da situação da produção de etanol e biodiesel no Brasil. Estud Av. 2010;24:223–53.
29. Larsen HC. Relationship of land values to warranted values, 1910-48. J Farm Econ. 1948;30:579–88.
30. Lima Filho RR, Aguiar GAM, Torres Junior AM. Mercado de terras: preços em alta, liquidez em baixa. Agroanalysis. 2013;33:19–22.
31. Magnan A, Sunley S. Farmland investment and financialization in Saskatchewan, 2003–2014: an empirical analysis of farmland transactions. J Rural Stud. 2017;49:92–103.
32. Novo AL, Jansen K, Slingerland M, Giller K. Biofuel, dairy production and beef in Brazil: competing claims on land use in São Paulo State. J Peasant Stud. 2010;37:769–92.
33. Novo A, Jansen K, Slingerland M. The sugarcane-biofuel expansion and dairy farmers' responses in Brazil. J Rural Stud. 2012;28:640–9.
34. Oliveira JT, Costa DN. Evolução recente do preço de terra no Brasil: 1966–74. Rev Econ Rural. 1977;15:259–76.
35. Parks R. Efficient estimation of a system of regression equations when disturbances are both serially and contemporaneously correlated. J Am Stat Assoc. 1967;62:500–9.
36. Pinheiro FA, Reydon BP. O preço da terra e a questão agrária: algumas evidências empíricas relevantes. Rev Econ Rural. 1981;19:5–15.
37. Plata LEA. Dinâmica do preço da terra rural no Brasil: Uma análise de cointegração. In: Reydon BP, Cornélio FNM, editors. Mercados de terras no Brasil: estrutura e dinâmica. Brasília: NEAD; 2006. p. 125–54.
38. Reinsel RD. Effect of seller financing on land prices. Agric Financ Rev. 1972;33:32–5.
39. Reinsel RD, Reinsel EI. The economics of asset values and current income in farming. Am J Agric Econ. 1979;61:1093–7.
40. Reydon BP, Fernandes VB, Telles TS. Land tenure in Brazil: the question of regulation and governance. Land Use Policy. 2015;42:509–16.
41. Reydon BP, Plata LEA. O plano real e o mercado de terras no Brasil: lições para a democratização do acesso à terra. In: Reydon BP, Cornélio FNM, editors. Mercados de terras no Brasil: estrutura e dinâmica. Brasília: NEAD; 2006. p. 267–84.
42. Reydon BP, Plata LEA, Sparovek G, Goldszmidt RGB, Telles TS. Determination and forecast of agricultural land prices. Nova Econ. 2014;24:389–408.
43. Rezende GC. Política agrícola, preço da terra e estrutura agrária. Rev Econ Rural. 1982;20:73–100.
44. Sauer S, Pereira Leite S. Agrarian structure, foreign investment in land, and land prices in Brazil. J Peasant Stud. 2012;39:873–98.
45. Sayad J. Preço da terra e mercados financeiros. Pesqui Planej Econ. 1977;7:623–62.
46. Sayad J. Especulação em terras rurais, efeitos sobre a produção agrícola e o novo ITR. Pesqui Planej Econ. 1982;12:87–108.
47. Schultz TW. A framework for land economics: the long view. J Farm Econ. 1951;33:204–15.
48. Scofield WH. Prevailing land market forces. J Farm Econ. 1957;39:1500–10.
49. Sousa AF. Terra gera alta rentabilidade. Agroanalysis. 2009;29:18–9.
50. Sparovek G, Barretto A, Berndes G, Martins S, Maule R. Environmental, land-use and economic implications of Brazilian sugarcane expansion 1996–2006. Mitig Adapt Strat Global. 2009;14:285–98.
51. Telles TS, Palludeto AWA, Reydon BP. Price movement in the Brazilian land market (1994–2010): an analysis in the light of post-Keynesian theory. Rev Econ Polit. 2016;36:109–29.

52. Traill B. An empirical model of the U.K. land market and the impact of price policy on land values and rents. Eur Rev Agric Econ. 1979;6:209–32.

53. Tweeten LG, Martin JE. A methodology for predicting U.S. farm real estate price variation. J Farm Econ. 1966;48:378–93.

54. Visser O. Running out of farmland? Investment discourses, unstable land values and the sluggishness of asset making. Agric Hum Values. 2017;34:185–98.

55. Weersink A, Clark S, Turvey CG, Sarker R. The effect of agricultural policy on farmland values. Land Econ. 1999;75:425–39.

56. Wooldridge JM. Econometric analysis of cross section and panel data. 1st ed. Cambridge: MIT Press; 2002.

57. Zhang W, Nickerson CJ. Housing market bust and farmland values: Identifying the changing influence of proximity to urban centers. Land Econ. 2015;91:605–26.

58. Zilli JB, Barros GSC, Bogoni NM. Precificação de terras de propriedades rurais em Cascavel–PR: Uma análise das opções reais. Teor Evid Econ. 2012;18:34–60.

Lessons for promotion of new agricultural technology: a case of Vijay wheat variety in Nepal

Krishna P. Timsina[1*], Yuga N. Ghimire[1], Devendra Gauchan[2], Sanjiv Subedi[3] and Surya P. Adhikari[4]

Abstract

Background: Wheat is the third important cereal after rice and maize in Nepal. Its yield suffers from several factors such as lack of reliable irrigation, inclement weather, incidence of disease and lack of improved technology. New virulence race Ug99 (Uganda-99) has threatened all available commercial wheat varieties around the world and Asia. Some of the popular varieties of wheat in Nepal are also getting susceptible to different diseases. Vijay is one of the recently released first Ug99-resistant improved wheat varieties for Terai region in Nepal. Nepal has put special emphasis on the seed production and diffusion of this variety to promote rapidly in the farmers' fields to mitigate the potential epidemics of newly emerging pathotype of stem rust (Ug99). This variety is being promoted by several organizations and seed companies' before and after its release. Therefore, this study was undertaken to know the seed supply of Vijay and identify the factors that are contributing for its commercialization.

Methods: Sunsari, Morang, Rupandehi, Banke and Kailali districts were selected purposely representing Eastern, Western, Mid-western and Far-Western Terai regions of Nepal, respectively. Seed companies, agro-vets (input dealers) and cooperatives involved in Vijay seed multiplication and distribution in the respective districts were selected for the purpose of the study. From the list of agro-vets in respective study sites, a total of 87 wheat seed selling agro-vets were selected randomly as a sample. The sample survey covered 44% of the target population in both categories (dealers and non-dealers of national seed company). Out of total samples, 40% were national seed company dealer agro-vets, while 60% were non-dealers. Multivariate regression analysis was used to find out different factors responsible for commercialization of Vijay seed.

Results: Results indicate that about 67% of the agro-vets were involved in Vijay seed trading and their average time of involvement was 2.65 years. Majority of the agro-vets perceived this variety performed better than other existing popular varieties such as Gautam, Bhrikuti, Aditya and Nepal 297. However, there was still gap in timely supply of the seed. Based on the estimated Certified-1 seed production in 2015/16, Vijay must cover 56.88% (11,943 ha) of total wheat area in Terai, but share of Vijay seed was about 22% of the total wheat seed sold by agro-vets in 2015/16. Nevertheless, it seems that this variety is getting popular in the study area compared to other improved wheat varieties in short time period. The factors influencing commercialization of Vijay were: agro-vets having dealership of national seed company, perception on comparative better performance of Vijay with Nepal 297, total quantity of wheat seed sold by agro-vets and total business transaction of the agro-vets. Moreover, other factors such as pre-release multiplication of Vijay variety by private seed companies and participation on seed related training have found positive contribution on its commercialization.

*Correspondence: krishnatimsina2000@gmail.com
[1] Socioeconomics and Agricultural Research Policy Division (SARPOD), Nepal Agricultural Research Council (NARC), GPO 5459, Khumaltar, Lalitpur, Nepal
Full list of author information is available at the end of the article

Conclusion: It seems that the current production of Vijay's breeder seed is more than enough to cultivate in total wheat area in Terai of Nepal if proper seed cycle is maintained. But this is not in reality. NARC should decentralized foundation seed to private actors such as private seed companies, national seed companies and community-based organizations to maximize the resource use efficiency in the seed cycle through introducing effective monitoring and technical backstopping mechanism from public sector. Therefore, the close coordination and commitment of the public and private seed companies, community seed groups, cooperatives and public extension agencies to multiply seeds in subsequent cycles is required. Moreover, awareness program about superiority of Vijay with other improved wheat varieties should be emphasized for the rapid commercialization of the Vijay seed which would be an instrumental to mitigate the potential epidemics of newly emerging pathotype of stem rust (Ug99) in the future.

Keywords: Vijay, Seed flow, Dealership, Commercialization, Business transaction

Background

Wheat (1,736,849 Mt.) is the third most important cereal after rice (4,299,079 Mt.) and maize (2,231,517 Mt.), contributing 20% of the total cereal production in Nepal [17]. Over 60% of wheat is produced in the Terai (plain) region, though they are also produced in the mid hills and high hills regions of Nepal. Wheat yields suffer from some factors such as lack of reliable irrigation, inclement weather, incidence of disease and lack of improved technology [21]. NWRP [21] reported stem rust disease of wheat was under control since last six decades in a global basis. New virulence race Ug99 (Uganda-99) has threatened future wheat production, and it has posed new problem since all available commercial wheat varieties around the world and Asia in particular are susceptible to this new biotype of stem rust. Some of the popular varieties of wheat in Nepal are getting susceptible to different disease, e.g., Nepal 297 become susceptible to new biotypes of leaf rust, Bhrikuti moderately susceptible to foliar blight [17]. Therefore, new varieties with high yield potential and resistant to major diseases are needed to replace old varieties from the growers to increase the production and maintain at least the present productivity level [22].

In Nepal, both formal and informal types of seed system are prevailing. The informal seed system has been playing a significant role in fulfilling seed demand of the farmers [30]. Gauchan et al. [9] have reported that the formal sector representing public, private and community led seed system is becoming prominent in recent years, even though nearly 90% of the seed supply comes from traditional individual farmer-led model representing mainly informal farmer-based seed system. Until 2017, NARC has released and recommended 43 wheat varieties for different agro-ecological domains, i.e., 26 for Terai and 17 for hills. But, 13 wheat varieties have been denotified and only 30 varieties are under cultivation "Appendix 1" (Official record at NWRP, January 2018). Moreover, it is reported that, NWRP has produced 23 metric tons of wheat breeder seeds and 4.17 metric tons

of nucleus seed of only 15 popular wheat varieties in 2015/16 due to shortage of land [23]. About 94% of wheat area was covered by modern varieties in the hills whereas it was 100% in the Terai plains [18]. Other studies at different districts indicate variation in coverage of modern varieties of wheat in Nepal which ranges from 65 to 95% in hilly districts and from 94 to 100% in different Terai districts [25, 31, 32, 35]. Gauchan [7] has reported more than 15 years age of most of commercially produced rice varieties in Nepal. Similarly, prevalence of older varieties with 12 years of adoption lags and 18 years of weighted varietal age in Nepal was reported by Gautam et al. [12] and Velasco et al. [36]. The adoption lags can be reduced to promote agricultural technologies using community-based organizations and private seed companies [1, 2, 24]. About 27% adoption gap of improved pigeon pea is reported in Malawi due to incomplete exposure to the improved pigeon pea varieties [26].

Vijay (BL 3063) is one of the recently released varieties in 2010 for Terai region in Nepal. It was the first Ug99-resistant improved wheat variety developed and released in Nepal [15]. This is also resistant to leaf rust and moderately susceptible to yellow rust and tolerant to spot blotch (HLB) and suitable for harvesting using combined thresher [8]. It has rapid grain filling trait under heat stress conditions having good bread and *chapati*-making quality [19]. A recent expert elicitation study conducted by NARC-CIMMYT study in 2014 indicated that Vijay is becoming one of the popular wheat varieties in Nepal [8]. Out of several varieties released and disseminated in Nepal in the last 50 years, Vijay is one of the top 5 most popular varieties adopted in Nepal with area coverage of 6% in 2014 (Table 1). NWRP has reported about 21% of estimated area covered by Vijay variety in Nepal based on the supply of breeder seed in 2014.

High-quality genetically pure breeder and foundation seeds are produced mainly by NARC research stations to use for multiplication of next generation of commercial seeds (certified and improved). Recently, private seed companies and some cooperatives are also authorized to

Table 1 Top five popular wheat varieties by agro-ecological domains and reasons for their popularity *Source*: **Gauchan and Dongol (2015); based on Expert Elicitation Workshop, Kathmandu (2014)**

S. no	Name of popular varieties	Domains or place name	% Area	Cause of popularity- or adopted attributes
1	Gautam	Terai	18.82	Resistant to foliar leaf blight, heat tolerant, stay green, wide adaptation, high yield
2	Nepal 297	Terai	17.40	Early maturity, good yielder even if later planted, terminal heat and hailstone tolerant, good chapatti quality
3	Bhrikuti	Terai	13.61	High yielding, drought tolerant, resistant to leaf and stripe rust
4	WK 1204	Mid hills	12.51	High yielder, Resistant to yellow rust, good bread quality
5	Vijay	Terai	5.80	Resistant to Ug99 rust, tolerant to terminal heat, suitable to harvest using combined harvester

produce source seeds (mainly foundation) for subsequent cycle of seed multiplication with close supervision and monitoring of seed certifying agencies. Source seed is the initial generation high-quality genetically pure (breeder and foundation) seed used to produce next generation commercial (e.g., certified, improved) seeds of the same variety [14]. About 44 tons of source seed of Vijay were produced by research stations in Nepal and private seed agencies in 2009–2010 wheat seasons before its release in 2010 September [3]. Nepal has put special emphasis for the seed production and diffusion of this variety to promote rapidly in the farmers' fields to mitigate the potential epidemics of newly emerging pathotype of stem rust (Ug99). This variety is also being promoted by several organizations and seed companies' after release. This variety has highly potential for commercialization. Most of the past studies focused on general types of study on wheat variety adoption and farm level impacts [27, 31, 32]. Therefore, this study was undertaken to assess the importance of seed system development for rapid up-scaling of new variety and identify the factors that are contributing to its commercialization.

Methods

Selection of study sites and sample

Command areas of the different research centers of Nepal Agricultural Research Council (NARC), viz. Bhairahawa (Rupandehi), Tarahara (Sunsari) and Nepalgunj (Banke), where Vijay variety was tested and being promoted by the NARC research stations and commodity programs were selected purposely for the study. Similarly, Sunsari, Morang, Rupandehi, Banke and Kailali districts were selected purposely representing Eastern, Western, Mid-western and Far-western Terai regions of Nepal as they represent the major wheat growing areas of the selected regions (Fig. 1). As this study focused on speed of seed flow and level of commercialization, agrovets (seed/input dealers) were selected as respondents rather than farmers. Seed companies, agro-vets (which

are input dealers dealing mainly with agricultural inputs such as seeds, fertilizers, pesticides and veterinary medicines in Nepal.) and cooperatives involved in Vijay seed multiplication and distribution in the command areas of above-mentioned offices of NARC were selected to track the seed flow. From the list of agro-vets in respective study sites, a total of 87 wheat seed seller agro-vets including national seed company dealers were selected randomly. In totality, 44% samples of target population in both categories were selected. Out of total samples, 40% were national seed company (NSC) dealer agro-vets and 60% agro-vets were non-dealer agro-vets. The summary of the site selection and sampling technique is presented in Table 2.

Techniques used in data collection and analysis

Desk reviews were undertaken to understand and document varietal development, release and registration process and policies, importance of Ug99 resistant variety and tracking source seed production and multiplication by institutions in Nepal. Different published and unpublished documents of NARC and other institutions were reviewed related to wheat varietal development and dissemination process (focusing on pre-release and after release seed multiplication), including institutional constraints to rapid seed flow in relation to Vijay variety. Institutional survey was conducted with different agro-vets and seed companies. Expert consultation meeting was done at each proposed districts with district agriculture development offices (DADO), plant breeders/researchers and policy makers at central level to know the issues and challenges associated with Vijay variety of wheat in Nepal. After collecting the data, data were compiled, reviewed and cleaned before final analysis for the accuracy of the results. This study used both descriptive as well as inferential statistics. Multivariate regression analysis was used to find out different factors responsible for commercialization of Vijay seed. The details of the model are given below:

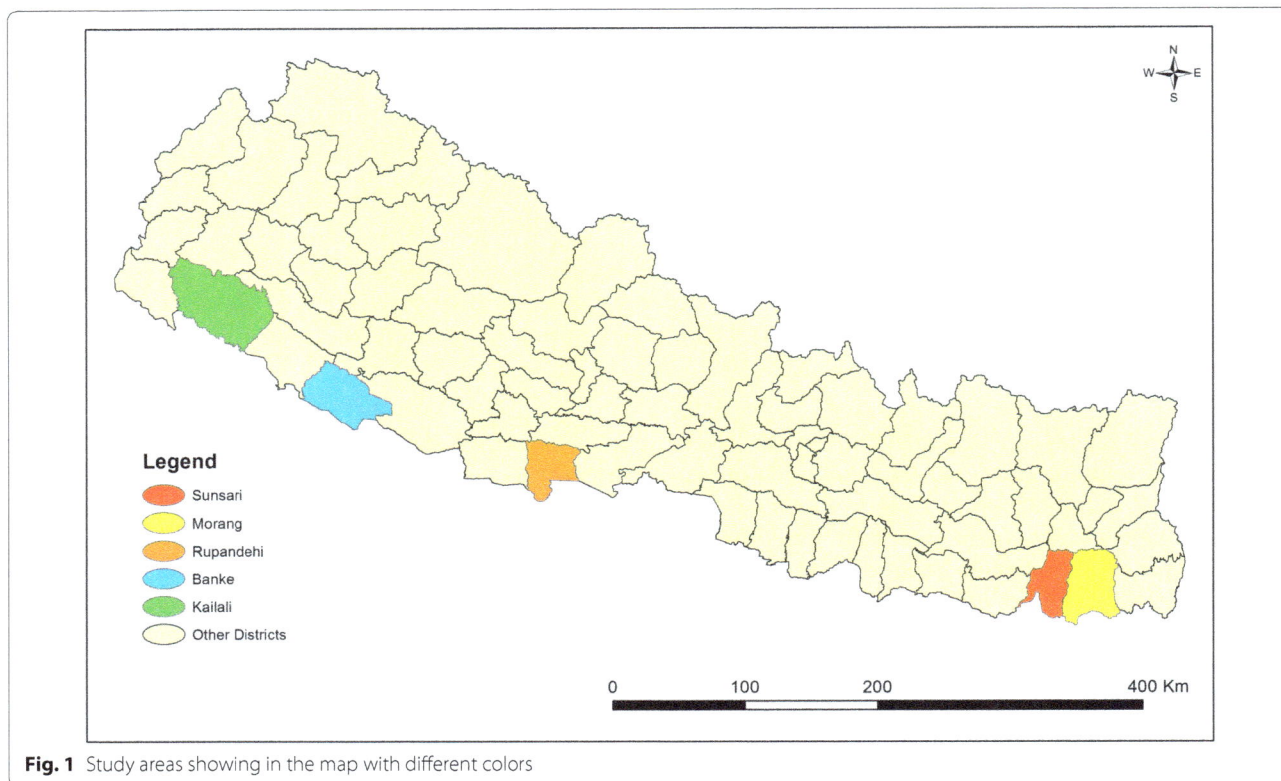

Fig. 1 Study areas showing in the map with different colors

Table 2 Details of NARC research stations, study area and sampling size and methods

Command area	Districts	Agro-vets				Seed companies	Professional	Methods
		Target population (no)		Sample size (no)				
		Wheat seller[a]	NSC dealer	Wheat seller[a]	NSC dealer			
Regional Agriculture Research Stations (RARs) Tarahara	Morang	15	16	7 (47)	6 (38)		1	Purposive and Simple random sampling
	Sunsari	15	10	11 (73)	8 (80)	1	2	Purposive and Simple random sampling
National wheat Research Program (NWRP)	Rupandehi	40	28	17 (43)	13 (46)	4	5	Purposive and Simple random sampling
Regional Agriculture Research Stations (RARs) Khajura	Banke	18	1	9 (50)	0 (0)	1	2	Purposive and Simple random sampling
	Kailali	35	19	10 (33)	6 (32)	1	2	Purposive and Simple random sampling
Total		123	74	54 (44)	33 (44)	7	12	

[a] Dealership holder Agro-vets were excluded. Total wheat seller agro-vets ($n = 54$) and dealers ($n = 33$). Professionals such as seed specialists, wheat scientists and planners were also interviewed in different research stations, DADO and Kathmandu at NARC, MOAD, DOA. Figure in parenthesis indicates sample percentage

$$TVS = \alpha + \beta_1 DNSC + \beta_2 DHQ + \beta_3 CVNL + \beta_4 TWS + \beta_5 AGP + \beta_6 DE + \beta_7 PT + \beta_8 DTV + \beta_9 PGV + \beta_{10} BT + \beta_{11} PSM$$

where TVS = Total qty. of Vijay variety sold (Own + NSC) in 2015/16 in kg as a proxy for level of up-scaling, DNSC = Dealer of National Seed Company (1 if yes, otherwise 0), DHQ = Distance from district headquarter (km), CVNL = Comparative performance of Vijay with Nepal 297 (1 if better than Nepal 297, otherwise 0), TWS = Total qty. of wheat seed (Own + NSC) sold in 2015/16 (kg), AGP = Age of the proprietor (year), DE = Duration of establishment (year), PT = participation on seed related training (1 if yes, 0 no), DTV = Duration of involvement in Vijay Trading (year), PGV = Perception on presence of required germination of Vijay (1 yes, 0 no), BT = Business Transaction (Total Qty. of cereal seed sold in 2014/15 (kg), PSM = Pre-release seed multiplication by private seed company (yes 1, 0 otherwise), α and $\beta_1 \ldots \beta_{11}$ = coefficient to be estimated.

Results and discussion

The details of socioeconomic and technical variables related to agro-vets surveyed are presented in Table 3. The education level of the agro-vets proprietor was more or less similar in the study area; on average, it was 11.82 years of education. Similar type of results was obtained in case of age of the proprietors. The wheat seller agro-vets were older (12.34 year) in RARS Tarahara area followed by NWRP area (9.37 year) and RARS Khajura area (9.32 year). The average transaction of seeds

and cereals seed in 2014/15 by each agro-vet was about 25.05 Mt and 22.45 Mt, respectively, in the study area. About 10.97 Mt of wheat seeds was sold by agro-vets in 2015/16, which includes both subsidized seed provided by government of Nepal (GoN) and non-subsidized seed. Average quantity of subsidized Vijay seed (4.28 Mt) sold by agro-vets in 2015/16 was higher than non-subsidized seed (2.79 Mt). This quantity was higher than the Vijay seed sold in the year 2014/15 in both categories. Majority of the agro-vet proprietors (90%) were received training related to seed production and marketing management. About 67% of the agro-vets were involved in Vijay seed trading, and their average time of involvement in Vijay trading was about 2.65 years. Seed company (SC) has been playing important role to create awareness about Vijay seed in the study area. About 85% of the agro-vets were selling Vijay seed in their own working single district, and they had mixed response about its availability on time, required quality and quantity. The Vijay variety is getting popular in the study area as more than 50% of the agro-vets had perceived that this variety performs better than other existing popular varieties such

Table 3 Description of socioeconomic and technical variables of sampled agro-vets *Source*: **Field Survey (2016); SC = Seed Company**

Description of variables	RARS Tarahara Area (N = 32)	RARS Khajura Area (N = 25)	NWRP Area (N = 30)	Overall (87)
Average qty. of total seed sold (Mt) in 2014/15	27.09	32.72	15.35	25.05
Average qty. of cereal seed sold (Mt) in 2014/15	26.68	32.26	14.42	24.45
Education level of respondents (Yrs.)	11.75	12.48	11.33	11.82
Year of Establishment of agro-vets (Yrs.)	12.34	9.32	9.37	10.45
Age of the respondents (Yrs.)	39.97	37.28	42.80	40.17
Avg. qty. of wheat seeds sold (Mt) in 2015/16	10.80	10.91	11.18	10.97
Avg. qty. of Vijay sold in 2015/16 (subsidized seed excluded)	.922	2.24	4.22	2.79
Avg. qty. of Vijay sold in 2015/16 (only subsidized seed)	2.22	8.50	1.60	4.28
Avg. qty. of Vijay sold in 2014/15 (subsidized seed excluded)	1.13	2.10	2.68	2.01
Avg. qty. of Vijay sold in 2014/15 (only subsidized seed)	2.80	5.80	1.67	2.34
% of Vijay selling on total wheat quantity (from	10.00	31.00	24.00	21.66
Participation on training related to seed (Yes %)	81.2	96.0	93.3	89.7
Time of involvement in Vijay seed trading (Yrs.)	2.74	2.48	2.74	2.65
Vijay seed selling status in 2015/16 (Yes %)	53.1	84.0	66.7	66.7
SC as a source of knowledge of Vijay (Yes %)	75.0	81.0	72.7	76.2
SC as a source of buying Vijay (Yes %)	50.0	95.0	50.0	65.5
Only one district as a selling destination of Vijay (Yes %)	70.0	86.7	94.7	86.4
Perception on availability of Vijay in required qty. (Yes %)	10.0	47.2	76.2	45.6
Perception on availability of Vijay on time (Yes %)	20.0	66.7	57.1	48.4
Perception on availability of Vijay in required quality (Yes %)	70.0	100.0	14.3	61.3
Perception on Vijay better than Gautam (Yes %)	55.0	62.5	84.2	67.3
Perception on Vijay better than Nepal 297 (Yes %)	30.0	71.4	57.9	53.3
Perception on Vijay better than Bhrikuti (Yes %)	–	100.0	63.2	72.0
Perception on Vijay better than Aditya (Yes %)	69.2	92.9	84.2	82.6

as Gautam, Bhrikuti, Aditya and Nepal 297. Gauchan and Timsina [11] reported cost–benefit analysis of Vijay (2.33) in comparison with existing popular varieties like Gautam (2.12), Bhrikuti (2.19) and Nepal 297 (2.6) indicates that it is economically at par or superior in some specific cases when it is grown in medium land type with irrigation. Other than economic benefits, farmers prefer Vijay due to its bold grain, attractiveness, natural maturity and good taste of bread/chapatti. Moreover, it performs well and high yielding in medium land as compared to low land [11, 29].

Commercialization prospective of Vijay seed

Public sector research organizations such as NWRP and other research centers produce breeder and foundation seeds, while NSC, seed companies, cooperatives and producer groups mainly produce and multiply certified or truthfully labeled seeds. In addition to Vijay seed multiplication, its distribution and commercialization are important aspects. So to find out the different factors that contribute to upscale, the Vijay seed marketed and factors determining extent of marketing was analyzed using ordinary least square (OLS) regression. Total qty. of Vijay variety sold (Own + NSC) in 2015/16 was used as dependent variable and different explanatory variables such as dealer of national seed company, distance from district head quarter (km), comparative performance of Vijay with Nepal 297, total qty. of wheat seed (Own + NSC) sold in 2015/16 (kg), age of the agro-vets proprietor (year), duration of agro-vets establishment (year), participation on seed related training (if yes 1), duration of involvement in Vijay seed trading (year), perception on presence of required germination of Vijay (if yes 1), total business transaction of the agro-vets (total qty. of seed sold in 2014/15 in kg) and pre-release seed multiplication by private seed company (if yes 1) were used as independent variables. To see the net effect of different variable, three models were run. Multicollinearity test was done among variable to increase the explanatory power of the model. The F test shows that all three models are best fitted. In our analysis, we used the standardized coefficients to see the real effect of variable.

Agro-vets having dealership of national seed company were used as independent variable in model 1, which is highly significant. The result shows 1 standard deviation (SD) increase in dealership of NSC, chances to increase in selling of Vijay seed by 0.426 SD compared to non-dealer agro-vets. This single variable covered the 18% variation of the model. In the last 3 years (2013/14–2015/16), there is increasing trend in the quantity of Vijay variety

allocated by NSC. It was 10% of total wheat variety in 2013/14 which was increased to 14% in 2015/16.

In the second model, we had used distance of agro-vets from district head quarter and comparative performance of Vijay with Nepal 297 as a control variable to determine the net effect of dealership. In this model, the effect of dealership on selling of Vijay seed is still significant; however, its explanatory power decreased by 0.11 SD. Moreover, negative relationship between distance of agro-vets from district head quarter and Vijay selling quantity is observed. Comparative performance of Vijay with Nepal 297 is also contributing significantly on selling of Vijay variety; it shows 1 SD increase in agro-vets who felt Vijay has better performance than Nepal 297, chance to increase in selling of Vijay Seed by 0.274 SD compared to agro-vets who felt worse performance of Vijay than Nepal 297. The second model is also best fitted, and the variation covered by the model is about 28%.

Out of 11 variables used in model three, only three variables, viz. comparative performance of Vijay with Nepal 297, total quantity of wheat sold and total business transaction of the agro-vets, are found significant (Table 4). The explanatory power of comparative performance of Vijay with Nepal 297 is reduced by 0.28 SD than model 2. The results show that 1 SD increase in quantity of total wheat seed sold by agro-vets response to increase in Vijay sold by 0.560 SD. Similarly, 1 SD increase in total quantity of seed sold by agro-vets, chance to decrease in quantity of Vijay sold by 0.365 SD. It means, among the wheat seeds selling, agro-vets have been giving priority for Vijay seed; however, agro-vets who had higher business transaction give lesser priority for Vijay. It may be due to higher profit margin in vegetables and hybrid rice seed compared to wheat seed [9].

Vijay seed multiplication and flow in the study area

The Vijay variety of wheat was released in 2010. However, the seed multiplication of this variety was started before its release. National Wheat Research Program (NWRP) of Nepal Agricultural Research Council (NARC) is producing only breeder seed of this variety. In totality, it had provided Vijay breeder seed to seven private seed companies and four different Regional/Agricultural Research Stations (R/ARS) of NARC few years before release of this variety. Among the different private seed companies, Kalika Seed Company from Rupandehi district had started to multiply 40 kg of breeder seed in 2007/8. It had produced foundation seed and then again multiply this foundation seed to C1 (Certified one) in own farm as well as in farmers field in contract production. In addition, different International Non-governmental organizations

Table 4 Factors contributing for Vijay seed up-scaling for its commercialization

Variables	Model 1		Model 2		Model 3	
	Standardized coefficients	P value	Standardized coefficients	P value	Standardized coefficients	P value
(Constant)	(1.858)	0.069	(.742)	0.461	(− 1.264)	0.213
Dealer of National Seed Company (1 yes, 0 No)	.426 (3.490)	0.001***	.415 (3.563)	0.001***	.116 (.764)	0.449
Distance of Agro-vets from DHQ (km)			− .162 (− 1.392)	0.170	− .029 (− .236)	0.814
Comparative performance of Vijay with Nepal 297(if Better than Nepal 297 1, otherwise 0)			.274 (.742)	0.022**	.246 (2.181)	0.034**
Total qty. of Wheat seed (Own +NSC) sold in 2015/16 (kg)					.560 (3.245)	0.002**
Age of the proprietor (Yr.)					.159 (1.258)	0.215
Duration of establishment (Yr.)					− .001 (− .010)	0.992
participation on seed related training (1 yes, 0 no)					.096 (.745)	0.460
Duration of involvement in Vijay Trading (Yr.)					− .032 (− .243)	0.809
Perception on presence of required germination of Vijay (1 yes, 0 no)					.004 (.037)	0.970
Business Transaction (total qty. of cereal seed sold in 2014/15 (kg)					− .365 (− 2.718)	0.009**
Pre-release seed multiplication by Private Seed Company (yes 1, 0 otherwise)					.070 (.401)	0.690
F value	12.18	0.001***	7.11	0.000***	4.39	0.000***
R square	.181		.287		.490	
Adjusted R square	.166		.247		.366	

*** and ** mean significant at 1% and 5% level of significance; Figure in parenthesis indicates the t value

(I/NGOs) and projects played important role to multiply this variety before its release. For example, out of 18, 486 kg seed produced by Kalika Private Seeds Company, about 11,986 kg seed was purchased by different organizations using USAID-Famine Seed Project funds for further pre-release seed multiplication in the 2009–2010 cycle [16]. Farmers were also actively participated in its multiplication through participatory varietal selection. NWRP had distributed about 3382 kg of breeder seed to different private seed companies and R/ARS for its multiplication before release of this variety.

After release of this variety, more private seed companies are attracted and different R/ARS of NARC have been involved to multiply this variety to commercialize it. Until 2016, about 15 private seed companies and 11 R/ARS of NARC have been involved to produce foundation seed of this variety. Moreover, national seed company and other I/NGOs have been involved to produce foundation seed to fulfill their program requirements. The C1 seed of Vijay variety has been produced through different modalities such as contract production of private seed companies with farmers, private Seed Companies' own farm, contract production of District Agriculture Development Offices (DADOs) with farmers group, community-based seed multiplication, contract production of agro-vets with farmer and production on national seed company's own farm "Appendix 2."

Until 2015/16, NWRP has produced 29 Mt of Vijay's breeder seed. One ton of breeder seed and 15 Mt of foundation seed is adequate to meet the 5% wheat area (36,000 ha) for Nepal if seed multiplication occurs regularly in the prescribed seed multiplication ratio (1:15) and prescribed seed cycle (breeder to foundation and foundation to certified seeds (C1 and C2) with proper planning and coordination [11]. Based on this assumption and availability of breeder seed, it is estimated that about 62 Mt foundation seed, 1710 Mt C1 seed and 21,498 Mt C2 seed available in the market which is more than enough to cover the total wheat area in Nepal. However, production of C2 seed was not in practice in the field even it is the component of seed cycle. Moreover, based on the estimated C1 seed in 2015/16, it must cover 56.88%

(11,943 ha) of total wheat area by Vijay variety in Terai "Appendix 3". It is found that Vijay wheat variety covers about 22% of the share on total wheat seed sold by agro-vets in 2015/16. An expert panel interview carried out for this study in the study areas expressed about 17% area covered by Vijay variety. Similarly, a national level expert elicitation conducted by CIMMYT-NARC collaborative study under Standing Panel Impact Assessment of the CGIAR (SIAC) in Nepal estimated 5.8% of the wheat area in Nepal in 2014 [8].

It shows inefficiency of the national R&D organization in the promotion and use of Vijay seed in the country. Similar result was reported by Ghimire et al. [13] and Timsina et al. [33, 34]. Little interaction and communication among various actors within the seed delivery chain resulted in a weak system that supplies less than 20% of seed requirement of farmers [4, 20]. Farmer seed networks can function efficiently in varietal diffusion [5]. Innovative platforms should be established in order to facilitate open communication and dialogue among all actors in the seed delivery system for its effective delivery [6]. Gauchan et al. [10] reported that well-planned strategies are required for further use of breeder's seeds and suggested to introduce mechanism of providing incentive and penalties for proper utilization of breeder to produce foundation seed by concerned agencies.

Proper planning and distribution of Vijay seed is required to create efficiency in resources use in the different research stations of NARC. If breeder seed is not available in all NARC stations, proper coordination and monitoring mechanism should be developed to produce breeder seed. It will also solve the current problem of land shortage to produce source seed in some of the research stations of NARC. Foundation seed production and supply should be decentralized to private actors such as private seed companies, national seed companies and community-based organization (CBOs) to maximize the resource use efficiency in the seed cycle. Therefore, the commitment of the public and private seed companies, community seed groups, cooperatives and public extension agencies to multiply seeds in subsequent cycles is required. The support for processing and storage, awareness about improved seed technologies and mechanization for foundation and certified seed is required at community level for community-based seed production [10]. SARPOD [28] has also emphasized to focus on production of breeder seed by NARC and foundation seed by private seed companies to capture their comparative

advantages and promote resource use efficiencies. In addition, public–private and community partnership in seed production and supply and developing network of diverse set of actors and institutions are required to meet the demand and supply of quality seed in required time. Gauchan [10] highlighted that strengthening breeding by public sector, seed multiplication by community sector and marketing by private sector are required for sustainable seed business in Nepal. The details of Vijay seed flow are presented in "Appendix 2."

Conclusions

Vijay is the first Ug99-resistant wheat variety developed and released in Nepal by NARC in 2010. However, the seed multiplication of this variety started before its release since 2007/8 through the involvement of private seed companies. Until 2010, about three Mt. of breeder and seven Mt. of Foundation seed had already been produced for its rapid dissemination before its release. About 67% of the agro-vets were involved in Vijay seed trading, and their average time of involvement was about 2.65 years. More than 50% of the agro-vets perceived that this variety performed better than other existing popular varieties such as Gautam, Bhrikuti, Aditya and Nepal 297. However, there is still mismatch between demand and supply of seed. Based on the availability of breeder seed of Vijay, it is estimated that about 62 Mt foundation seed, 1710 Mt. C1 seed and 21,498 Mt. C2 seed available in the market in 2015/16, which is more than enough to cover the total wheat area in Nepal. However, production of C2 seed was not in practice recently in the field even it is the component of seed cycle. Moreover, based on the estimated C1 seed in 2015/16, it must cover 56.88% (11,943 ha) of total wheat area by Vijay variety in Terai. But it has been found that Vijay seed shared about 22% on total wheat seed sold by agro-vets in 2015/16 and actual farm level use and adoption is estimated to be much lower. Regression analysis showed that presence of dealership of agro-vets with national seed company, relative performance of Vijay with Nepal 297, quantity of wheat seed sold by agro-vets were found to have significant positive contribution on Vijay seed sold. Similarly, total business transaction of the agro-vets was found to have significant negative contribution on Vijay seed selling. Among the wheat seeds, agro-vets has been giving priority to Vijay seed; however, agro-vets who had higher business transaction gave lesser priority to Vijay. It may be due to higher profit margin in vegetables and hybrid rice seed compared to wheat seed. In addition, other factors such as pre-release multiplication of this variety by private seed companies and participation on seed related training have found positive contribution on the commercialization of Vijay variety. This implies that future crop varietal improvement research program should focus on pre-release seed multiplication and fast-track release to rapidly promote commercialization of the crop varieties in the short period.

Moreover, from an analysis, we can conclude that the proper planning and distribution of Vijay seed is required to create resource use efficiency in the different research stations of NARC. NARC as a public research agency with expertise in plant breeding should focus only on breeder seed production, while foundation seed should be decentralized to private actors to increase resource use efficiency in the seed cycle. Therefore, close coordination and commitment from the private seed companies, national seed companies and community-based organization (CBOs) such as community seed groups, cooperatives and public extension agencies is required to multiply seeds in subsequent cycles. Furthermore, awareness program comparing good varietal traits of Vijay with other improved wheat varieties should be emphasized to accelerate the dissemination of Vijay seed rapidly in the country which would be an instrumental to mitigate the potential epidemics of newly emerging pathotype of stem rust (Ug99) in the future.

Abbreviations

CBOs: community-based organizations; DADO: District Agriculture Development Office; GoN: Government of Nepal; I/NGOs: International/Non-governmental organizations; NARC: Nepal Agricultural Research Council; NWRP: National Wheat Research Program; PSC: Private seed company; NSC: National seed company; FG: Farmers Group; RARS: Regional Agricultural Research Station; SARPOD: Socioeconomics and Agricultural Research Policy Division; SC: Seed company; SD: Standard deviation.

Authors' contributions

KPT was the lead investigator and the initiator of the study also responsible for literature search and write-up and finalization of the manuscript. YNG and DG were responsible for the overall study design and provided critical feedback on the manuscript. SS and SPA were responsible for filed study and drafting manuscript. All authors read and approved the final manuscript.

Author details

[1] Socioeconomics and Agricultural Research Policy Division (SARPOD), Nepal Agricultural Research Council (NARC), GPO 5459, Khumaltar, Lalitpur, Nepal. [2] Bioversity International Nepal Office, National Agriculture Genetic Resource Centre, Khumaltar, Lalitpur, Nepal. [3] Regional Agricultural Research Station (RARS, Nepalgunj), Nepal Agricultural Research Council (NARC), GPO 5459, Khumaltar, Lalitpur, Nepal. [4] Regional Agricultural Research Station (RARS, Tarahara), Nepal Agricultural Research Council (NARC), GPO 5459, Khumaltar, Lalitpur, Nepal.

Acknowledgements

We are grateful for the financial support from NARC for enabling the collection of data. We thank all respondents who were participated in the study. We would like to thank Mr. Jeevan Lamichhane for his suggestions.

Competing interests

The authors declare that they have no competing interests.

Funding

The funding was provided by Nepal Agricultural Research Council (NARC).

References

1. Abebaw D, Haile MG. The impact of cooperatives on agricultural technology adoption: empirical evidence from Ethiopia. Food Policy. 2003;38(1):82–91.
2. Ainembabazi JH, van Asten P, Vanlauwe B, Ouma E, Blomme G, Birachi EA, Nguezet PMD, Mignouna DB, Manyong VM. Improving the speed of adoption of agricultural technologies and farm performance through farmer groups: evidence from the Great Lakes region of Africa (Article). Agric Econ. 2017;48(2):241–59.
3. Bhatta MR, Joshi AK, Gautam NR, Pokharel DN. Accelerating the adoption of Ug99 resistant wheat cultivars in Nepal. Poster presented during 8th International Wheat Conference and BGRI-Workshop in St. Petersburg, Russia, May 30th–4th June; 2010.
4. Cleaver KM. A strategy to develop agriculture in Sub-Saharan Africa and a focus for the World Bank. World Bank technical paper-Africa technical department series 2013, Washington DC, USA. 1993. http://dx.doi.org/10.1596/0-8213-2420-9
5. David S, Louise S. Improving technology delivery mechanisms: lesson from bean seed systems research in East and Central Africa. Agric Hum Values. 1999;16:381–8.
6. Etwire E, Ariyawardana A, Mortlock MY. Seed delivery systems and farm characteristics influencing the improved seed uptake by smallholders in Northern Ghana. Sustain Agric Res. 2016;5(2):27–40.
7. Gauchan D. Pattern of adoption and farm level diffusion of modern rice varieties in Nepal. In: Bhandari DR, Khanal MP, Joshi BK, Acharya P, Ghimire KH, editors. Rice science and technology in Nepal (MN Paudel. Khumaltar: Published by Crop Development Directorate (CDD), Hariharbhawan and Agronomy Society of Nepal (ASoN); 2017. p. 639–44.
8. Gauchan D, Dongol D. Country report for wheat expert elicitation study, CIMMYT NARC Collaborative Study for Varietal Release and Adoption Data (2014). Strengthening Impact Assessment in CGIAR (SIAC) Project. Nepal Agricultural Research Council (NARC), Kathmandu, Nepal; 2015.
9. Gauchan D, Thapa Magar DB, Gautam S. Marketing of hybrid rice seed in Nepal: recent evidence from field survey. Socio- Economics and Agricultural Research Policy Division (SARPOD), Nepal Agricultural Research Council, Khumaltar, Lalitpur, Nepal; 2014.
10. Gauchan D, Singh S, Tripathi BP, Singh US. Policy workshop on development of Seed-Net in Nepal. A synthesis report of workshop on "Seed-Net Development" jointly held by IRRI and NARC on 23 December, 2011, Hotel Mountain, Kathmandu, Socioeconomics and Agricultural Research Policy Division, Nepal Agricultural Research Council, Nepal; 2014.
11. Gauchan D, Timsina KP. Tracking seed flow, varietal adoption and initial farm level impact of Improved UG99 Resistant Variety (BL 3063) in Nepal Terai. Technical report Submitted to CIMMYT, Kathmandu, Nepal; 2012.
12. Gautam S, Panta HK, Yelasco ML, Ghimire YN, Gauchan D, Pandey S. Tracking of improved rice varieties in Nepal. Country report for rice component of the project "Tracking of improved varieties in South Asia". Monograph. International Rice Research Institute, Philippines; 2013.
13. Ghimire YN, Pokharel TP, Khadka R. Lessons learned from the experiences of the scaling-up programme of the Nepal Agricultural Research Council. In: Proceedings of a workshop on uptake pathways and scaling up of agricultural technologies to enhance the livelihoods of Nepalese Farmers. Hill Agriculture Research Project (HARP); 2003. p. 137–144.
14. Government of Nepal (GoN). National seed vision (seed sector development strategy 2013–2025). Government of Nepal, Ministry of Agricultural Development, National seed Board, Seed Quality Control Center; 2013.
15. International Maize and Wheat Improvement Center (CIMMYT). Wheat-global alliance for improving food security and the livelihood of the resource-poor in the developing world. A Draft proposal Submitted by CIMMYT and ICARDA, to the CGIAR Consortium Board; 2010.

16. Joshi AK, Azab M, Mosaad M, Moselhy M, Osmanzai M, Gelalcha S, Bedada G, Bhatta MR, Hakim A, Malaker PK, Haque ME, Tiwari TP, Majid A, Jalal Kamali MR, Bishaw Z, Singh RP, Payne T, Braun HJ. Delivering rust resistant wheat to farmers: a step towards increased food security. Euphytica. 2010;179:187–96.

17. Ministry of Agriculture Development (MOAD). Statistical information on Nepalese Agriculture 2015/2016. Agri Statistics Section, Monitoring, Evaluation and Statistics Division, Ministry of Agricultural Development, Singha Durbar, Kathmandu, Nepal; 2017.

18. Ministry of Agriculture Development/Department of Agriculture (MOAD/DOA). *Krisi Diary*. Ministry of Agriculture and Cooperative/Department of Agriculture/Agriculture information and Communication centre, Hariharbhawan, Lalitpur; 2016.

19. National Wheat Research Program (NWRP). Annual report 2010/11. Published by NWRP, Bhairahawa, Rupandehi; 2011.

20. Niangado, O. Varietal developmental and seed system in West Africa: challenges and opportunities. Second Africa Rice Congress. Bamako, Mali, 2010. Retrieved from http://www.africarice.org/workshop/ARC/0p1%20 Niangado%20fin.pdf.

21. National Wheat Research Program (NWRP). Annual report 2008/09 to 2009/10. Published by NWRP, Bhairahawa, Rupandehi; 2010.

22. Nepal Agricultural Research Council (NARC); NARC's Strategic Vision for Agriculture Research (2011–2030); 2010.

23. Nepal Agricultural Research Council (NARC). Seed compilation report. Published by Monitoring and Evaluation Division, Nepal Agricultural Research Council, Singha Durbar Plaza, Kathmandu, Nepal; 2016.

24. Poolsawas S, Napasintuwong O. Farmer innovativeness and hybrid maize diffusion in Thailand. J Int Agric Ext Educ. 2013;20(2):50–63.

25. Shrestha HK, Manandhar HK, Regmi PP. Variety development cost versus variety adoption in major cereals in Nepal. Nepal J Sci Technol. 2012;13(1):7–15.

26. Simtowe F, Asfaw S, Abate T. Determinants of agricultural technology adoption under partial population awareness: the case of pigeon pea in Malawi. Agric Food Econ. 2016;4:7.

27. Socio-Economics and Agricultural Research Policy Division (SARPOD). Annual report (2011/12). Published by SARPOD, Khumaltar, Lalitpur, Nepal; 2012.

28. Socio-Economics and Agricultural Research Policy Division (SARPOD). Annual report (2012/13). Published by SARPOD, Khumaltar, Lalitpur, Nepal; 2013.

29. Socio- Economics and Agricultural Research Policy Division (SARPOD). Annual report (2014/15). Published by SARPOD, Khumaltar, Lalitpur, Nepal; 2015.

30. Thapa M, Acharya LP, Thapa B. Existing seed policies, seed regulatory frameworks and quality assurance systems in Nepal: ways forward. In: proceeding of the Fourth National Seed Seminar held from 19–20 June, 2008. Published by Government of Nepal, Ministry of Agriculture Cooperatives, National Seed Board, Hariharbhawan, Pulchowk, Lalitpur; 2008. p. 6l–76.

31. Timsina KP, Gairhe S, ThapaMagar DB, Ghimire YN, Gauchan D, Padhyoti Y. On farm research is a viable means of technology verification, dissemination and adoption: a case of wheat research in Nepal. Agron J Nepal. Published by Agronomy Society of Nepal (ASoN) and Crop Development Directorate (CDD), Department of Agriculture (DoA), Kathmandu. 2016; 4:9–24.

32. Timsina KP, Gairhe S, ThapaMagar DB, Ghimire YN, Gauchan D, Padhyoti Y. Effectiveness of on Farm research: Evidence of wheat research in Nepal. Government of Nepal, Nepal Agricultural research Council (NARC), Socio-economics and Agricultural Policy Research Division, Lalitpur, Nepal; 2016.

33. Timsina KP, Bastakoti RC, Shivakoti GP. Achieving strategic fits in onion seed supply chain in Nepal. J Agribus Dev Emerg Econ. 2016;6(2):127–49.

34. Timsina KP, Shivakoti GP, Bradford KJ. Supply situation of vegetable seeds in Nepal: an analysis from policy perspective. J Nepal Hortic. 2015;2015(10):26–36.

35. Timsina KP, Shrestha KP, Pandey S. Factors affecting adoption of new modern varieties of Rice in eastern Terai of Nepal. In: The Proceedings of 4th Society of Agricultural Scientist-Nepal (SAS-N) conference held at Lalitpur 4–6 April, 2012. Published by Nepal Agricultural Research Council (NARC) & Society of Agricultural Scientists (SAS-N), Nepal; 2012. p. 48–54.

36. Velasco ML, Tsusaka TW, Yamano T. Tracking of improved varieties in South Asia (TRIVSA). TRIVSA Synthesis Report on Rice. International Rice Research Institute (IRRI). Los Banos, Philippines; 2013.

## 3

Millets: a solution to agrarian and nutritional challenges

Ashwani Kumar[1,2]* , Vidisha Tomer[2], Amarjeet Kaur[1], Vikas Kumar[2] and Kritika Gupta[2]

Abstract

World is facing agrarian as well as nutritional challenges. Agricultural lands with irrigation facilities have been exploited to maximum, and hence we need to focus on dry lands to further increase grain production. Owing to low fertility, utilization of dry lands to produce sufficient quality grains is a big challenge. Millets as climate change compliant crops score highly over other grains like wheat and rice in terms of marginal growing conditions and high nutritional value. These nutri-cereals abode vitamins, minerals, essential fatty acids, phyto-chemicals and antioxidants that can help to eradicate the plethora of nutritional deficiency diseases. Millets cultivation can keep dry lands productive and ensure future food and nutritional security.

Keywords: Millets, Dry lands, Nutrition, Nutri-cereals, Micronutrient deficiency

Background

Progress in scientific knowledge and technological innovations have led mankind into yet another stage of modern civilization. Application of novel research strategies into fundamental and translational research has brought an all-round development. In agriculture, strategized technological innovations, viz. development and selection of high yielding variety, use of synthetic fertilizers and pesticides, mechanization and irrigation facilities, have resulted in sufficient availability of food. Estimated global cereal production was 2605 million tons in 2016 and was forecasted to be 2597 million tons in 2017 [1]. Several short-sighted measures have enhanced productivity but have undermined sustainability and are eroding the very capacity of resource base leading to nutrient deficient saline soil and lowering water beds. In addition, changing climatic conditions have further hastened the vulnerability of farmers towards declining crop production. Dry lands constitute 40% of the global land surface and are home for about 1/3rd of the global population. These low fertile soils are predicted to elevate up to 50–56% in 2100 AD, and 78% of dry land expansion is expected to occur in developing countries [2–4]. According to the report of World Bank, hunger is a challenge for 815 million people worldwide [5]. The spate of farmer's suicides in an agriculture-based country like India has reached to an average of 52 deaths/day, and reports of farmers selling their blood to earn a livelihood in drought-hit region of the country depict the severity of the agrarian crisis [6].

Sustainable crop substitutes are needed to meet the world hunger (cereal demand) and to improve income of farmers. Role of millets cannot be ignored for achieving sustainable means for nutritional security (Fig. 1). International crops research institute for the semi-arid tropics (ICRISAT) is focusing on increasing the productivity of millets and has included finger millet (*Eleucine corcana*) as sixth mandatory crop [3, 4]. Millets abode vital nutrients and the protein content of millets grains are considered to be equal or superior in comparison to wheat (*Triticum aestivum*), rice (*Oryza sativa*), maize (*Zea mays*) and sorghum (*Sorghum bicolor*) grains [7]. The role of millets in designing the modern foods like multigrain and gluten-free cereal products is well known [8]. Due to the richness of millets in polyphenols and other biological active compounds, they are also considered to impart role in lowering rate of fat absorption, slow release of

*Correspondence: ashwanichandel480@gmail.com
[1] Food Science and Technology, Punjab Agricultural University, Ludhiana, Punjab 141004, India
Full list of author information is available at the end of the article

Millets: an approach for sustainable agriculture and healthy world

Food Security	Nutritional Security	Safety from diseases	Economic security
• Sustainable food source for combating hunger in changing world climate • Resistant to climatic stress, pests and diseases	• Rich in micronutrients like calcium, iron, zinc, iodine etc. • Rich in bioactive compounds • Better amino acid profile	• Gluten free: a substitute for wheat in celiac diseases • Low GI: a good food for diabetic persons • Can help to combat cardiovascular diseases, anaemia, calcium deficiency etc.	• Climate resilient crop • Sustainable income source for farmers • Low investment needed for production • Value addition can lead to economic gains

Fig. 1 Benefits of millets in a nutshell

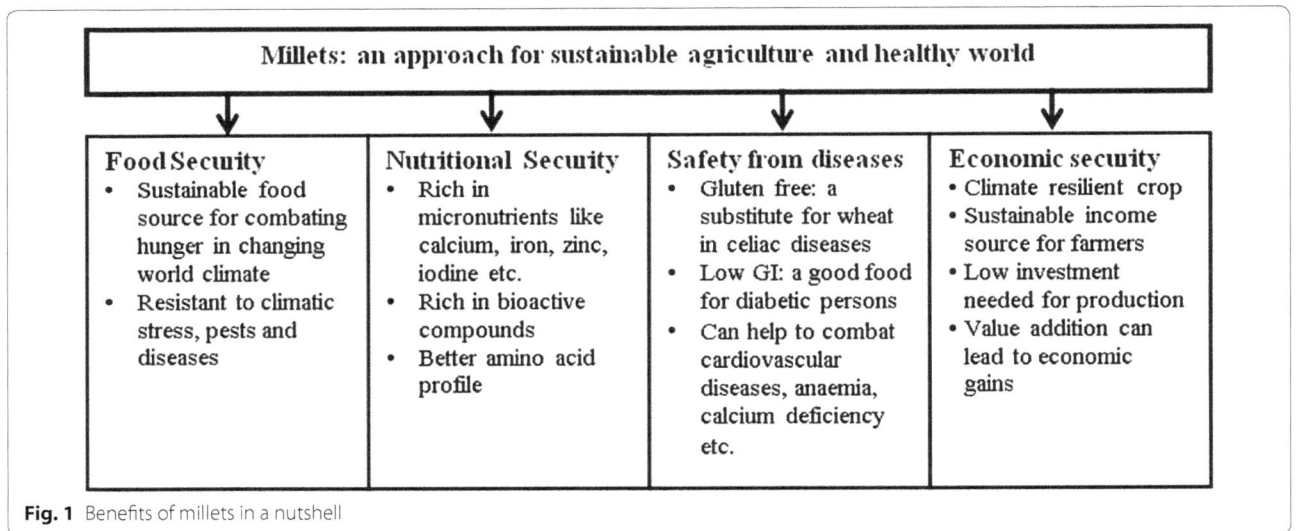

sugars (low glycaemic index) and thus reducing risk of heart disease, diabetes and high blood pressure. Due to increased awareness regarding the health promoting profile of millets, inclination towards their consumption has been observed. Present review envisages the agrarian requirements, nutritional information and health benefits imparted by these grains. Review also explores the millet-based products made traditionally along with the latest researches conducted worldwide.

Agrarian importance of millets

The demand of food will increase proportionately with growth in world population. At present about 50% of world's total calorie intake is derived directly from cereals [9]. Rice, wheat and maize have emerged as the major staple cereals with a lesser extent of sorghum and millets. Sharma [10] reported that an increase in the areas of crops with intense water requirements like rice, sugarcane (*Saccharum officinarum*) and cotton (*Gossypium*) has resulted in the increase in 0.009% in the distance between the ground level and ground water table and this loss is approximately equivalent to a loss of 7191 L of ground water per hectare. There is a lesser possibility of increasing the production of major staple cereals as the world is already facing the challenges of increase in dry lands and deepening of ground water level [3, 10]. According to the report of the National Rainfed Area Authority (NRAA) even after realizing the full irrigation potential, about half of the net sown area will continue to remain rainfed [11]. This alarms the need of shifting to the alternative of current cereal staples.

Millets cultivation can be a solution to this problem as these can grow on shallow, low fertile soils with a pH of soil ranging from acidic 4.5 to basic soils with pH of

8.0 [12]. Millets can be a good alternative to wheat especially on acidic soils. Rice is very sensitive to saline soils and has poor growth and yield on a soil having salinity higher than 3dS/m [13]. On the other hand, millets like pearl millet (*Pennisetum glaucum*) and finger millet can grow up to a soil salinity of 11–12 dS/m. Millets have a low water requirement both in terms of the growing period and overall water requirement during growth. The rainfall requirement of certain millets like pearl millet and proso millet (*Panicum miliaceum*) is as low as 20 cm, which is several folds lower than the rice, which requires an average rainfall of 120–140 cm [13]. Most of the millets mature in 60–90 days after sowing which makes them a water saving crop. Barnyard millet (*Echinochloa frumentacea*) has the least maturation time of 45–70 days among millets, which is half to the rice maturation (120–140 days) time [14]. Millets fall under the group of C4 cereals. C4 cereals take more carbon dioxide from the atmosphere and convert it to oxygen, have high efficiency of water use, require low input and hence are more environment friendly. Thus, millets can help to phase out climatic uncertainties, reducing atmospheric carbon dioxide, and can contribute in mitigating the climate change. The major millets and their growing conditions in comparison to the staple cereals, i.e. rice and wheat, are tabulated in Table 1.

Scientific interventions in terms of the use of molecular biomarkers, sequence information, creation of mapping populations and mutant have led to the development and release of high yielding varieties of millets throughout the world [22, 29]. Newly developed hybrids are resistant to diseases and has increased per hectare production as compared to their parent varieties [29, 30]. Millets have abundant natural diversity, and the release of new hybrids

Table 1 Optimum agrarian conditions for major cereals and millets

Crop	Scientific name	Optimum soil type	Height range	Temperature	pH	Soil salinity (dS/m)	Rainfall required	Maturity time (days)	References
Rice	*Oryza sativa*	Heavy to sandy loam	Sea level up to 2500 m	21–37 °C	6.5 to 8.5	Less than 3.0 dS/m	Range 100–300 cm Average 120–140 cm	100–160	[13]
Wheat	*Triticum aestivum* L.	Light clay or heavy loam	Sea level to 2500 m	Range 1.3–35 °C Average 15.5 °C	6.0 to 7.0	6.0 dS/m	Range 30–100 cm	90–125	[15, 16]
Sorghum	*Sorghum bicolor*	Clay loamy soils to shallow soils	Sea level to 3000 m	Range 7–30 °C Average 26–30 ℃	5.0–8.0	4–6 dS/m	40–100 cm	90–120	[17, 18]
Pearl Millet	*Pennisetum glacum*	Loamy soils, shallow soils, soils with clay, clay loam and sandy loam texture	Sea level to 2700 m	30–34 °C *can grow up to 46 °C	6.0–7.0 *can grow up to 8.0 pH	11–12 dS/m *yields are economically well up to ECe 8dS/m	20–60 cm	60–70	[19]
Finger millet	*Eleusine coracana*	Rich loam to poor upland shallow soils	Sea level to 2300 m	26–29 °C *lower productivity below 20 °C	4.5 to 7.5	11–12 dS/m	50–60 cm	90–120	[20]
Proso millet	*Panicum miliaceum*	Sandy loam, slightly acidic, saline, low fertility soils	1200–3500 m above sea level	20–30 °C	5.5 to 6.5	–	20–50 cm	60–90	[21]
Foxtail Millet	*Setaria italica* L.	Sandy to loamy soils	Sea level to 2000 m	Range 5–35 °C Average 16–25 °C	5.5–7.0	6 dS/m	30–70 cm	75–90	[22–24]
Barnyard Millet	*Echinochloa, E. frumentacea* (Indian barnyard millet) and *E. esculenta* (Japanese barnyard millet),	Medium to heavy soils	Sea level to 2000 m	Range 15–33 °C Average 27–33 °C	4.6–7.4	3–5 dS/m	–	45–70	[25, 26]
Kodo millet	*Paspalum scrobiculatum* L.	Fertile to marginal soils	Up to 1500 m	25–27 °C	–	–	800–1200 mm	100–140	[27]
Little Millet	*Panicum sumatrense*	–	Up to 2100 m	–	–	–	–	80–85	[27, 28]

increases this variation by several folds. For example, pearl millet has approximately 140 species or subspecies belonging to the genus *Pennisetum* [31] and further maintenance of the gene bank accessions has increased this number to 65,400. The primary global collection of pearl millet is at ICRISAT with 33% of the world's gene bank accessions. The largest gene accessions for finger millet, i.e. approximately 27% of the world's total 35,400 accessions, are maintained by Bureau for Plant Genetic Resources, India. Chinese Institute of Crop Germplasm Resources (ICGR) maintains world's 56% of the accessions of foxtail millet (*Setaria italica*), while National Institute of Agrobiological Sciences in Japan maintains the largest proso millet accessions collection with 33% of the world's approximately 17,600 genebank accessions [32]. In addition to the improvement in varieties, the advancements in the post-harvest operations of millets have eased their processing. In past, due to the lack of suitable machinery, traditional methods like pounding, winnowing, etc., were used for the decortication of millet grains. These methods were labour intensive, and hence, the production of edible millets was limited [33]. Millet-specific threshers, decorticator, destoners and polishers have been designed by intervention of government agencies as well as private companies. These recent developments in post-harvest operations of millets have eased their processing and have paved way for utilization of millets in the development of food products. The cultivation of millets can provide an overall solution to the existing agrarian challenges and can prove a milestone in achieving United Nations commitment to end malnutrition in all its forms by 2030 [34].

Nutritional importance

World is in the clinch of several health disorders and chronical diseases. As per 2016 Global Nutrition report, 44% population of 129 countries (countries with available data) experience very serious levels of undernutrition, adult overweight and obesity [35]. A nutrient imbalanced diet is responsible for most of these diseases. According to the estimates of United Nations Food and Agriculture Organization, about 795 million people (10.9% of world population in 2015) were reported undernourished. While on the other hand more than 1.9 billion (39% of world's population) adults ≥ 18 years of age were overweight and further 13% were reported to be obese [36, 37]. The average body mass index (BMI) of the world's population was reported to be 24 kg/m^2 in 2014 which is above the WHO standards for optimum health (21 to 23 kg/m^2) [38]. Obesity-related complications like cardiovascular diseases and diabetes have already been declared as epidemic by the world health organization. India is the home of world's largest undernourished

population. About 194.6 million people, i.e. 15.2% of total population of India, are undernourished. According to the 2017 Global Hunger Index report, India ranked 100th among 119 countries. The score of India is even poorer than Nepal, Sri Lanka and Bangladesh [39]. Protein energy malnutrition (PEM) was reported to result in 4,69,000 deaths with 84,000 deaths from the deficiency of other vital nutrients such as iron, iodine and vitamin A [40]. Obesity is also a major health concern in India with the prevalence rate of 11% in men and 15% in women. Status of malnutrition in world and India is presented in Table 2. Millets secure sixth position in terms of world agricultural production of cereal grains and are still a staple food in many regions of world. These are rich source of many vital nutrients and hence, promise an additional advantage for combating nutrient deficiencies in the third world countries.

Macronutrients

Millets are nutritionally similar or superior to major cereal grains. The additional benefits of the millets like gluten-free proteins, high fibre content, low glycaemic index and richness in bioactive compounds made them a suitable health food [27]. The average carbohydrates content of millets varies from 56.88 to 72.97 g/100 g. Least carbohydrate content has been reported in barnyard millet [8, 46]. The protein content of all the millets is comparable to each other with an average protein content of 10 to 11%, except finger millet, which has been reported to contain protein in the range of 4.76 to 11.70 g/100 g in different studies [47–49]. Finger millet protein is rich in essential amino acids like methionine, valine and lysine, and of the total amino acids present, 44.7% are essential amino acids [50]. This content is higher than the required 33.9% essential amino acids in FAO reference protein [51]. The mean value of protein reported from different studies depicts that proso millet has the highest protein content among millets (Table 3). The protein present in proso millet is comparable to wheat, but the amount of essential amino acids like leucine, isoleucine and thiamine is much higher in proso millet. The lipid content of millets as a group is comparable to that of wheat and rice (2.0% in wheat and 2.7% in rice) and ranges from 1.43 to 6 g/100 g. Among millets, the least lipid content has been reported in finger millet while the highest lipid content has been reported in pearl millet [46, 49, 52]. Millets are richest source of fibres, i.e. crude fibre as well as dietary fibre. Barnyard millet is the richest source of crude fibre with an average content of 12.8 g/100 g [8]. The highest dietary fibre content, i.e. 38% and 37%, has been reported to be in little millet (*Panicum sumatrense*) and kodo millet (*Paspalum scrobiculatum*), respectively. This content is 785% higher than rice and wheat; this make millets

Table 2 Status of malnutrition in world and India

	Agencies/studies	References
World		
2 billion people suffer from micronutrient malnutrition 800 million people suffer from calorie deficiency 2 million adults are overweight One in 12 adults has type 2 diabetes 159 million children under age 5 are stunted (too short for their age) 50 million children under age 5 are wasted (Less weight for their height) 41 million children under age 5 are overweight (More weight for their height)	United Nations International Children's Emergency Fund, International Food Policy Research Institute	[34, 35]
India		
60 million children underweight (highest in world) 30% low birth weight babies 75% pre-school children suffer from iron deficiency anaemia 85% districts have endemic iodine deficiency	World bank	[41]
35.7% of children under five are underweight; 58.4% of children between 6 and 59 months are anaemic; 53% of (non-pregnant) women are anaemic;	National Family Health Survey 2015–2016	[42]
Global hunger index score = 31.4 (serious hunger situation) 21% of children in India suffers from wasting Ranked 34th among leading countries with a serious hunger situation Ranked third behind only Afghanistan and Pakistan (In south Asia)	Global hunger index 2017	[39]
17.3% stunted 15.1% wasted 29.4% children underweight 9.4% severely underweight	Rapid Survey on Children 2013–2014 (subjects—pre-school children)	[43]
51 million people suffer from diabetes which is expected to increase to 79.4 million by 2030 (the increasing consumption of highly polished rice grains and decreasing consumption of coarse cereals contributes to this trend)	Kaveeshvar and Cornwall	[44]
18.5% children overweight 5.3% children obese	Misra et al.	[45]

Table 3 Proximate composition of different millets (per 100 g) [7, 8, 46, 48, 57–59]

Millet type	Carbohydrates	Protein	Fat	Crude fibre	Ash	Calorific value
Rice	82.86 ± 7.53	4.99 ± 1.38	1.90 ± 1.03	1.63 ± 0.42	0.99 ± 0.42	369 ± 27.82
Wheat	69.88 ± 1.66	13.78 ± 1.40	2.81 ± 0.18	1.77 ± 0.15	1.63 ± 0.26	438 ± 1.75
Sorghum	72.97 ± 2.25	10.82 ± 2.45	3.23 ± 1.60	1.97 ± 0.35	1.70 ± 0.66	329.0
Pearl millet	69.10 ± 1.52	11.4 ± 0.8	4.87 ± 0.12	2.0 ± 0.55	2.13 ± 0.21	363.0
Foxtail millet	67.30 ± 5.70	11.34 ± 0.91	3.33 ± 0.76	8.23 ± 1.66	3.37 ± 0.12	352 ± 1.41
Finger millet	71.52 ± 3.59	7.44 ± 0.87	1.43 ± 0.12	3.60	2.63 ± 0.06	334 ± 2.82
Barnyard millet	56.88 ± 6.86	10.76 ± 1.11	3.53 ± 1.19	12.8 ± 2.4	4.30 ± 0.26	300.0
Proso millet	67.09 ± 4.79	11.74 ± 0.86	3.09 ± 1.18	8.47 ± 3.4	2.73 ± 0.72	352.5 ± 1.62
Kodo millet	63.82 ± 7.94	9.94 ± 1.6	3.03 ± 1.03	8.20 ± 2.3	2.83 ± 0.40	349.5 ± 4.95

low glycaemic foods and hence a good choice for diabetic patients. In vitro studies of the soluble polysaccharides of finger millet (arabinose and xylose mainly) have proved them to be potent prebiotics and also possess wound dressing potential [53, 54]. This resistant starch contributes towards dietary fibre, which acts as a prebiotic and hence enhances the health benefits of the millets [55]. Resistant starch also helps in the production of desirable metabolites such as short-chain fatty acids in the colon, especially butyrate, which helps to stabilize colonic cell proliferation as a preventive mechanism for colon cancer [56]. Table 3 gives the mean and standard deviation of the macronutrient content of millets as reported by various researchers.

Micronutrients

The minerals and vitamins are known as micronutrients as they are required in very small quantities. Minerals play an important role in the building of bones, clotting of blood, sending and receiving signals, keeping normal heart beat, cell energy production, transportation of oxygen, metabolize and synthesize fats and proteins, act as co-enzymes, provide immunity to the body and help nervous system work properly [60]. The mineral content in millets ranges from 1.7 to 4.3 g/100 g, which is several folds higher than the staple cereals like wheat (1.5%) and rice (0.6%). Calcium and iron deficiency is highly prevalent in India [61], and a large chunk of adult population is suffering from osteoporosis. Calcium content of finger millet is about eight times higher than wheat and being the richest source of calcium (348 mg/100 g) it has the ability to prevent osteoporosis. Barnyard millet and pearl millet are the rich source of iron, and their consumption can meet the iron requirement of pregnant women suffering from anaemia. The iron content of barnyard millet is 17.47 mg/100 g which is only 10 mg lower than the required daily value. Foxtail millet contains highest content of zinc (4.1 mg/100 g) among all millets and is also a good source of iron (2.7 mg/100 g) [57]. These nutrients, i.e. zinc and iron, play an important role in enhancing the immunity. Millets are also good source of β-carotene and B-vitamins especially riboflavin, niacin and folic acid. The thiamine and niacin content of millets is comparable to that of rice and wheat. The highest thiamine content in millets, i.e. 0.60 mg/100 g, is found in foxtail millet. Riboflavin content of the millets is several folds higher than the staple cereals, and barnyard millet (4.20 mg/100 g) has the highest content of riboflavin followed by foxtail

millet (1.65 mg/100 g) and pearl millet (1.48 mg/100 g). The detail of micronutrient content of millets has been discussed in Table 4. The incorporation of millets in the diet can help to eradicate nutritional deficiencies. Platel [62] has proposed for the use of millet flour as a vehicle for iron and zinc fortification in India.

Phenolic compounds

Phenolic compounds form a very large group of compounds containing the phenol functional group as a fundamental component. Conveniently, these may be classified into phenolic acids, flavonoids and tannins. Phenolic acids are further sub-classified as hydroxybenzoic acids, hydroxycinnamic acids, hydroxyphenylacetic acids and hydroxyphenylpropanoic acids. Chandrasekara and Shahidi [63] determined and characterized the free, hydrolyzed (esterified and etherified) and bound phenolic compounds in millets by HPLC–DAD-ESI-MS^n. The highest amounts of hydroxybenzoic acid derivatives (62.2 µg/g) and flavonoids (1896 µg/g) were found in the soluble fraction of finger millet. Little millet (173 µg/g) and foxtail millet (171 µg/g) had the highest amount of hydroxycinnamic acid and their derivate in soluble form. The highest contribution to the total phenol content is in the form of the insoluble bound phenolics attached to the cell wall. Flavonoids are more prevalent in free form. Major phenols identified and quantified by different researchers are given in Table 5. Millets phenols are reported to have antioxidant, anti-mutagenic, anti-oestrogenic, anti-inflammatory, antiviral effects and platelet aggregation inhibitory activity [64]. Total antioxidant capacity of finger, little, foxtail and proso millets is higher due to their high total carotenoid and tocopherol content which varied from 78 to 366 and 1.3 to 4.0 mg/100 g, respectively, in different millet varieties [65]. The beneficial effect of phenolics in diabetes is due to partial inhibition of amylase and α-glucosidase during enzymatic hydrolysis of complex carbohydrates and delays the

Table 4 Comparison of micronutrients content of millets with the staple cereals (mg/100 g) [7, 8, 46, 48, 57–59]

Cereal grain	Calcium	Iron	Phosphorus	Zinc	Thiamine	Niacin	Riboflavin
Rice	0.12 ± 0.07	1.25 ± 0.78	0.52 ± 0.02	0.5	0.50 ± 0.13	5.56 ± 1.76	0.06 ± 0.02
Wheat	43.41 ± 3.69	5.24 ± 0.80	357.74 ± 26.54	2.9	0.44 ± 0.05	4.31 ± 1.00	0.10 ± 0.01
Sorghum	35.23 ± 7.42	5.29 ± 1.28	266.30 ± 32.3	3.01 ± 0.89	0.28	5.19	0.05
Pearl millet	35 ± 8.9	10.3 ± 7.0	339	–	0.30 ± 0.1	1.11 ± 1.3	1.48 ± 1.9
Foxtail millet	31 ± 11	3.5 ± 1.2	300	60.6	0.60	0.55 ± 0.6	1.65 ± 2.2
Finger millet	348 ± 3.5	4.27 ± 0.6	250	36.6 ± 3.7	0.40 ± 0.1	0.80 ± 0.9	0.60 ± 0.7
Barnyard millet	18.33 ± 6.0	17.47 ± 2.0	–	57.45 ± 1.9	0.33	0.10	4.20
Proso millet	10 ± 3.5	2.2 ± 1.2	200	–	0.41	4.54	0.28
Kodo millet	32.33 ± 4.6	3.17 ± 1.3	300	32.7 ± 2.2	0.15	0.09	2.0

Table 5 Phenolic compound content (µg/g defatted meal) in different types of millets (Adapted from Chandrasekara and Shahidi [63]) (content of phenolic compounds in bound form)

Phenolic compound	Pearl	Finger	Proso	Foxtail	Barnyard[c]	Kodo
Hydroxybenzoic acid and derivatives						
Methyl vanillate	19.8	–	–	–	–	–
Protocatechuic acid	11.8[a],	23.1[a], 48.2	69.7	10.2	–	39.7
p-Hydroxybenzoic acid	22[a]	8.9[a], 1.7	55.4	14.6[a], 5.63	–	10.5
Vanillic	16.3[a], 7.08	15.2[a],	85.8	87.1[a], 22.1	–	40.1
Syringic	17.3[a]	7.7[a]	–	93.6[a]	–	–
Gentisic acid	96.3[a]	61.5[a]	–	21.5[a]	–	–
Hydroxycinnamic acid and derivatives						
Caffeic acid	21.3[a]	16.6[a], 11	–	10.6[a], 34	–	276
p-Coumaric acid	268.9[a], 53.5	36	1188	2133.7[a], 848	–	767
Trans-ferulic acid	637	331	332	631	–	1844
Cis-ferulic acid	81.5	65.3	18.6	101	–	100
8,8'-Aryl ferulic acid	–	–	–	19.6	–	94.8
5,5'-Di ferulic acid	57	11.8	5.44	62.2	–	173
Flavonoids[b]	7.1	1896	1.9	169	–	179

[a] values are taken from Dykes and Rooney [65] (expressed as µg phenolic acid/mg samples)

[b] Content of phenolic compounds in soluble fraction of millet grains

[c] Data not available

absorption of glucose, which ultimately controls the postprandial blood glucose levels.

Other health benefits

Sireesha et al. [66] has demonstrated the anti-hyperglycaemic and anti-lipidemic activities of the aqueous extract of foxtail millet (*Setaria italica*) in streptozotocin-induced diabetic rats. In the study, they reported the dose of 300 mg of *Setaria italica* seed aqueous extract per kilo gram (kg) body weight produced a significant fall (70%) in blood glucose in diabetic rats after 6 h of administration of the extract. They also found lower levels of triglycerides, total LDL (low-density lipoproteins) and VLDL (very low-density lipoproteins) cholesterol and an increase in the levels of HDL (high-density lipoproteins) cholesterol in diabetic treated rats compared to those in diabetic untreated rats which demonstrates the hypolipidemic effect of aqueous extract. Choi et al. [67] studied the effect of dietary protein of Korean foxtail millet and found its importance in increasing insulin sensitivity and cholesterol metabolism. A remarkable reduction in insulin concentration of the rats fed with foxtail millet was demonstrated by this experiment. Lee et al. [68] investigated the effect of millet consumption on lipid levels and C-reactive protein concentration; it was found that hyperlipidemic rats fed with foxtail millet had decreased levels of triglycerides, which was in contrast to its previous researches [67]. Levels of C reactive protein, which is an indicator of inflammation, were also found to

decrease in foxtail millet fed rats. Aqueous and ethanolic extracts of kodo millet have been reported to produce a dose-dependent fall in fasting blood glucose [69]. Further millets are gluten free and might have anti-carcinogenic properties [65]. The health benefits of millets in a nutshell are given in Table 6.

Effects of processing on millets

Processing of millets decreases the anti-nutritional factors in millets and improves the bio-accessibility of nutrients. Many processing methods have been used traditionally like roasting/popping, soaking, germination and fermentation [80]. All these methods have been reported to have a significant impact on the nutritional value of the grain. Malting of millets improves access to nutrients and has been reported to increase the bio-accessibility of iron by 300% and of manganese by 17% [81]. The anti-nutritional factors decreased significantly with an increase in germination time due to hydrolytic activity of the enzyme phytase that increases during germination. The phytate content of millets can be reduced by germination as during the germination the hydrolysis of phytate phosphorus into inositol monophosphate takes place which contributes to the decrease in phytic acid. The tannins are also leached during soaking and germination of grains, and hence it results in the reduction in tannins [82, 83]. Boiling and pressure cooking also result in reduction in tannins. Fermentation is known to reduce the anti-nutritional factors and hence improves

Table 6 Benefits of millets in a nutshell

Disease	Functional factor	Mechanism of action	References
PEM	Optimum carbohydrate and high quality protein	Sustainable crop option in arid and semi-arid regions	[8]
Micronutrient deficiencies	High content of Iron, iodine, zinc, calcium, magnesium and other micronutrients compared to other cereals	Inclusion of millets in diet Bio-fortification of staple cereals	[70, 71]
Obesity	Dietary fibre	Controls release of carbohydrates Soluble fibre leads to highly viscous intestinal contents that possess gelling properties and could delay the intestinal absorption of carbohydrates Low glycaemic index	[72]
Diabetes	Dietary fibre	Slow glucose release and low glycaemic load	[72]
	Protein concentrates rich in antioxidants	Seed coat phenolics act as inhibitors which decrease postprandial hyperglycaemia by blocking the action of complex carbohydrate hydrolyzing enzymes (amylase, alpha-glucosidase); increase in adinopectin concentration may improve insulin sensitivity	[71]
Cardiovascular diseases	Protein concentrate of foxtail millet	Elevated levels of adinopectin which protects cardiovascular tissues by: (1) Inhibition of pro-inflammatory and hypertrophic response (2) Stimulation of endothelial cell responses	[67]
	Administration of proso/foxtail millet	Reducing plasma triglycerides, LDL through improved cholesterol metabolism Lower C reactive protein: a marker of inflammation and a stronger predictor of cardiovascular events in clinical applications	[68]
	Phenolic extracts from seven millet varieties (kodo, finger proso, foxtail, little and pearl millet	Kodo millet exhibited higher inhibition to lipid peroxidation, analogous to butylated hydroxyanisole at 200 ppm	[63]
Cancer	Phenolic extracts from seven millet varieties (kodo, finger proso, foxtail, little and pearl millet	Inhibition of lipid peroxidation in liposomes, singlet oxygen quenching and inhibition of DNA scission Millet extracts inhibited H-29 cell proliferation in the range of 28–100% after 4 days of administration	[63]
	35 kDa protein FMBP extracted from foxtail millet bran extract	FMBP, homologous to peroxidase suppress colon cancer cell growth through: (1) Induction of G1 phase arrest (2) Loss of mitochondrial trans-membrane potential resulting in caspase-dependent apoptosis in colon cancer cells	[73]
Inflammation and wound healing	Antioxidants: 50 g of finger millet per 100 g feed in diabetic and non-diabetic rats	Enhances dermal wound healing process in diabetes with oxidative stress-mediated modulation of inflammation	[74]
	Administration of proso/foxtail millet	Lower C reactive protein	[68]
Ageing	Antioxidant: Methanolic extract of finger millet	Inhibit glycation and cross-linking of collagen Scavange free radicals in protection against ageing	[69]
Anti-microbial activity	Protein extracts, polyphenols	Anti-fungal and antibacterial activity: active against Bacillus cereus, Aspergillus niger	[75]
	Seed coat phenolic extract	Loss of fungal functionality by: (1) Oxidation of microbial membranes and cell components by the free radicals (2) Inactivation of enzymes due to irreversible complex formation with nucleophilic amino acids (3) Complex formation of phenolic compounds with biopolymers such as proteins, polysaccharides and metal ions making them unavailable to micro-organisms	[76]
Ocular diseases and disorders	Polyphenols, flavanoids: Wistar rats maintained on 5% finger millet seed coat matter (SCM) for 6 weeks	(1) Direct scavenging of reactive oxygen species (ROS), anti-apoptotic activity, and phase 2 induction (2) Inhibiting nitric oxide (NO) production (3) Inhibiting certain enzymes responsible for the production of superoxide anions (xanthine oxidase and protein kinase C) (4) Prevents the accumulation of sorbitol by inhibiting aldose reductase by non-competitive inhibition and reduce the risk of diabetes-induced cataract diseases	[77, 78]
Coeliac Disease	Protein of all millets	Absence of gluten in millet protein prevents coeliac disease and related complications	[79]

the protein digestibility. Irradiation has also shown inhibitory effect against anti-nutrients, and it enhances the protein digestibility [84]. Extrusion cooking or high temperature short time (HTST) processing has been reported to reduce anti-nutrients like phytates, tannins and increase bioavailability of minerals [52].

Millet-based contemporary foods

Nutritional quality and drought-resistant properties of millets have drawn attention of various research agencies all over the world and have increased focus to improve the millet varieties and to enhance their use in processed food products. A schematic diagram for the preparation of composite foods from millets is shown in Fig. 2. Some of the research work carried on the utilization of millet crops is discussed in this section.

Cookies

Shadang and Jaganathan [85] formulated the bakery products like biscuits, cakes and cookies using foxtail millet, finger millet, proso millet and pearl millet added with wheat flour. For biscuit and cake, the ratios of 10:90, 20:80 and 30:70 were selected, whereas for cookies, the flours were used in the ratios of 15:85, 20:80 and 25:75, respectively. The sensory evaluation of their products revealed that the combinations of all the three levels were well acceptable for the three products. Rai et al. [86] utilized alternate flours/meals based on rice (*Oryza sativa*), maize (*Zea mays*), sorghum (*Sorghum vulgare*) and pearl millet (*Pennisetum glaucum*) for the preparation of gluten-free cookies. Their study revealed that the combination of pearl millet and sorghum flour had the maximum sensory scores followed by the cookies prepared from rice and sorghum, maize and pearl millet, rice and pearl millet and control cookies. Best pasting properties were obtained from blends of maize and pearl millet followed by pearl millet and sorghum flour. However, maximum yield was obtained in control (wheat) cookies, i.e. 186.8%, while cookies prepared from rice and maize had the highest spread ratio. The cookies prepared from blend of pearl millet and sorghum was nutritionally rich and had higher fat, protein, ash and calorific values.

Surekha et al. [87] prepared the barnyard millet flour-based cookies with three different variations, viz. plain, pulse and vegetable. Their research findings indicated that among the three treatments, pulse cookies (90% barnyard millet flour + 10% soybean and green gram flour) had the highest (85%) overall acceptability followed by vegetable cookies (90% barnyard millet flour + 10% dehydrated carrots) with 80% overall acceptability with the least acceptability of 73.33% plain barnyard millet varieties. These cookies had a significant increase in

macronutrient and micronutrient composition as compared to simple wheat flour-based cookies.

Bread

Ballolli et al. [88] prepared bread using varying concentrations of wheat flour and foxtail millet. It was found that wheat flour can be successfully replaced with foxtail millet flour up to 50% without significant effect on flavour and overall acceptability. However, the scores for colour, texture and appearance were reduced as compared to controlled sample. Addition of foxtail millet also resulted in a slight increase in the total protein and mineral content in comparison to the control bread.

Biscuits

Anju and Sarita [89] prepared biscuits using foxtail and barnyard millet. In the recipe, refined wheat flour was replaced to 45% with millet flour and all other ingredients like hydrogenated fat, eggs, baking powder and curd were same as the standard process for the biscuit making. The sensory evaluation of millet-based biscuits revealed a good overall acceptability and had a higher content of crude fibre, total ash and total dietary fibre as compared to refined wheat flour biscuits. Biscuits from foxtail millet flour had the lowest glycaemic index (GI) of 50.8 compared to 68 for biscuits from barnyard millet flour and refined wheat flour.

Pearl millet flour-based sweet, salty and cheese biscuits were prepared and reported by Sehgal and Kwatra [90]. Different blends containing pearl millet flour (40–80%), refined wheat flour (10–50%) and green gram flour (10%) were prepared. The sweet and salty biscuits prepared from refined wheat flour, blanched pearl millet and green gram were nutritionally sound as compared to biscuits prepared from wheat flour alone but had higher anti-nutrient (polyphenol and phytic acid) content.

Saha et al. [91] prepared biscuits from composite flour containing finger millet and wheat flour in the ratio of 60:40 and 70:30 (*w/w*). The hardness of biscuit dough was more in blend of 60:40 than 70:30 levels. An increase in adhesiveness and resistance of biscuit dough was found with the increasing levels of wheat flour. But the blend of 70:30 showed more breaking strength and expansion of biscuit after baking in comparison to blend of 60:40.

Snack foods

Dhumal et al. [92] developed potato and barnyard millet-based oil free, microwave puffed ready-to-eat fasting foods. Barnyard millet flour and potato mash, i.e. 50:50, 55:45 and 60:40, were prepared in three proportions and was steamed for 10, 15 and 20 min. Appropriate cold extrudates were obtained from mixture of barn-

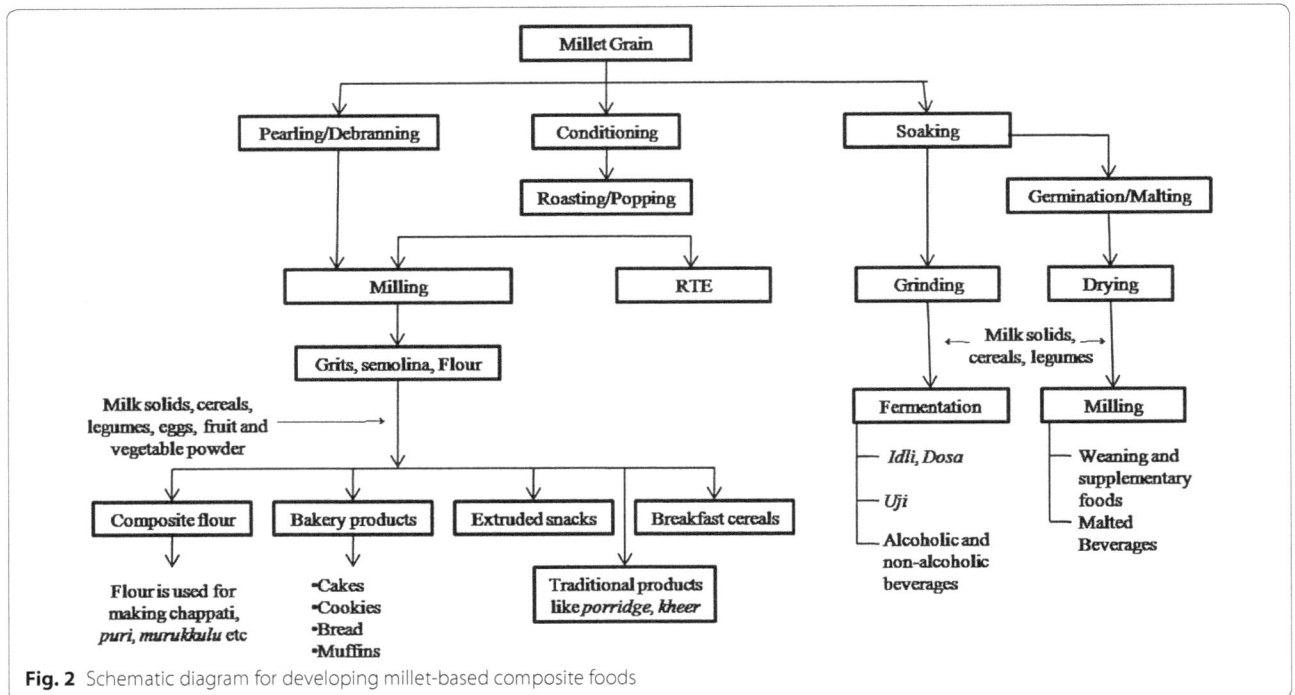

Fig. 2 Schematic diagram for developing millet-based composite foods

yard millet flour and potato mash (55:45) after steaming the dough rolled in 50 mm thickness in kitchen pressure cooker (1 kg/cm^2 pressure) for 15 min. The cold extrudates prepared after steaming for 10 min were very white while that prepared after steaming for 20 min were brown in colour.

A Barnyard millet-based ready-to-eat snack food was prepared by Jaybhaye and Srivastav [80], in which barnyard millet, potato mash and tapioca powder was used in the proportion of 60:37:3. The dough was formed into thin rectangular-shaped, steam-cooked cold extrudate samples and was puffed with HTST puffing process at optimum temperature and time (238 °C/39.35 s). The puffed product had a moisture content of 9% and an expansion ratio of 2.05. After puffing, the product was oven-toasted at 116.26 °C for 20–23 min.

Multigrain flour

Kamaraddi and Shanthakumar [93] prepared multigrain flour by incorporating various millet flours. They concluded that the substitution of wheat flour with 10–20% of millet flour was possible. They optimized 10% substitution of finger, foxtail and little millet. The proso millet can be replaced to a level of 15% and barnyard millet up to 20%. The further increase in millet content resulted in a lower gluten content, sedimentation value, loaf volume of dough and decreased content of proteins in some flours as compared to wheat flour. The addition of millet also changed the colour of crumb from creamish white

to dull brown. An increase in protein, ash and fat content was observed on addition of some millet flours.

Traditional millet-based products

Millets has been used for the purpose of food and feed from ancient times and has been a staple food particularly in diets of African and Asian people. These are consumed as flat bread, porridge, roasted and alcoholic and non-alcoholic beverages (Fig. 2). Millet porridge is a traditional food in Indian, Russian, German and Chinese cuisines. Millets are also used to replace commonly used cereals in local dishes like *idli, puttu, adai, dosa*, etc. Other traditional products like *baddis, halwa, burfi, papad* with added millet are also reported [68, 94, 95].

Non-alcoholic products
Appalu

Appalu is made from a mixture of pearl millet flour and Bengal gram flour. Spices like sesame seeds, carom seeds, chilli powder and salt are added and kneaded into dough. Then, the dough is divided into small balls and flattened into round shape. These are then fried and served hot.

Samaipayasam

The word *samai* means little millet while *payasam* means kheer. For preparation of *samaipayasam*, roasted groundnuts, fennel and jaggery are ground into a fine powder separately. Little millet is added to boiling water by constantly stirring. After the flour is stirred in, jaggery solution is added and the mixture is cooked for a few minutes on low flame. This dish is served hot. This recipe is also made with other millets instead of little millet [96].

Korramurukulu

This crispy savoury Indian snack is prepared from a mixture of foxtail millet flour, Bengal gram flour. To this, small amount of spices like cumin seeds, chilli powder, sesame seeds and salt are added and formed into stiff dough with the help of water. The dough is placed in the hand extruder and murukus extruded are deep-fried until these turn brown [97].

Alcoholic beverages
Sur

It is a finger millet-based (*Eleucine coracana*) fermented beverage mostly prepared in the Lug valley of Kullu; Bhangal, Luharti of Kangra district, Balh valley, Barot valley of district Mandi and regions of Sirmour, Himachal Pradesh, India [98, 99]. A mixture (inocula) of roasted barley and local herbs known as '*dhaeli*' is used to carry out fermentation. The millet flour is kneaded with water to make dough and left in a container for 7–8 days for natural fermentation. The fermented flour is half baked into *rotis*, cut to pieces and cooled. Then, the *roti* pieces and powdered *dhaeli* with sufficient amount of water and jaggery are put into smoke-treated earthen pots and allowed to ferment for 10 days by covering the pot. After the completion of fermentation, liquid is filtered and stored in specially designed earthen pots, sealed air tight from the top. The product has been reported to have 5–10% of alcohol [99].

Madua

Madua is among the most popular finger-millet-based beverage prepared in Arunachal Pradesh. The millet is roasted for 30 min followed by cooling and cooking until soft. The softened grains are mixed with starter culture and allowed to ferment in a perforated basket covered with *Ekam* leaves for 4–7 days. After completion of fermentation, hot water is poured from top and collected in a container. The collected liquid is known as *madua*. A good quality *madua* is golden yellow in colour, sweet in taste and has good alcoholic flavour. *Themsing, rakshi, mingri* and *lohpani* are other finger millet-based alcoholic beverages produced and consumed in Arunachal Pradesh, India [100].

Oshikundu

Oshikundu is a traditional cereal-based sour–sweet beverage of Namibia. It exits in both alcoholic and non-alcoholic form. It is brewed from pearl millet (*Pennisetum glaucum*) meal locally known as *mahangu*, malted sorghum (*Sorghum bicolor*), bran and water. Brewing of *oshikundu* is a household practice by rural women for their daily household consumption and for sale in the open markets in some towns of northern Namibia. The production process involves the addition of boiled water to *mahangu* meal, and the mixture is left to cool to room temperature with occasional stirring. Malted sorghum meal and bran are then added to the mixture. The step of bran addition is optional depending on the availability and preference of using bran in brewing. After the preparation of mixture, some amount of previously fermented *oshikundu* is added. The final mixture is then diluted with water depending on the amount of starting material used and desired volume of the final product. The mixture is then left to ferment at room temperature for an average one and half hour after which *oshikundu* is ready to drink. The alcohol is produced by the yeast fermentation of malt sorghum. It is a perishable beverage with a shelf life of less than 6 hours and is drunk on the same day [101].

Koozh

Koozh is another fermented beverage made with millet flour and rice and consumed mainly by ethnic communities in Tamil Nadu, India [102]. It is mainly prepared using finger millet (*Eleucine corcana*); however, use of pearl millet has been reported in other places. The preparatory steps of *koozh* involve two fermentation stages. The process starts with grinding of the millet to flour, mixing with subsequent water to make slurry and left this to ferment overnight. On the second day, broken rice (20% by weight of millet) is cooked in excess water, into which the overnight fermented millet slurry is mixed and cooked to make a thick porridge called *noyee*. The fermentation of this porridge for 24 h results in *kali*, a semisolid porridge to which the required amount of potable water was added (1:6 w/v) and hand-mixed with salt to prepare *koozh*.

Acceptability of millet-based products

The effect of addition of millets on the sensory acceptability of food products is scanty. Some researchers have reported the increased acceptability of the products on addition of millets, and literature on the decreased acceptability is also available. Florence et al. [103] reported high sensory acceptability in pearl millet-based cookies. Okpala et al. [104] reported a sensory acceptability of 7.1 on a scale of 9.0 points for 100% sorghum-based cookies. The acceptability was increased to 7.2 when a blend of cocoyam flour, fermented sorghum flour and germinated pigeon pea flour was used in the ratio of 66.6:16.7:16.7, respectively. In a study based on extruded products prepared from sorghum flour, corn flour, whey protein isolate and defatted soy flour, decreased accept-ability was reported with increased content of sorghum [105]. The use of millets as a blend with other cereals, pulses or legume has been reported to have an increase in overall acceptability of the product [104–106]. In addition to sensory aspects, the presence of anti-nutritional factors like phytic acid, tannins and phenols limits the use of millets as food [105]. High content of phytic acid was reported in the biscuits prepared using pearl millet [104]. Similar results have been reported by Mbithi-Mwikya et al. [50] in composite mix developed from unprocessed finger millet, kidney beans, peanuts and mango puree. The products were reported to be unfit for the infant consumption due to the presence of phytic acid, trypsin inhibitor and tannins content. However, the processing methods like roasting, malting, germination and soaking have been reported to reduce the anti-nutritional content [82, 105].

Conclusions

Millets can easily thrive in extreme conditions like drought, and some wild varieties can even prevail in flooded areas and swampy grounds. These have low gly-caemic index, abode gluten-free protein and are rich in minerals (calcium, iron, copper, magnesium, etc.), B-vita-mins and antioxidants. These extraordinary traits make them nutritious and climate change compliant crops. These can not only serve as an income crop for farmers but also improve the health of the community as a whole. Existing limitations, i.e. the presence of anti-nutritional factors and low sensory acceptability of millet-based products, can be overcome by the scientific interventions. The anti-nutritional factors can be inactivated by processing methods like cooking, roasting, germination and fermentation. The sensory acceptability of millet-based products can be enhanced by mixing millet flours with other flours of high acceptability and preparing composite foods. The use of millets in commercial/pack-aged food will encourage farmers to grow millets and will open new opportunities and revitalize the farmers. The inclusion of millet-based foods in international, national and state-level feeding programs will help to overcome the existing nutrient deficiencies of protein, calcium and iron in developing countries.

Authors' contributions
AK and VT carried out a major part of the literature review, drafted the manuscript and are equally first author. A Kaur co-authored, supervised the manuscript preparation and helped to finalize the manuscript. VK and KG carried out literature review for selected sections and helped to revise the manuscript. All authors read and approved the final manuscript.

Author details
[1] Food Science and Technology, Punjab Agricultural University, Ludhiana, Punjab 141004, India. [2] Food Technology and Nutrition, School of Agriculture, Lovely Professional University, Phagwara, Punjab 144411, India.

Acknowledgements
The authors are thankful to Department of Food Science and Technology, PAU, Ludhiana and Lovely Professional University for providing the necessary facili-ties, which were used for the preparation of manuscript.

Competing interests
The authors declare that they have no competing interests.

Funding
This research work has no funding.

References
1. FAO. World food situation; 2017. http://www.fao.org/worldfoodsitua-tion/csdb/en/. Accessed 25 Jul 2017.
2. Huang J, Haipeng YH, Xiaodan GX, Wang G, Guo R. Accelerated dryland expansion under climate change. Nat Clim Change. 2016;6:166–71.
3. ICRISAT. Small Millets. http://www.icrisat.org/homepage. Accessed 16 Apr 2017.
4. ICRISAT. ICRISAT adds finger millet as its 6th mandate crop; 2015. http://www.icrisat.org/newsroom/news-releases/nr-2015/ICRISAT-adds-finger-millet-6th-mandate-crop.pdf. Accessed 23 May 2017.
5. World Bank. Agriculture and Food; 2017. http://www.worldbank.org/en/topic/agriculture/overview. Accessed 22 Mar 2018.
6. Sharma D. More than make in India, Jaitley needs to focus on farm in India. The wire; 2016. https://thewire.in/22520/budgeting-for-agricul-ture-and-revitalising-the-economy/. Accessed 23 Jan 2018.
7. Kajuna. STAR MILLET: Post-harvest operations. Food and Agricultural Organization, United Nations; 2001. http://www.fao.org/3/a-av009e.pdf. Accessed 27 Oct 2017.
8. Saleh ASM, Zhang Q, Chen J, Shen Q. Millet grains: nutritional quality, processing, and potential health benefits. Compr Rev Food Sci Food Saf. 2013;12:281–95.
9. Awika JM. Major cereal grains production and use around the world. In: Advances in cereal science: implications to food processing and health promotion. American Chemical Society; 2011. p. 1–13.
10. Sharma CP. Overdraft in India's water banks: studying the effect of production of water intensive crops on ground water depletion. Master Thesis—Georgetown University, Washington DC; 2016.
11. National Rainfed Area Authority (NRRA); 2012. http://www.indiaenvi-ronmentportal.org.in/category/28905/publisher/national-rainfed-area-authority/. Accessed 4 Jan 2018.
12. Gangaiah. Agronomy—Kharif Crops Finger Millet; 2008. http://nsdl.nis-cair.res.in/jspui/bitstream/123456789/527/1/Millets%20(Sorghum,%20Pearl%20Millet,%20Finger%20Millet)%20-%20%20Formatted.pdf. Accessed 23 Mar 2018.
13. International Rice Research Institute (IRRI). Steps to successful rice pro-duction. Rice production manual, Los Banos (Phillipines); 2015. www.knowledgebank.irri.org. Accessed 28 Jun 2017.
14. Hulse JH, Laing EM, Pearson OE. Sorghum and the millets. New York: Their composition and nutritional value. Academic Press; 1980.
15. Food and Agricultural Organization (FAO). Crop salt tolerance data. Agricultural drainage management in arid and semi-arid crops; 1985. http://www.fao.org/docrep/005/y4263e/y4263e0e.htm. Accessed 23 May 2017.
16. Farmers portal; 2017. http://farmer.gov.in/imagedefault/pestanddiseas-escrops/wheat.pdf. Accessed 15 Jun 2017.
17. Dixon GE. Sorghum. In: Doggett H, editor. London: Longmans, Green; 1970. p. 403.
18. Fageria NK, Baligar VC, Jones CA. Growth and mineral nutrition of field crops. Boca Raton: CRC Press; 2010.
19. Hannaway DB, Larson C. Forage fact sheet: pearl millet (Pennisetum americanum). Corvallis: Oregon State University; 2004.
20. Satish L, Rathinapriya P, Rency SA, Ceasar SA, Prathibha M, Pandian S, Kumar R, Ramesh M. Effect of salinity stress on finger millet (Eleusine coracana (L.) Gaertn): histochemical and morphological analysis of coleoptile and coleorhizae. FLORA. 2016;222:111–20.
21. Habiyaremye C, Matanguihan JB, D'Alpoim Guedes J, Ganjyal GM, Whiteman MR, Kidwell KK, Murphy KM. Proso millet (Panicum miliaceum L.) and its potential for cultivation in the Pacific Northwest, US: a review. Front. Plant Sci. 2017;7:1961.

22. Brink M. *Setaria italica* (L.) P. Beauv. Record from Protabase. In: Brink M, Belay G, editors. PROTA (Plant Resources of Tropical Africa/Ressources-végétales de l'Afrique tropicale), Wageningen, Netherlands; 2006.

23. Hariprassana K. Foxtail millet, *Setaria italica* (L.) P. Beauv. In: Jagananth PV, editor. Millets and sorghum: biology and genetic improvement. Wiley: New York; 2017. p. 112–48.

24. Krishnamurthy L, Upadhyaya HD, Gowda CLL, Kashiwagi JR, Singh Purushothaman S, Vadez V. Large variation for salinity tolerance in the core collection of foxtail millet (*Setaria italica* L.) germplasm. Crop Pasture Sci. 2014;65(4):353–61.

25. Farrell W. Plant guide for billion-dollar grass (*Echinochloa frumentacea*), USDA-Natural Resources Conservation Service; 2011.

26. Mitchell WA. Japanese millet (*Echinochloa crus galli var. frumentacea*). Sect. 7.1.6, US Army Corps of Engineers Wildlife Resources Management Manual. Technical Report EL-89-13. Department of Defense Natural Resources Program. US Army Engineer Waterways Exp. Stat., Vicksburg, Missisipi; 1989.

27. Kannan SM, Thooyavathy RA, Kariyapa RT, Subramanian K, Vijayalakshmi K. Seed production techniques for cereals and millets. In: Vijayalakshmi K, editor. Seed node of the revitalizing rainfed agriculture network Centre for Indian knowledge systems (CIICS). 2013. p. 1–39. http://www.ciks.org/downloads/seeds/5.%20Seed%20Production%20Techniques%20for%20Cereals%20and%20Millets.pdf. Accessed 29 Dec 2017.

28. Dayakar RB, Bhaskarachary K, Arlene Christina GD, Sudha Devi G, Tonapi A. Nutritional and health benefits of millets. ICAR. Indian Institute of Millets Research (IIMR) Rajendranagar, Hyderabad, 2017; p. 112.

29. Joel A, Kumaravadivel N, Nirmalakumari A, Senthil N Mohanasundaram K, Raveendran T, Mallikavangamudi V. A high yielding Finger millet variety CO (Ra) 14. Madras Agric J. 2005;92:375–80.

30. ICAR- Indian Institute of Millets Research. Millets annual report 2016–17; 2017. http://millets.res.in/annual_report/ar16-17.pdf. Accessed 27 Mar 2018.

31. Recommended package of practices: pearl millet. http://millets.res.in/technologies/Recommended_package_of_practices-Pearl_millet.pdf. Accessed 26 Mar 2018.

32. State of diversity of major and minor crops (Appendix 4) http://www.fao.org/docrep/013/i1500e/i1500e14.pdf. Accessed 27 Mar 2018.

33. Balasubramanian S, Vishwanathan R, Sharma R. Post-harvest processing of millets: an appraisal. Agric Eng Today. 2007;31(2):18–23.

34. United Nations Children Fund. From promise to impact: ending malnutrition by 2030; 2016 https://data.unicef.org/wp-content/uploads/2016/06/130565-1.pdf. Accessed 24 Mar 2018.

35. IFPRI. Global Nutrition Report: Malnutrition Becoming the "New Normal" Across the Globe; 2016. https://www.ifpri.org/news-release/global-nutrition-report-malnutrition-becoming-%E2%80%9Cnew-normal%E2%80%9D-across-globe. Accessed 24 Mar 2018.

36. FAO. The State of Food Insecurity in the World 2015. Food and Agricultural Organization of the United Nations; 2015. http://www.fao.org/worldfoodsituation/csdb/en/. Accessed 15 Mar 2017.

37. WHO. Obesity and overweight Fact sheet N°311"; 2015. http://www.who.int/mediacentre/factsheets/fs311/en/. Accessed 23 Jan 2018.

38. WHO. Overweight and obesity. Global Health Observatory (GHO) data; 2016. http://www.who.int/gho/ncd/risk_factors/obesity_text/en/. Accessed 23 Jan 2018.

39. Von Grebmer K, Bernstein J, Hossain N, Brown T, Prasai N, Yohannes Y. 2017 global hunger index: The inequalities of hunger. International Food Policy Research Institute. 2017. http://www.globalhungerindex.org/pdf/en/2017.pdf. Accessed 21 Jan 2018.

40. Lozano R, Naghavi M, Foreman K. Global and regional mortality from 235 causes of death for 20 age groups in 1990 and 2010: a systematic analysis for the Global Burden of Disease Study 2010. Lancet. 2012;380:2095–128.

41. Gragnolatia M, Shekarb M, Gupta MC, Bredenkampd C, Lee Y. India's undernourished children: A call for reform and action, health, nutrition and population (HNP) discussion paper. Washington DC: World Bank; 2005.

42. National Family Health Survey. National Family Health Survey 2015–16: India. Mumbai, India: International Institute for Population Sciences; 2017. http://rch ips.org/NFHS/factsheet_NFHS-4.shtml. Accessed 28 Mar 2018.

43. GoI. Ministry of Women and Child Development Government of India Rapid Survey on Children 2013–2014 India Factsheet; 2015.

44. Kaveeshvar SA, Cornwall J. The current state of diabetes mellitus in India. Australas Med J. 2014 7(1):45–8.

45. Misra A, Shah P, Goel K, Hazra DK, Gupta R, Seth P, Tallikoti P, Mohan I, Bhargava R, Bajaj S, Madan J, Gulati S, Bhardwaj S, Sharma R, Gupta N, Pandey RM. The high burden of obesity and abdominal obesity in urban Indian schoolchildren: a multicentric study of 38,296 children. Annal Nutrit Metab. 2011;58(3):203–11.

46. Leder I. Sorghum and Millets. Cultivated plants, primarily as food sources. In: Gyargy F, editor. Encyclopedia of life support systems, UNESCO, Eolss Publishers, Oxford; 2004.

47. Baebeau WE, Hi u KW. Plant foods. Human Nutrition. 1993;43(2):97–104.

48. Panghal A, Khatkar BS, Singh U. Cereal proteins and their role in food industry. Indian Food Indus. 2006;25(5):58–62.

49. Singh P, Raghuvanshi SR. Finger millet for food and nutritional security. Afr J Food Sci. 2012;6:77–84.

50. Mbithi-Mwikya S, Ooghe W, Van Camp J, Nagundi D, Huyghebaert A. Amino acid profile after sprouting, Autoclaving and lactic acid fermentation of finger millet (*Elusine coracana*) and kidney beans (*Phaseolus vulgaris* L.). J Agric Food Chem. 2000;48:3081–5.

51. FAO. Amino Acid Scoring Pattern. In Protein quality evaluation, FAO/WHO Food and Nutrition Paper, Italy. 1991;12–24.

52. Nirmala M. Subbarao MVSST, Muralikrishna G. Carbohydrates and their degrading enzymes from native and malted finger millet (Ragi, *Eleusine corcana*, Indaf-15). Food Chem. 2000;69:175–80.

53. Manisseri C, Gudipati M. Prebiotic Activity of Purified Xylobiose obtained from *Ragi* (*Eleusine coracana*, Indaf-15) bran. Indian J Med Microbiol. 2012;52(2):251–7.

54. Mathanghi SK, Sudha K. Functional and phytochemical properties of finger millet (*Eleusine coracana* L.) for health. Int J Pharm Chem Biol Sci. 2012;2(4):431–8.

55. Shobana S, Malleshi NG. Preparation and functional properties of decorticated finger millet (*Eleusine coracana*). J Food Engg. 2007;79:529–38.

56. Englyst HN, Kingman SM, Cummings JH. Classification and measurement of nutritionally important starch fractions. Eur J Clin Nutr. 1992;46(2):33–50.

57. Chandel G, Kumar M, Dubey M, Kumar M. Nutritional properties of minor millets: neglected cereals with potentials to combat malnutrition. Curr Sci. 2014;107(7):1109–11.

58. Millet Network of India. Millets: future of food and farming; 2016. http://www.swaraj.org/shikshanter/millets.pdf2016. Accessed 3 March 2017.

59. Pontieri P, Troisi J, Fiore RD, Bean SR, Roemer E, Boffa A, Giudice AD, Pizzolante G, Al fano P, Giucice LD. Mineral contents in grains of seven food grade sorghum hybrids grown in mediterranean environment. J Crop Sci. 2014;8(11):1550–9.

60. Soetan KO, Olaiya CO, Oyewole OE. The importance of mineral elements for humans, domestic animals and plants—a review. Afr J Food Sci. 2010;4(5):200–22.

61. Aggarwal V, Seth A, Aneja S. Sharma B, Sonkar P, Singh S, Marwaha RK. Role of calcium deficiency in development of nutritional rickets in Indian children: a case control study. J Clin Endocrinol Metab. 2012;97(10):3461–6.

62. Platel K. Millet flours as a vehicle for fortification with iron and zinc. In: Preedy VR, Srirajaskanthan R, Patel VB, editors. Handbook of food fortification and health, eds. New York: Springer; 2013. p. 115–23.

63. Chandrasekara A, Shahidi F. Determination of antioxidant activity in free and hydrolyzed fractions of millet grains and characterization of their phenolic profiles by HPLC-DAD-ESI-MSn. J Funct Foods. 2011;3:144–58.

64. Devi PB, Vijayabharathi R, Sathyabama S, Malleshi NG, Priyadarisini VB. Health benefits of finger millet (*Eleusine coracana* L.) polyphenols and dietary fiber: a review. J Food Sci Technol. 2014;51(6):1021–40.

65. Dykes L, Rooney LW. Review sorghum and millet phenols and antioxidants. J Cereal Sci. 2006;44:236–51.

66. Sireesha Y, Kasetti RB, Nabi SA, Swapna S, Apparao C. Antihyperglycemic and hypolipidemic activities of Setaria italica seeds in STZ diabetic rats. Pathophysiology. 2011;18:159–64.

67. Choi YY, Osada Y. Ito, Nagasawa T, Choi MR, Nishizawa N. Effects of dietary protein of Korean foxtail millet on plasma adiponectin, HDL-cholesterol, and insulin levels in genetically type 2 diabetic mice. Biosci Biotechnol Biochem. 2005;69(1):31–7.

68. Lee SH, Chung IM, Cha YS, Park Y. Millet consumption decreased serum concentration of triglyceride and C-reactive protein but not oxidative status in hyperlipidemic rats. Nutri Res. 2010;30(4):290–6.

69. Hegde PS, Chandrakasan G, Chandra TS. Inhibition of collagen glycation and crosslinking in vitro by methanolic extracts of finger millet (Eleusine coracana) and kodo millet (Paspalum scrobiculatum). J Nutr Biochem. 2002;13:517–21.

70. Shashi BK, Sharan S, Hittalamani S, Shankar AG, Nagarathna TK. Micronutrient composition, antinutritional factors and bioaccessibility of iron in different finger millet (Eleusine coracana) genotypes. Karnataka J Agric Sci. 2007;20(3):583–5.

71. Shobana S, Sreerama YN, Malleshi NG. Composition and enzyme inhibitory properties of finger millet (Eleusine coracana L.) seed coat phenolics: mode of inhibition of a-glucosidase and a-amylase. Food Chem. 2009;115:1268–73.

72. Jenkins DJA, Jenkins MA, Wolever TMS, Taylor RH, Ghafari H. Slow release carbohydrate: mechanism of action of viscous fibers. J Clin Gastroenterol. 1986;1:237–41.

73. Shan S, Li Z, Newton IP, Zhao C, Li Z, Guo M. A novel protein extracted from foxtail millet bran displays anticarcinogenic effects in human colon cancer cells. Toxicol Lett. 2014;227(2):129–38.

74. Rajasekaran NS, Nithya M, Rose C, Chandra TS. The effect of finger millet feeding on the early responses during the process of wound healing in diabetic rats. Biochem Biophys Acta. 2004;1689:90–201.

75. Viswanath V, Urooj A, Malleshi NG. Evaluation of antioxidant and antimicrobial properties of finger millet polyphenols (Eleusine coracana). Food Chem. 2009;11:340–6.

76. Siwela M, Taylor JR, de Milliano WA, Dudu KG. Influence of phenolics in finger millet on grain and malt fungal load, and malt quality. Food Chem. 2010;121:443–9.

77. Chethan S. Finger millet (Eleusine coracana) seed polyphenols and their nutraceutical potential. Mysore: Thesis—Doctorate of Philosophy. University of Mysore; 2008.

78. Harsha MR, Platel K, Srinivasan K, Malleshi NG. Amelioration of hyperglycemia and its associated complications by finger millet (Eleusine coracana) seed coat matter in streptozotocin induced diabetic rats. Br J Nutr. 2010;104:1787–95.

79. Jnawali P, Kumar V, Tanwar B. Celiac disease: overview and considerations for development of gluten-free foods. Food Sci and Hum Wellness. 2016;5(4):169–76.

80. Jaybhaye RV, Srivastav PP. Development of barnyard millet ready-to-eat snack food: part II. Food Sci. 2015;6(2):285–91.

81. Platel K, Eipeson SW, Srinivasan K. Bioaccessible mineral content of malted finger millet (Eleusine coracana), Wheat (Triticum aestivum), and Barley (Hordeum vulgare). J Agric Food Chem. 2010;58:8100–3.

82. Handa V, Kumar V, Panghal A, Suri S, Kaur J. Effect of soaking and germination on physicochemical and functional attributes of horsegram flour. J Food Sci Technol. 2017;54(13):4229–39.

83. Hussain I, Uddin MB, Aziz MG. Optimization of anti-nutritional factors from germinated wheat and mung bean by response surface methodology. Food Res Int J. 2011;18:957–63.

84. Pushparaj FS, Urooj A. Influence of processing on dietary fiber, tannin and in vitro protein digestibility of pearl millet. Food Nutri Sci. 2011;2:895–900.

85. Shadang C, Jaganathan D. Development and standardisation of formulated baked products using millets. Int J Res Appl Nat Soc Sci. 2014;2:75–8.

86. Rai S, Kaur A, Singh B. Quality characteristics of gluten free cookies prepared from different flour combinations. J Food Sci Technol. 2014;51:785–9.

87. Surekha N, Ravikumar SN, Mythri S, Devi R. Barnyard Millet (Echinochloa Frumentacea Link) cookies: development, value addition, consumer acceptability, nutritional and shelf life evaluation. IOSR J Environ Sci Toxicol Food Technol. 2013;7:1–10.

88. Ballolli U, Malagi U, Yenagi N, Orsat V, Garipey Y. Development and quality evaluation of foxtail millet [Setaria italica (L.)] incorporated breads. Karnataka J Agric Sci. 2014;27(1):52–5.

89. Anju T, Sarita S. Suitability of foxtail millet (Setaria italica) and barnyard millet (Echinochloa frumentacea) for development of low glycemic index biscuits. Malays J Nutr. 2010;16(3):361–8.

90. Sehgal A, Kwatra A. Use of pearl millet and green gram flours in biscuits and their sensory and nutritional quality. J Food Sci Technol. 2007;44(5):536–8.

91. Saha S, Gupta A, Singh SRK, Bharti N, Singh KP, Mahajan V, Gupta HS. Compositional and varietal influence of finger millet flour on rheological properties of dough and quality of biscuit. LWT—J Food Sci Technol. 2010;44:616–21.

92. Dhumal CV, Pardeshi IL, Sutar PP, Jayabhaye RV. Development of potato and barnyard millet based ready to eat (RTE) fasting food. J Ready Eat Food. 2014;1(1):11–7.

93. Kamaraddi V, Shanthakumar G. Effect of incorporation of small millet flour to wheat flour on chemical, rheological and bread characteristics. In: Recent trends in millet processing and utilization. CCS Hisar Agricultural University. 2003. p. 74–81.

94. Malathi D, Thilagavathi T, Sindhumathi G. Traditional recipes from Banyard millet. Coimbtore: Postharvest Technology Centre, Agricultural Engineering College and Research Institute, Tamil Nadu Agricultural University; 2012.

95. Malathi D, Thilagavathi T, Sindhumathi G. Traditional recipes from kodo millet. Coimbtore: Postharvest Technology Centre, Agricultural Engineering College and Research Institute, Tamil Nadu Agricultural University; 2012.

96. Hema P. Samaipayasam/sama kheer/little millet porridge; 2017. https://www.mylittlemoppet.com/samai-payasam-sama-kheer-little-millet-porridge/. Accessed 20 Jan 2018.

97. Deccan Development Society. http://ddsindia.com/www/pdf/Millet%20Cook%20Book.pdf. Accessed 20 Jan 2018.

98. Joshi VK, Kumar A, Thakur NS. Technology of preparation and consumption pattern of traditional alcoholic beverage 'Sur' of Himachal Pradesh. Int J Food and Ferment Technol. 2015;5(1):75–82.

99. Kumar A. Refinement of the traditional sur production in Himachal Pradesh, Thesis—Master in Food Technology. Dr. YS Parmar University of Horticulture and Forestry, Nauni, Solan; 2013.

100. Shrivastava K, Greeshma AG, Shrivastava B. Biotechnology in action—a process technology of alcoholic beverages is practices by different tribes of Arunachal Pradesh, North East India. Indian J Trad Knowl. 2012;11:81–9.

101. Werner E, Youssef A, Kahaka G. Survey on indigenous knowledge and household processing methods of oshikundu; a cereal-based fermented beverage from Oshana, Oshikoto. Ohangwena and Omusati Regions in Namibia. MRC: University of Namibia, Windhoek, Namibia; 2012.

102. Ilango S, Antony U. Assessment of the microbiological quality of koozh, a fermented millet beverage. African J Microbiol Res. 2014;8(3):308–12.

103. Florence SP, Urooj A, Asha MR, Rajiv J. Sensory, physical and nutritional qualities of cookies prepared from pearl millet (Pennisetum typhoideum). J Food Process Tech. 2014;5(10):1.

104. Okpala L, Okoli E, Udensi E. Physico-chemical and sensory properties of cookies made from blends of germinated pigeon pea, fermented sorghum, and cocoyam flours. Food Sci Nutri. 2013;1(1):8–14.

105. Devi NL, Shobha S, Tang X, Shaur SA, Dogan H, Alavi S. Development of protein-rich sorghum-based expanded snacks using extrusion technology. Int J Food Prop. 2013;16(2):263–76.

106. Deshpande HW, Poshadri A. Physical and sensory characteristics of extruded snacks prepared from Foxtail millet based composite flours. Int Food Res J. 2011;18(2):751–6.

## 4

Small farmers' preferences for weather index insurance: insights from Kenya

Kenneth W. Sibiko[1]* , Prakashan C. Veettil[2] and Matin Qaim[3]

Abstract

Background: Smallholder farmers in developing countries are particularly vulnerable to climate shocks but often lack access to agricultural insurance. Weather index insurance (WII) could reduce some of the problems associated with traditional, indemnity-based insurance programs, but uptake has been lower than expected. One reason is that WII contracts are not yet sufficiently tailored to the needs and preferences of smallholder farmers. This study combines survey and choice-experimental data from Kenya to analyze the experience with an existing WII program and how specific changes in the contractual design might encourage uptake.

Results: Many smallholders struggle with fully understanding the functioning of the program, which undermines their confidence. Regular provision of relevant rainfall measurements and thresholds would significantly increase farmers' willingness to pay for WII. Mechanisms to reduce basis risk are also positively valued by farmers, although not to the same extent as higher levels of transparency. Finally, offering contracts to small groups rather than individual farmers could increase insurance uptake.

Conclusions: Better training on WII and regular communication are needed. Group contracts may help to reduce transaction costs. Farmer groups can also be important platforms for learning about complex innovations, including novel risk transfer products. These concrete results are specific to Kenya; however, they provide some broader policy-relevant insights into typical issues of WII in a small-farm context.

Keywords: Climate risk, Smallholder farmers, Crop insurance, Discrete choice experiment, Africa

Background

Climate change will affect agricultural production through higher mean temperatures and more frequent weather extremes [1, 2]. Higher variability in crop yields and food prices may increase poverty and food insecurity, especially in developing countries [3, 4]. Smallholder farmers, who make up a large share of the world's poor and undernourished people, could suffer the most [5]. Often located in the tropics and subtropics, smallholders are particularly vulnerable to climate shocks, and they are usually also ill-equipped to cope with risks [6]. After severe weather events, small farm households often end up selling productive assets to smooth consumption [7].

Frequent weather extremes are also associated with risk-avoidance strategies, such as low uptakes of productivity-enhancing inputs and technologies [8]. Thus, climate shocks can cause and perpetuate poverty traps in the small-farm sector. Agricultural insurance could help, but is literally non-existent in most developing countries due to institutional constraints, including high transaction costs and issues of moral hazard and adverse selection [9–11].

Weather index insurance (WII) is a relatively new type of financial risk transfer product, which could help to overcome some of the problems with traditional insurance schemes [12, 13]. Unlike indemnity-based crop insurance, where an insured farmer receives compensation for the verifiable loss at the end of the growing season, WII makes claim payments based on the realization of an objectively measured weather variable (e.g., rainfall) that is correlated with production losses [14, 15]. Neither the insured farmer nor the insurer can easily manipulate

*Correspondence: kenwaluse@gmail.com
[1] Department of Agricultural Economics and Rural Development, School of Agriculture and Food Security, Maseno University, P.O. Box Private Bag, Maseno, Kenya
Full list of author information is available at the end of the article

rainfall measurements, which reduces issues of information asymmetry. Moreover, instead of reducing effort to increase chances of compensation, farmers with WII actually have an incentive to make the best farming decisions [13]. In comparison with traditional insurance, WII is less expensive to administer, which can lead to more affordable contracts and faster payments to farmers, who often need the funds for timely planting in the subsequent season [16].

Despite these potential benefits, voluntary uptake of index insurance products is much lower than was initially anticipated [17]. Importantly, poor households, who are particularly risk-averse and could benefit most from novel micro-insurance products, were found to be hesitant in adopting WII, unless when premiums are subsidized or bundled with other benefits, such that insurance becomes quasi-compulsory [18, 19]. This mismatch between anticipated and actual demand among smallholder farmers is attributed to liquidity constraints during planting time, limited trust, and lack of insurance experience [20–22]. Others cite basis risk or the residual risk that often remains with the index insurance holder as a major issue [14, 23–26]. Several field experimental studies were undertaken to better understand farmers' insurance demand and its determinants [27–29]. However, farmers' preferences and willingness to pay for specific attributes of WII contracts have rarely been analyzed. Such knowledge could help to better adjust WII contracts and policies to the needs of smallholder farmers in different contexts. Here, we address this knowledge gap by using data from smallholder farmers in Kenya.

It would be interesting to observe how farmers actually respond to certain changes in the contractual design of a WII scheme. However, observational data with suitable variations in insurance contracts are not available. As an alternative, choice experiments can be conducted to analyze peoples' preferences for hypothetical contract features that are not (yet) observable in the market. A few studies used choice experiments to examine farmer attitudes toward WII in developed countries, such as Germany and Finland [30, 31]. Two recent studies applied this method to estimate farmers' willingness to pay for WII in Ethiopia and Bangladesh [32, 33]. The study by [32] found that apart from the premium charged and expected pay-outs, demand for WII was also determined by perceived frequency of droughts and the type of institutions involved in WII distribution. On the other hand, [33] analyzed gender disparities in preferences for WII. They showed that female farmers were more insurance averse, mainly due to differences in financial literacy, and the level of trust toward insurance providers. We add to this choice-experimental literature by analyzing more explicitly how farmers might react to changes

in WII contracts aimed at reducing typical issues in a smallholder context. In particular, we study possible mechanisms to reduce basis risk and to increase farmers' confidence in WII products.

A typical problem that contributes to low confidence in WII is that farmers often do not fully understand when exactly a payment is triggered [12, 14, 25]. Even when the rainfall threshold is clearly stated in the contract, this refers to a weather station located at some distance to the farm, so the insured farmer is usually not perfectly informed. A larger network of weather stations to decrease the mean distance to farms may be one mechanism to reduce basis risk. Another mechanism to improve confidence is regular communication of the weather data recorded at relevant stations. Transparent communication could also help to reduce farmers' distrust in the insurance provider. While some experimental evidence on the importance of trust in micro-insurance uptake exists [21, 33, 34], the specific influence of insurer transparency on WII demand has never been researched. We use contract features related to distance and regular communication in our choice experiment.

In addition, we analyze the possible role of insurance contracts with farmer groups instead of individual farmers. Group contracts are being proposed as a potential mechanism to increase WII uptake in the small-farm sector [10, 12, 35]. Farmer groups could influence demand for WII through several pathways. First, groups can help to reduce transaction costs. Second, groups can be efficient channels for disseminating information about innovative technologies and products [36, 37]. Third, and related to the previous point, groups may provide a learning platform that increases farmers' confidence in trying out unfamiliar insurance products [38]. Finally, farmer groups often involve networks that interact in various social dimensions and have norms on how to internalize idiosyncratic risks of their members [39]. Against this background, group WII contracts that help to mitigate covariate weather risks could have interesting complementary effects [10, 12, 35, 40]. Empirical evidence on the effect of group contracts on farmers' willingness to adopt WII is scarce. A few studies have confirmed a positive influence of informal risk-sharing networks [41, 42]. Others suggest that group dynamics and possible distrust toward other members might actually make group insurance less attractive than individual contracts [43, 44].

The purpose of this paper is to analyze farmers' preferences for WII, to estimate the average willingness to pay for proposed contract attributes, and to understand causes of heterogeneity in farmer preferences, so as to suggest ways of improving insurance uptake. Our analysis builds on a survey and choice experiment carried out with smallholder farmers in Kenya. Farmers in Kenya already had the opportunity to gain first-hand experience with WII contracts. Since 2009, the so-called *Kilimo Salama* Program has provided index-based crop insurance products in various parts of the country. We briefly describe this existing program in the next section, before presenting and discussing details of the methodological approach and results.

Weather index insurance in Kenya

Crop production in Kenya takes place mostly under rain-fed conditions, with weather fluctuations having a great impact on productivity [45]. Well-designed WII contracts could therefore be beneficial for development given such production uncertainties. Several pilot projects to introduce WII have been implemented with technical support from the World Bank and other development agencies. *Kilimo Salama*, which was launched by the Syngenta Foundation for Sustainable Agriculture, is the most widely known and successful out of these projects [46]. *Kilimo Salama* was started in 2009 as a small initiative with only 200 farmers. By 2013, the project covered close to 200,000 farmers in Kenya, Rwanda, and Tanzania, with a total sum insured of 12.3 million US dollars [47, 48]. While this growth within a few years is impressive, it cannot mask the fact that up till now only a small fraction of farmers has actually adopted WII. In 2014, *Kilimo Salama* transitioned into a commercial business under the new name 'Agriculture and Climate Risk Enterprise' (ACRE). In this study, we stick to the old name because this is better known in the literature.

Kilimo Salama offers rainfall index insurance products that cover farmers against drought and excess rain. As is common for weather-based insurance schemes, *Kilimo Salama* relies on data from automated weather stations to monitor local rainfall. Farmers are allowed to choose the station that best represents their farm conditions. Initially, the contracts were designed for maize and wheat, but more recently products for other crops have also been developed [47]. Contracts are sold for a crop season divided into three phases (early growth, flowering, and grain filling), which vary in duration and rainfall thresholds. Contracts are location-specific, and threshold (or strike) levels reflect the minimum agronomic requirements for normal plant growth during each particular phase. If the cumulative rainfall in a given phase falls below the threshold (for drought) or exceeds the threshold (for excess rain), a pay-out is triggered for all farmers holding a contract with reference to the particular weather station. The pay-out amount is calculated per millimeter of rainfall below (or above) the strike level and increases proportionally up to the maximum pay-out. However, as we learned through our survey, farmers are rarely aware of the exact details of the pay-out function,

even when they purchase an insurance contract. At the end of the contract period, the sum of triggered pay-outs over the three phases is sent to farmers through mobile money transfer. This is different from traditional indemnity-based crop insurance programs, where the insurer has to physically visit the farm to assess individual crop damage.

One important element for the smooth functioning of *Kilimo Salama* is the existence of a vibrant mobile money network (M-*PESA*) that facilitates farmers' access to various financial services [48, 49]. In many cases, farmers purchase WII linked to agricultural loans; in the event of unfavorable weather conditions, the insurer compensates the credit institution, which then writes off the loans of affected farmers. *Kilimo Salama* also offers input insurance through local input dealers. In that case, the insurance premium is included in the price of purchased inputs.

In 2011, *Kilimo Salama*-plus was launched, which offers the option to either only insure the cost of the inputs at a lower premium or the value of the output at a higher premium. Both options are offered through local input dealers on behalf of the insurer. The dealers have technical equipment to directly transmit purchase information to an administrative server, which also automatically triggers pay-outs to farmers via M-*PESA*. To keep our choice experiment simple and easy to understand for farmers, our hypothetical contracts build on the output-based insurance option, as is explained in more detail below.

Materials and methods
The farm survey

This study builds on data from a choice experiment and a socio-economic survey of farm households in Kenya. Primary data were collected in 2014 among smallholder farmers in Embu County, Kenya. Embu was chosen because WII initiatives have been implemented in that area for more than 5 years [50]. This ensured farmers' familiarity with this type of insurance. Farmers in Embu are predominantly small-scale, and uncertainty about the timing and amount of rainfall is a serious issue in this part of Kenya [51].

The farm households to be surveyed were selected using a stratified sampling procedure. At first, we purposively selected Embu-East sub-county, which had a relatively high number of farmers insured under *Kilimo Salama*. However, even in Embu-East insurance coverage was below 10%. Embu-East has two administrative divisions (Kyeni and Runyenjes); within each division, we randomly selected three sub-locations (smallest administrative units). In each of the six sub-locations, we interviewed all farmers that were insured at the time of the

Table 1 Attributes of WII contracts used in the choice experiment

Attribute	Attribute levels					
Premium rate	2%	5%	7%	10%	15%	20%
Strike level	− 10%	− 20%	− 40%	±10%	±20%	±40%
Distance to weather station	5 km (ward radius)		20 km (district radius)		50 km (county radius)	
Transparency	Weekly text messages and radio broadcast of recorded rainfall		No text message or radio broadcast of recorded rainfall			
Contracted party	Individual farmer		Small group (10 farmers)		Large group (100 farmers)	

survey or had purchased an insurance contract in previous years. These farmers were identified through lists provided by *Kilimo Salama* field staff. Overall, we surveyed 152 "ever-insured" farmers. In addition, we randomly selected 234 non-insured farmers in the same six sub-locations, resulting in a total sample size of 386.[1] While we deliberately over-sampled insured farmers, the two sub-samples are representative for "ever-insured" and non-insured farmers in Embu-East.

The survey involved face-to-face interviews, which were administered with the help of a small team of local enumerators. The enumerators were students from Egerton University that we hired and trained for this research. The survey instrument included a structured questionnaire to capture socio-economic data at farm and household level, including risk preferences, past experiences with weather shocks, and attitudes toward the existing WII contracts. In addition, each sample farmer participated in a carefully designed choice experiment. In this choice experiment, farmers were asked to make selections between various hypothetical WII insurance options to better understand possible responses to contract changes. Details of the choice experiment are explained in the following.

Discrete choice experiment

We developed and used a discrete choice experiment (DCE) to evaluate subjective preferences of farmers for WII contracts. In particular, we want to assess how farmers value specific contract attributes and trade-off between different attribute levels, which is not possible with other common preference elicitation methods such as contingent valuation [52]. The theoretical basis for DCEs is Lancaster's consumer choice theory, which postulates that an individual derives utility from the different attributes of a good [54]. DCEs are also consistent

with random utility theory, which suggests that, given a finite set of alternatives, a rational individual will always prefer the alternative that yields the highest utility [52]. DCEs are frequently applied in agriculture and environmental valuation to study consumer and producer preferences in multi-attribute choice problems [55–58]. But, as explained, choice-experimental methods have not yet been widely used to analyze farmer preferences for WII.

Experimental design

For designing the DCE, we first identified contract attributes of possible interest in the WII context through a review of the relevant literature [25, 30, 40, 59, 60]. Then, we carried out focus group discussions with farmers in Kenya and also consulted local insurance agents and agricultural extension officers to narrow down the list of possible attributes to those most meaningful in a smallholder context. In order not to overburden participants in the experiment , we eventually decided to use five contract attributes, as shown in Table 1.

"Premium rate" is the fee charged for insurance coverage. This is expressed as a percentage of the maximum pay-out (expected value of harvest per acre), irrespective of the type of crop cultivated. In the existing WII contracts, premium rates are calculated based on the historical frequency and severity of certain weather events plus a small percentage loading for implementation costs. A simple example, severe droughts in Kenya occur every 7–10 years, so the average premium charged in existing contracts is about 10% (assuming a maximum pay-out). Yet, since the rates are adjusted to local weather conditions where shocks may occur more or less often, premium rates in the *Kilimo Salama* Program range from 5 to 25% depending on the location [47]. We included six levels ranging from 2 to 20% in the DCE, in order to predict farmers' responsiveness to changing prices.

Apart from the premium rate, which is treated as numerical, all the other attributes were effects-coded, thus ensuring that the effect of reference levels is not correlated with the intercept [61]. "Strike level," refers to the percentage deviation in rainfall at which the index triggers a pay-out to the insurance holder in a particular

---

[1] Sample size of 386 was determined using a formula by [53]: $n = \frac{Z^2 pq}{d^2}$; where $Z = 1.96$, $p = 0.5$ is the proportion of population that is likely to have taken up WII, $q = (1 − p)$, and $d = 0.5$ is the acceptable margin of error at 95% confidence level.

phase of the crop season. We chose to include six levels, where a negative sign (e.g., -20%) refers to drought contracts, and double signs (e.g., $\pm20\%$) refer to contracts that insure both drought and excess rainfall. Strike levels indicate the magnitude of loss (mild, moderate, or severe) that farmers have to personally manage before a pay-out is triggered. Strike levels also determine how frequently insured farmers will receive compensation over the years. Higher levels (say 40% rainfall deviation) decrease the probability of compensation, hence making insurance contracts more affordable. But this also reduces eligibility and frequency of payments, since payments will only be triggered by rare but extremely severe losses [16, 18]. The tick size (i.e., the payment per millimeter of rainfall deviation) was not varied across attribute levels.

The third attribute is distance from the farm to the weather station, which we use as a proxy for basis risk. With shorter distances, pay-outs will be more closely correlated with actual yield losses on the farm [60]. Distance also signifies the radius of the insurance zone. Insured farmers within this zone pay the same premium rate and receive pay-outs at the same time [13]. For this attribute, we considered three levels as shown in Table 1.

The fourth attribute relates to insurer transparency. In two attribute levels, we differentiate between transparent and non-transparent contracts, referring to the weather information provided to farmers. For transparent contracts, insured farmers would receive weekly text messages from the insurer, summarizing rainfall measurements at the reference weather station, required measurements for a pay-out, and whether a threshold for pay-out has actually been reached in that phase. This information would be publicly verifiable, by comparing with radio broadcasts about local weather facilitated by the national meteorological department. In the *Kilimo Salama* Program, such information is currently not provided to farmers, but the proposed intervention would be technically feasible without much extra cost.

The last attribute refers to the "contracted party," which allows us to analyze farmer preferences for individual versus group contracts. Currently, *Kilimo Salama* sells contracts only to individuals. As explained, group contracts may potentially be attractive for farmers to reduce transaction costs and benefit from mutual learning and broader risk-sharing arrangements. But the effectiveness of groups may depend on group size [36, 62]. Hence, we distinguish between small groups (10 members) and large groups (100 members) in different attribute levels. In Kenya, a minimum of 10 members is required for a group to be legally registered.

The next step in the DCE design was to come up with meaningful choice alternatives from varying combinations of attributes and attribute levels. The generic nature of the research problem prompted the use of an unlabeled experiment [63]. A full factorial design based on the five attributes and associated attribute levels gives a total of 648 ($2^1 \times 3^2 \times 6^2$) possible combinations. Using SAS macros [64], we developed 12 generic choice sets with a calculated D-efficiency of 0.79. To prevent fatigue and resulting inefficiency in answering, these 12 choice sets were randomly divided into three blocks, and only one of these blocks was randomly assigned to each participating farmer. That is, each farmer participated in four choice sets by choosing one out of three hypothetical insurance contracts. Every choice set also included a "no-insurance" opt-out choice, which farmers could select when none of the contract choices was satisfactory to them. This design makes it possible to interpret welfare effects resulting from the proposed contract modifications [55].

Prior to presenting the choice sets, the different attributes and attribute levels were explained to farmers in their local language. The choice cards also had shortened texts and pictorial representations of the attribute levels to facilitate understanding. An example of a choice set presented to farmers is shown in Fig. 1.

Econometric Model

The choice data were analyzed using mixed logit (ML), a popular model in discrete choice analysis [65]. ML has several advantages over standard logit models. First, it allows utility parameters to vary over decision-makers rather than being fixed, hence accommodating for preference heterogeneity in the sample. Second, it relaxes the independence from irrelevant alternatives (IIA) assumption in standard logit models. In our case, Hausman specification tests showed that the IIA assumption was violated, so that the standard logit model would have produced biased estimates. Third, ML allows for correlation of unobserved factors over choice situations. In our experiment, each farmer responded to four choice sets, meaning that individual-specific characteristics did not vary. Correlation over choice sets could also occur due to learning effects or fatigue among respondents [63, 65].

The ML models were run in STATA using a maximum simulated likelihood estimator [66]. We assumed a lognormal distribution for the premium rate attribute, allowing us to restrict the coefficient sign to be negative (rational farmers will always prefer a lower premium, holding other things constant) while still being able to account for preference heterogeneity [67]. The coefficients for the non-monetary attributes were assumed to be independent and normally distributed because the direction of preferences could not be determined prior to estimation.

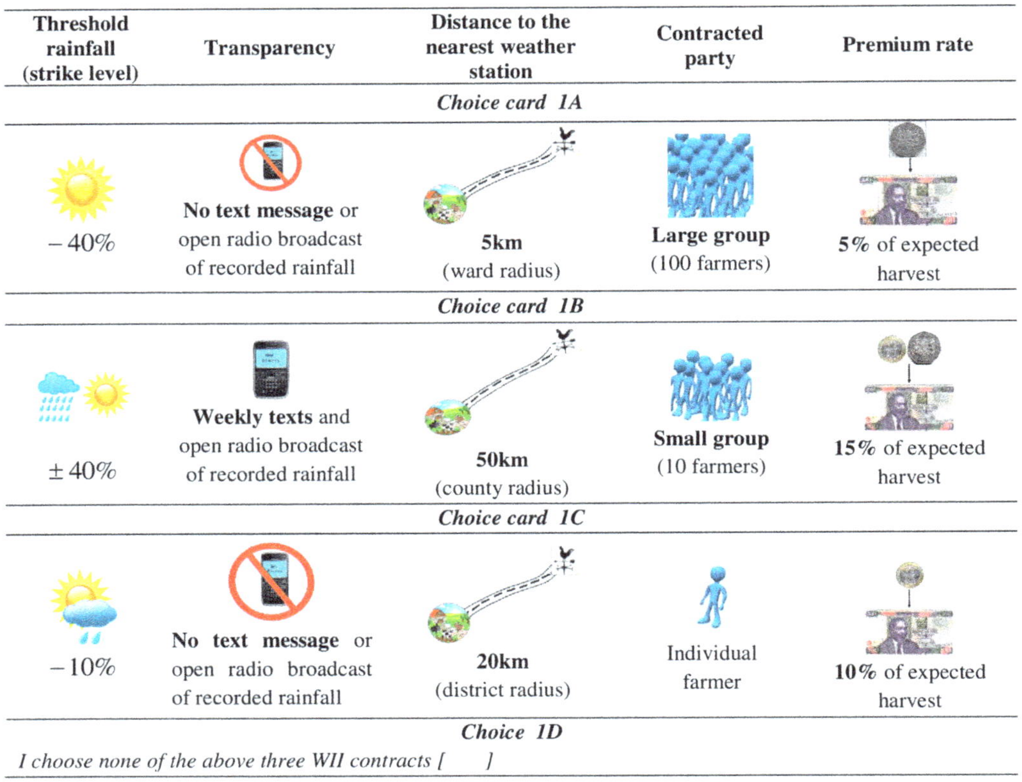

Fig. 1 Example of a choice set

We start by first specifying a main-effects model, assuming preference heterogeneity for all attributes. The simplified empirical model is expressed as:

$$y_{nt} = \alpha_n + \beta_n p_{nt} + \gamma'_n x'_{nt} \tag{1}$$

where y_{nt} is a binary variable that takes a value of one if farmer n chooses a WII contract in choice scenario t, and zero otherwise. α is an alternative specific constant (ASC), and β and γ' are parameters to be estimated for the premium rate (p_{nt}) and other contract attributes (x'_{nt}), respectively. The ASC captures the average effect of unobserved factors on utility [65]. In our specification, the ASC is defined such that it tells us how farmers value the no-contract option when observed factors are controlled for. That is, a negative ASC coefficient reveals a negative general attitude toward the non-contract option (a positive preference for WII contracts) and vice versa.

Next, we add interaction terms to analyze the influence of farmer-specific characteristics on contract preferences and thus better understand causes of preference heterogeneity. These extended models are specified as follows:

$$y_{nt} = \alpha_n + \beta_n p_{nt} + \gamma'_n x'_{nt} + \lambda'_n (\text{ASC} \times z'_n) \tag{3}$$

where WII_n^{2014}, WII_n^{2013}, and WII_n^{before} are dummy variables that take a value of one if the household had last purchased WII in 2014, 2013, or any previous year, respectively. Thus, we can evaluate the influence of previous contract experience and drop-out on current contract preferences. In Eq. (3), z'_n is a vector of socioeconomic factors that are expected to influence farmers' demand for WII.

Finally, by working out the total derivative of utility (U_{nt}) with respect to changes in the premium rate and other contract attributes $[\text{d}U_{nt} = \beta_n \text{d}p + \gamma'_n \text{d}x']$ and setting this expression equal to zero, we can solve for:

$$\frac{\text{d}p}{\text{d}x_k} = -\frac{\gamma_{nk}}{\beta_n} = \text{WTP}_{nk} \tag{4}$$

which is the marginal willingness to pay (WTP) of farmer n for a change in attribute x_k [63]. Given that the premium rate is log-normally distributed, we use the median

$$y_{nt} = \alpha_n + \beta_n p_{nt} + \gamma'_n x'_{nt} + \delta_{n1}(\text{ASC} \times \text{WII}_n^{2014}) + \delta_{n2}(\text{ASC} \times \text{WII}_n^{2013}) + \delta_{n3}(\text{ASC} \times \text{WII}_n^{before}) \tag{2}$$

Table 2 Socio-economic characteristics of sample farmers

Variables	Full sample ($n = 386$)	Ever-insured ($n = 152$)	Non-insured ($n = 234$)
Male household head (%)	67.9 (46.8)	58.6 (49.4)	73.9*** (44.0)
Education of farmer (years)	8.2 (4.0)	8.2 (4.0)	8.2 (4.0)
Age of farmer (years)	52.1 (14.6)	53.7 (13.1)	51.1* (15.4)
Farming experience (years)	26.8 (16.2)	29.1 (15.7)	25.3** (16.4)
Household size (persons)	4.6 (1.9)	4.6 (2.0)	4.5 (1.9)
Farming as primary occupation (%)	92.0 (27.2)	94.7 (22.3)	90.2*** (29.8)
Off-farm secondary occupation (%)	33.4 (47.2)	29.0 (45.4)	36.3*** (48.1)
Farm size (acres)	2.1 (1.9)	2.4 (2.4)	1.9*** (1.5)
Land title (%)	66.4 (47.2)	74.7 (43.5)	61.0*** (48.8)
Share of farm income (%)	67.2 (32.7)	72.0 (31.1)	64.1** (33.4)
Total annual income ('000 Ksh)	185.5 (370.3)	156.9 (214.3)	204.0 (442.6)
Share of land under maize (%)	46.1 (17.5)	49.5 (18.7)	43.9*** (16.3)
Received WII training in 2013 (%)	41.2 (49.3)	61.8 (48.7)	27.8*** (44.9)
WII trainings in 2013 (number of contacts)	2.3 (8.5)	3.5 (10.5)	1.6** (6.8)
Group membership (%)	88.1 (32.4)	90.8 (29.0)	86.3 (34.4)
Years in group	11.2 (12.8)	14.2 (13.4)	9.2*** (12.1)
Access to farming loan (%)	20.2 (40.2)	23.0 (42.2)	18.4 (38.8)
Farming loan received in 2013 ('000 Ksh)	22.4 (60.0)	15.2 (23.0)	28.2 (78.1)
Satisfaction with WII (1 = very dissatisfied, 5 = very satisfied)	3.4 (0.5)	3.5 (0.6)	3.3*** (0.4)
Distance to weather stations (km)	43.6 (12.5)	44.8 (12.6)	42.8 (12.4)
Risk preference (1 = risk-averse, 5 = neutral, 10 = loving)	6.75 (2.87)	6.82 (2.90)	6.71 (2.86)

Mean values are shown with standard deviations in parentheses

***, **, * indicate difference in means between sub-samples is statistically significant at 1, 5, and 10% level, respectively

parameter which is less sensitive than the mean [68]. The median estimate for the premium rate is calculated as $-\exp(\beta_n)$ [66].

Results and discussion

We first introduce sample descriptive statistics and farmers' experience with the existing *Kilimo Salama* insurance scheme, before presenting and discussing results from the model estimates with the choice-experimental data.

Socio-economic characteristics

Table 2 presents descriptive statistics of socio-economic characteristics for the full sample of farmers, as well as separately for the sub-samples of ever-insured and non-insured farmers. Overall, sample farmers from Embu County are typical smallholders with an average farm size of around two acres. Statistically significant differences between the sub-samples are observed for sex, age, farming experience, and occupation of the household head. Female-headed households are more likely to purchase insurance than male-headed households mainly due to differences in their willingness/ability to take risk. Women tend to make less risky investment choices and are more vulnerable to

weather-related risks [33]; hence, they would have a stronger demand for WII compared to men. Furthermore, insured farmers are older and more experienced than their non-insured colleagues, and they derive a larger share of their income from farming. This suggests that, to some extent, insurance may be a substitute for income diversification, which otherwise tends to be a common strategy to cope with risk. Farmers with access to WII training and those who have been organized in farmer groups for a longer period of time are also more likely to purchase insurance.

Farmers were also asked how willing they are to take risks in their farming decisions using a scale of 1 = very risk-averse to 10 = very risk-loving. This direct question about farmers' perception of their risk behavior is an alternative to more comprehensive lotteries that can also be used to elicit risk attitudes. In [69], it is argued that farmers sometimes overstate their risk preference (understate their risk aversion) when asked directly, but that in terms of comparing relative risk attitudes, answers to direct questions are equally reliable as lotteries. The last row in Table 2 reveals that average risk preferences are indeed relatively high. However, as the same question was used for all sample farmers, relative comparisons should be in order. Interestingly, we do not observe

Table 3 Farmers' experience with agricultural shocks during the past five years

Agricultural risks	Farmers affected (%)	Frequency (past 5 years)	Severity (scale: 1–4)
Input price hike	88.6	4.2 (1.2)	3.1 (0.7)
Output price drop	85.2	4.1 (1.2)	3.0 (0.7)
Drought/insufficient rain	85.2	3.0 (1.2)	3.0 (0.7)
Pests and diseases	84.7	3.8 (1.6)	2.9 (0.8)
Excess rain	23.6	1.7 (1.1)	3.0 (0.8)
Frost	22.5	3.3 (1.6)	2.4 (0.8)
Hail storms	9.6	1.4 (0.8)	2.0 (1.1)
Wildlife problem	6.5	3.5 (1.8)	2.8 (0.8)

Mean values are shown with standard deviations in parentheses

a statistically significant difference in risk attitudes between ever-insured and non-insured farmers.

Table 3 outlines the main agricultural risks encountered by farmers in the study area. A 5-year recall period was used to enhance reliability in respondents' answers. In addition to asking respondents about the frequency of events, they also had to rate the severity of shocks based on experienced losses, using a four-point Likert scale (1 = no effect, 2 = mild, 3 = severe, 4 = very severe). Over 80% of the farmers were affected by input and output price shocks, drought, and crop pests during the last 5 years. Other weather-related shocks, such as excess rain, frost, and hailstorms, were more localized, and also occurred less often.

Farmers' experience with existing WII

We now look at experiences with the existing WII in the *Kilimo Salama* Program, based on farmers' responses to the survey questions. Table 4 shows that the number of insured farmers has increased since 2009, when WII started as a small pilot project. However, the number of insured farmers has not further increased since 2012, and has actually fallen in 2014. Similarly, the number and

share of insured farmers who received payments have declined since 2012. The lower share of farmers paid in 2013 may possibly have contributed to lower insurance purchase in 2014. Yet, the majority (62%) of all ever-insured farmers has been compensated at least once since the start of the program.

Farmers' responses reveal that actual insurance payments do not always coincide with their own assessment of yield losses. Differences may be due to basis risk, but they contribute to a lower level of confidence from the farmers' point of view. Figure 2 illustrates that in the early years of the WII program, several farmers had received pay-outs without having experienced significant yield losses. It is possible that the insurance program paid more generously in the beginning to encourage more farmers to participate in subsequent years. However, in 2013, when many farmers experienced crop losses due to low rainfall, the index failed to trigger a pay-out. As indicated above, this may have contributed to lower insurance uptake in 2014.

Some of the ever-insured farmers purchased insurance in several years, others only in 1 year. The average number of years that farmers in this sub-sample were insured is 2.2 (out of the 6 years considered). Dropping-out is not uncommon, indicating that not all farmers are fully satisfied with their WII experience. In the survey, we assessed the farmers' level of satisfaction, using a list of 22 statements. Farmers were asked whether they agree or

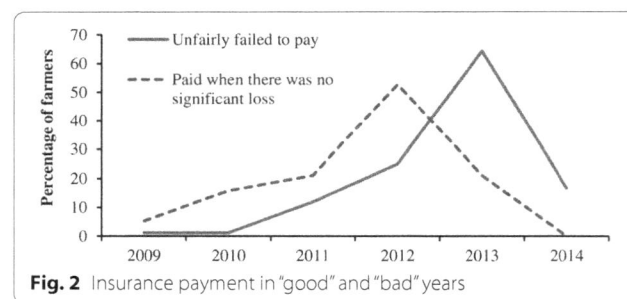

Fig. 2 Insurance payment in "good" and "bad" years

Table 4 Number of farmers who purchased WII and received payment (2009–2014)

Year	Farmers who purchased WII (number)	Farmers who received payment (number)	Share of insured farmers who received payment (%)
2009	14	4	2.6
2010	35	16	10.5
2011	54	23	15.1
2012	88	55	36.2
2013	86	26	17.1
2014	60	1	0.7
Overall (in any year)	152	94	61.8

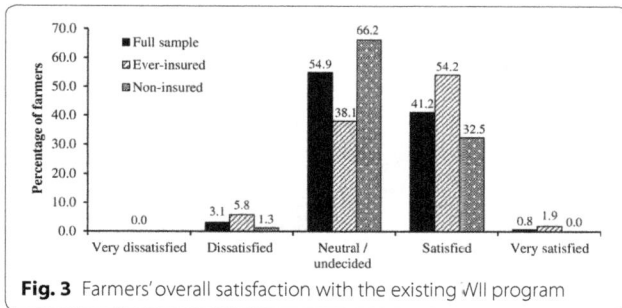

Fig. 3 Farmers' overall satisfaction with the existing WII program

disagree with each statement based on a five-point Likert scale ranging from "completely disagree" to "completely agree." As most farmers in our sample were aware of WII and had some opinion, the same questions were asked to all respondents, not only those who had ever purchased insurance themselves. Out of the 22 responses for each farmer, we calculated mean satisfaction levels, as summarized in Fig. 3. The majority of the farmers are in the "neutral" category, meaning that they are neither particularly satisfied nor dissatisfied with the WII program. Yet, further disaggregation shows that mean levels of satisfaction are higher among those who had ever purchased insurance themselves than among the non-insured. Overall, this analysis suggests that most farmers have neutral or positive attitudes toward WII in general, but that there is scope for further improvement in the insurance products.

Choice-experimental results

We now present and discuss results from the DCE. Model (1) in Table 5 shows the ML estimates of the main-effects model without interaction terms. Most of the mean parameters are statistically significant with expected signs, suggesting that the chosen contract attributes are relevant for farmers in this context. Most of the standard deviation parameters, which are shown in the lower part of Table 5, are significant as well, pointing at considerable preference heterogeneity.

In model (1), the mean parameter for the ASC is negative, suggesting that farmers have a positive general attitude toward WII contracts. The premium rate coefficient is negative, meaning that farmers prefer lower-priced over higher-priced insurance contracts, holding other contract attributes constant. For the strike level, − 10% is the reference against which the other coefficients can be compared. The coefficient for − 20% is not statistically significant. However, the coefficient for − 40% is statistically significant. The negative sign indicates that farmers have a preference for pay-outs already starting at lower absolute threshold levels. The coefficients for ± 20

and ± 40% are positive and significant, suggesting that farmers value insurance that covers excess rainfall in addition to drought. The coefficient for ± 10% is insignificant, which may be due to the fact that excess rain in moderate dimensions is often less harmful for crop yields. More heavy excess rain, however, can be quite damaging, as farmers' responses in Table 3 above have shown. This is in line with the estimation results in Table 5.

The positive and significant coefficient for insurer transparency reveals a strong farmer preference for receiving regular text messages about rainfall measurements as part of the insurance contract. This result confirms that information transparency and regular communication can increase farmers' confidence in WII products, which was also pointed out by [34]. Concerning distance to the weather station, where 5 km is the reference; the negative and significant coefficient for the 50 km alternative shows that farmers prefer shorter distances that are associated with lower basis risk. Currently, the average distance to the weather stations in our sample of farmers is 44 km (Table 1). The estimation results suggest that insurance uptake could be higher with more weather stations installed. Previous research also showed that reducing basis risk can be an important way of increasing the attractiveness of WII contracts [14, 24].

Regarding group insurance, results in Table 5 show that small-group contracts are more likely to be chosen over individual contracts, whereas large-group contracts have a lower probability of being chosen. This implies that offering group contracts could motivate more farmers to take up WII, which is consistent with recent findings from Tanzania and Ethiopia [38, 42]. However, it also becomes evident that structural aspects such as group size matter, as larger groups may be associated with lower levels of group cohesion [36].

Models with interaction effects

To explain possible sources of preference heterogeneity, we added interaction terms as additional covariates, as was explained above. Results of these extended model estimates are shown in models (2) and (3) of Table 5. We concentrate the discussion on the coefficients of the interaction terms. In model (2), ASC is interacted with actual insurance uptake in the past. The insignificant coefficients for the interactions with WII uptake in 2013 and 2014 suggest that recent adopters and non-adopters of insurance contracts have similar preferences. However, the interaction with WII uptake before 2013 is positive and significant, meaning that earlier adopters who then dropped out have less positive attitudes toward insurance contracts. This is plausible, as their decision to drop out

Table 5 Estimated model results for weather index insurance preferences

Variables	Model (1)		Model (2)		Model (3)	
Mean parameters						
ASC (1 = no insurance)	− 7.54***	(1.72)	− 9.53***	(2.46)	4.25	(3.93)
Premium rate (%)	− 5.39***	(2.01)	− 5.26***	(1.86)	− 5.50***	(1.86)
Strike level						
− 20%	− 0.16	(0.11)	− 0.16	(0.11)	− 0.15	(0.11)
− 40%	− 0.44***	(0.11)	− 0.43***	(0.11)	− 0.45***	(0.12)
± 10%	− 0.01	(0.10)	− 0.01	(0.10)	− 0.01	(0.10)
± 20%	0.18*	(0.10)	0.17*	(0.10)	0.17*	(0.10)
± 40%	0.20*	(0.11)	0.19*	(0.11)	0.21*	(0.11)
Transparency (1 = weekly texts)	0.85***	(0.08)	0.86***	(0.08)	0.86***	(0.08)
Distance to station						
50 km	− 0.27***	(0.07)	− 0.26***	(0.07)	− 0.27***	(0.07)
20 km	0.08	(0.06)	0.07	(0.07)	0.08	(0.06)
Contracted party						
Small group	0.25***	(0.08)	0.26***	(0.08)	0.25***	(0.08)
Large group	− 0.33***	(0.09)	− 0.33***	(0.09)	− 0.33***	(0.09)
WII^{2014} × ASC			− 1.23	(1.51)		
WII^{2013} × ASC			− 1.89	(1.60)		
WII^{before} × ASC			2.67*	(1.49)		
Satisfaction (scale: 1–5) × ASC					− 2.15*	(1.12)
Risk-attitude (scale:1–10) × ASC					− 0.51***	(0.20)
Received WII training × ASC					− 3.22**	(1.54)
Education × ASC					0.0004	(0.13)
Female × Education × ASC					0.30**	(0.12)
Group membership × ASC					− 2.64**	(1.25)
Off-farm occupation × ASC					− 2.61	(3.21)
Larger farm(1 = if ≥ $\bar{x}$) × ASC					1.93*	(1.14)
Std. deviation parameters						
ASC	3.76***	(0.88)	5.03***	(1.32)	3.66***	(0.95)
Premium rate (%)	4.29***	(0.83)	0.20	(0.31)	0.16	(0.37)
Strike level						
− 20%	0.15	(0.37)	0.39	(0.25)	0.42*	(0.23)
− 40%	0.38	(0.24)	0.05	(0.25)	0.04	(0.25)
± 10%	0.07	(0.24)	0.21	(0.21)	0.10	(0.25)
± 20%	0.13	(0.24)	0.09	(0.38)	0.13	(0.26)
± 40%	0.15	(0.25)	0.62***	(0.09)	0.60***	(0.09)
Transparency (1 = weekly texts)	0.60***	(0.09)	0.45***	(0.11)	0.49***	(0.11)
Distance to station						
50 km	0.48***	(0.11)	0.30**	(0.13)	0.24	(0.17)
20 km	0.22	(0.16)	0.79***	(0.12)	0.81***	(0.12)
Contracted party						
Small group	0.78***	(0.12)	1.04***	(0.13)	1.03***	(0.12)
Large group	1.00***	(0.12)	4.33***	(0.79)	4.33***	(0.83)
Log likelihood	− 1490.76		− 1489.72		− 1474.11	
Chi-squared	279.12***		278.75***		240.55***	

The number of observations in all three models is 6176. Coefficient estimates are shown with standard errors in parenthesis. The reference values for the effects-coded contract attributes are − 10% strike level, no text message, 5 km distance, and individual contract

***, **, * indicate statistically significant at the 1, 5, and 10% level, respectively

from the existing WII program was probably related to not being fully satisfied.

The results in model (3) confirm that levels of satisfaction with the existing insurance program determine farmer attitudes: higher levels of satisfaction contribute to a higher general preference for WII. Somewhat surprising is the negative coefficient for the ASC interaction with risk attitudes, which implies that risk-loving farmers have more positive attitudes toward WII. One would usually expect the opposite, namely that risk-averse farmers have a stronger preference for crop insurance. We interpret this result as another sign that not all farmers are fully confident with the functioning of WII contracts. Given the lack of transparency regarding rainfall measurements and pay-out triggers, risk-averse farmers may not feel properly insured against weather shocks. Some may even consider WII as a kind of gamble on random weather outcomes. This is consistent with previous studies showing that risk-averse farmers are often less likely to adopt WII [20, 22, 70].

Limited confidence may also be related to the complexity of WII, especially for smallholder farmers who are often unfamiliar with formal insurance products [71]. The other interaction terms in model (3) confirm the important role of training and learning. Farmers who received training as part of the *Kilimo Salama* Program have more positive attitudes toward WII. Furthermore, membership in a farmer group, which can serve as a learning platform for innovations, affects attitudes toward WII in a positive way.

Finally, we were interested in the role of farm size. To analyze possible heterogeneity between smaller and larger farms, we created a dummy variable that takes a value one if a particular farm is above the mean farm size in the sample. The positive and significant coefficient for the interaction of this dummy with ASC reveals that smaller farms have a higher preference for WII. Since farm size is an indicator of wealth, it is possible that smaller farms would be more interested in insurance because they lack the economic muscle to individually cope with weather shocks. However, this is a welcome finding, as it demonstrates the potential of properly designed WII products to benefit smallholder farmers. This potential is not yet fully realized.

Willingness to pay (WTP)

Based on the estimates in model (1), we calculated farmers' WTP for WII contracts and for changes in particular contract attributes. We used individual-specific coefficients to obtain WTP point estimates for the farmers in our sample [63]. Results are presented in Table 6. We only show results for attribute levels with significant

Table 6 Marginal willingness to pay for WII attributes

Variables	Mean (%)	Std. deviation	Lower CI	Upper CI
ASC	7.56	1.90	7.37	7.75
Transparency	0.79	0.34	0.76	0.83
Strike level				
−40%	−0.41	0.09	−0.42	−0.40
±20%	0.16	0.03	0.16	0.17
±40%	0.18	0.03	0.18	0.19
Large group contract	−0.31	0.66	−0.38	−0.24
Small group contract	0.23	0.46	0.19	0.28
Distance (50 km)	−0.24	0.20	−0.26	−0.22

Confidence intervals (CI) refer to the 95% level. Willingness to pay (WTP) was calculated by dividing individual-specific coefficients for attribute level by the median of the premium rate coefficient. For the ASC, the coefficient was multiplied by −1 to obtain the WTP for insurance in general. WTP is only shown for attribute levels with significant coefficient estimates in model (1) of Table 5

coefficient estimates. For the ASC, we multiplied the coefficient estimates by −1 because we are interested in the WTP for insurance, not for the no-insurance option. On average, farmers are willing to pay about 7.6% of their expected harvest for a WII contract. As mentioned, the actual price varies by location, but the average premium rate in the *Kilimo Salama* Program is 10%. Moderate premium reductions could probably increase insurance uptake significantly. The mean estimate also suggests that contracts priced at 20 or 25%, as observed in some locations, are way above what the average farmer is willing and able to pay for WII.

The WTP estimates for the different attribute levels can be interpreted as increments over the base value of insurance. That is, the mean WTP for a contract with transparent communication of weather data through weekly text messages would be 7.56 + 0.79 = 8.35% of the expected harvest. The point estimates for the different attribute levels are all highly significant but quite small in magnitude, which may be due to the assumed lognormal distribution of the premium rate variable. However, even if the marginal WTP for the attribute levels was underestimated, relative comparisons should still be in order because the same calculation methods were used for all attributes. The highest marginal WTP is observed for the transparency attribute. Transparency also seems to be more important than distance to the weather station. Even though farmers are willing to pay less for contracts with reference stations further away from their farm, the WTP comparison suggests that transparent communication and information provision may have a larger effect on insurance uptake than investing in additional weather stations to reduce basis risk.

Concerning the other attributes, farmers are willing to pay 0.41 percentage points less for contracts that only start paying at a rainfall threshold level of −40%. For contracts that are also covering excess rainfall, farmers are willing to pay more, but the additional WTP is relatively small. Comparing values across attributes we learn that—at least for this study area—focusing on drought risk with a lower absolute strike level is more valuable for farmers than covering additional risks, as the existing *Kilimo Salama* Program does. Finally, the estimates show that large-group contracts would only be chosen over individual contracts if the premium was 0.31 percentage points lower, whereas small-group contracts would result in a 0.23 percentage point higher WTP.

Conclusions

Weather index insurance (WII) could reduce the high transaction costs involved in traditional, indemnity-based crop insurance programs and could therefore be of particular relevance for smallholder farmers in developing countries. However, the uptake of WII in the small-farm sector has been relatively low up till now. One reason is probably that WII contracts are not sufficiently tailored to the needs and preferences of smallholder farmers. Improved contractual design might help toward more widespread insurance uptake. In this study, we have contributed to the knowledge base focusing on the situation of smallholder farmers in Kenya. We have combined farm survey and choice-experimental data to analyze the experience with an existing WII program and to better understand how hypothetical changes in the insurance contracts might improve the situation.

While the existing WII program in Kenya was launched in 2009, the number of participating farmers has remained relatively low. Several farmers also decided to discontinue their insurance contracts after one or 2 years of participation. One issue is that the insurance contracts are too expensive from the farmers' point of view. Our analysis has shown that farmers' mean willingness to pay is about 25% lower than the average premium rate charged by the insurance provider. Lower premium rates could probably contribute to increased insurance uptake.

Beyond the premium rate, we identified several other contract attributes that seem to be critical. Many farmers struggle with fully understanding the functioning of WII contracts and when exactly pay-outs are triggered. The resulting uncertainty undermines farmers' confidence and thus lowers their demand for insurance. Risk-averse farmers in particular were found to have a low preference for WII contracts, even though they are actually the main target group of insurance products. Our estimates suggest that better training and communication could increase farmers' confidence and thus insurance uptake.

Transparent provision of relevant rainfall measurements and thresholds—for instance through regular text messages—could significantly increase farmers' willingness to pay for WII. Mechanisms to reduce basis risk are also valued by farmers, although not to the same extent as higher levels of transparency. Improving communication may therefore be more important for WII providers than investing into additional weather stations in order to reduce basis risk. Offering contracts to farmer groups rather than individuals was also found to be a promising avenue for wider insurance uptake. Group contracts could help to reduce transaction costs. Furthermore, farmer groups can be important platforms for learning about complex innovations, including novel risk transfer products. For this, however, group sizes should be relatively small, as larger groups often lack the necessary cohesion.

We caution that the results are specific to Kenya and that choice-experimental data may be subject to hypothetical bias. Hence, the exact estimates should not be generalized and over-interpreted. However, the findings still provide interesting insights into typical issues of WII design in a small-farm context. Given that smallholder farmers are particularly vulnerable to climate shocks, improving their access to crop insurance is of high policy relevance. More research is needed to further add to the knowledge base about suitable contractual designs in particular situations.

Authors' contributions

All authors (KWS, PCV, and MQ) conceptualized and designed the survey. KWS analyzed data. KWS and MQ wrote the paper. All authors read and approved the final manuscript.

Author details

[1] Department of Agricultural Economics and Rural Development, School of Agriculture and Food Security, Maseno University, P.O. Box Private Bag, Maseno, Kenya. [2] International Rice Research Institute, New Delhi, India. [3] Department of Agricultural Economics and Rural Development, University of Goettingen, 37073 Göttingen, Germany.

Acknowledgements

The authors thank local extension officers in Embu, Kenya, for their assistance during the farm survey. We are also grateful to Professors and Doctorate students at the Department of Agricultural Economics, University of Goettingen, for their useful comments during the development of this paper.

Competing interests

The authors declare that they have no competing interests.

Ethics approval and consent to participate

Although we interviewed Kenyan farmers to generate data used to in this study, participation in the farm survey and choice experiment did not involve any risk for farmers. Hence, the study was not subject to institutional review board approval at the University of Goettingen, as was confirmed by the University's Ethics Commission. In Kenya, we obtained clearance from the ministries of agriculture and education before collecting the data. Prior to the interviews, farmers were informed about the purpose of the research and asked for their verbal consent to participate. We did not ask for written consent, because many of the respondents were not familiar with formal paper work. It was clarified that the information collected would be treated confidentially, analyzed anonymously, and only used for purposes of this research.

Funding

This research was undertaken with financial support from the Kenyan National Commission for Science, Technology, and Innovation (NACOSTI) and the German Research Foundation (DFG).

References

1. Daryanto S, Wang LX, Jacinthe PA. Global synthesis of drought effects on maize and wheat production. PLoS ONE. 2016;11(1):e0156362.
2. Lesk C, Rowhani P, Ramankutty N. Influence of extreme weather disasters on global crop production. Nature. 2016;529:84–7.
3. Wheeler T, von Braun J. Climate change impacts on global food security. Science. 2013;341(6145):508–13.
4. Brown ME, Kshirsagar V. Weather and international price shocks on food prices in the developing world. Global Environ Change. 2015;35:31–40.
5. World Bank. World development report: development and climate change. Washington, DC: The World Bank; 2010.
6. Vermeulen SJ, Campbell BM, Ingram JSI. Climate change and food systems. Annu Rev Environ Res. 2012;37:195–222.
7. Carter MR, Barrett CB. The economics of poverty traps and persistent poverty: an asset-based approach. J Dev Stud. 2006;42(2):178–99.
8. Dercon S, Christiaensen L. Consumption risk, technology adoption, and poverty traps: evidence from Ethiopia. J Dev Econ. 2011;96(2):159–73.
9. Hazell PBR, Hess U. Drought insurance for agricultural development and food security in dryland areas. Food Sec. 2010;2(4):395–405.
10. De Janvry A, Dequiedt V, Sadoulet E. The demand for insurance against common shocks. J Dev Econ. 2014;106:227–38.
11. Jensen ND, Barrett CB. Agricultural index insurance for development. Appl Econ Perspect Policy. 2016. https://doi.org/10.1093/aepp/ppw022.
12. Barnett BJ, Mahul O. Weather index insurance for agriculture and rural areas in lower-income countries. Am J Agric Econ. 2007;89(5):1241–7.
13. IFAD. The potential for scale and sustainability in weather index insurance for agriculture and rural livelihoods. Rome: International Fund for Agricultural Development and World Food Program; 2010.
14. Musshoff O, Odening M, Xu W. Management of climate risks in agriculture: will weather derivatives permeate? Appl Econ. 2011;43(9):1067–77.
15. World Bank. Weather index insurance for agriculture: guidance for development practitioners. Washington, DC: The World Bank; 2011.
16. Rao KN. Index based crop insurance. Agric Agric Sci Procedia. 2010;1:193–203.
17. Binswanger-Mkhize HP. Is there too much hype about index-based agricultural insurance? J Dev Stud. 2012;48(2):187–200.
18. Clarke DJ, Mahul O, Rao KN, Verma N. Weather based crop insurance in India. Policy research working paper no. 5985, Washington DC: The World Bank; 2012.
19. Miranda MJ, Farrin K. Index insurance for developing countries. Appl Econ Perspect Policy. 2012;34(3):391–427.
20. Giné X, Townsend R, Vickery J. Patterns of rainfall insurance participation in rural India. World Bank Econ Rev. 2008;22(3):539–66.
21. Cole S, Giné X, Tobacman J, Topalova P, Townsend R, Vickery J. Barriers to household risk management: evidence from India. Am Econ J Appl Econ. 2013;5(1):104–35.
22. Hill RV, Hoddinott J, Kumar N. Adoption of weather-index insurance: learning from willingness to pay among a panel of households in rural Ethiopia. Agric Econ. 2013;44(4–5):385–98.
23. Breustedt G, Bokusheva R, Heidelbach O. Evaluating the potential of index insurance schemes to reduce crop yield risk in an arid region. J Agric Econ. 2008;59(2):312–28.
24. Norton MT, Turvey C, Osgood D. Quantifying spatial basis risk for weather index insurance. J Risk Finance. 2013;14(1):20–34.
25. Elabed G, Bellemare MF, Carter MR, Guirkinger C. Managing basis risk with multiscale index insurance. Agric Econ. 2013;44(4–5):419–31.
26. Jensen ND, Barrett CB, Mude AG. Index insurance quality and basis risk: evidence from northern Kenya. Am J Agric Econ. 2016;98(5):1450–69.
27. Carter MR, Barrett CB, Boucher S, Chantarat S, Galarza F, McPeak J, et al. Insuring the never before insured: explaining index insurance through financial education games. BASIS brief 2008–07, Madison: University of Wisconsin; 2008.
28. Norton M, Osgood D, Madajewicz M, Holthaus E, Peterson N, Diro R, et al. Evidence of demand for index insurance: experimental games and commercial transactions in Ethiopia. J Dev Stud. 2014;50(5):630–48.
29. Takahashi K, Ikegami M, Sheahan M, Barrett CB. Experimental evidence on the drivers of index-based livestock insurance demand in Southern Ethiopia. World Dev. 2016;78:324–40.
30. Liebe U, Maart SC, Musshoff O, Stubbe P. Risk management on farms: an analysis of the acceptance of weather insurance using discrete choice experiments. Ger J Agric Econ. 2012;61(2):63–79.
31. Liesivaara P, Myyrä S. Willingness to pay for agricultural crop insurance in the northern EU. Agric Finance Rev. 2014;74(4):539–54.
32. Castellani D, Viganò L, Tamre B. A discrete choice analysis of smallholder farmers' preferences and willingness to pay for weather derivatives: evidence from Ethiopia. J Appl Bus Res. 2014;30(6):1671–92.
33. Akter S, Krupnik TJ, Rossi F, Khanam F. The influence of gender and product design on farmers' preferences for weather-indexed crop insurance. Global Environ Change. 2016;38:217–29.
34. Patt A, Peterson N, Carter M, Velez M, Hess U, Suarez P. Making index insurance attractive to farmers. Mitig Adapt Strateg Glob Change. 2009;14(8):737–53.
35. Pacheco JM, Santos FC, Levin SA. Evolutionary dynamics of collective index insurance. J Math Biol. 2016;72(4):997–1010.
36. Fischer E, Qaim M. Smallholder farmers and collective action: what determines the intensity of participation. J Agric Econ. 2014;65(3):683–702.
37. Wollni M, Fischer E. Member deliveries in collective marketing relationships: evidence from coffee cooperatives in Costa Rica. Euro Rev Agric Econ. 2015;42(2):287–314.
38. Traerup SLM. Informal networks and resilience to climate change impacts: a collective approach to index insurance. Global Environ Change. 2012;22(1):255–67.
39. Townsend RM. Consumption insurance: an evaluation of risk-bearing systems in low-income economies. J Econ Perspect. 1995;9(3):83–102.
40. Delpierre M, Boucher S. The impact of index-based insurance on informal risk-sharing networks. Selected paper at the Agricultural and Applied Economics Association Annual Meeting, Washington, DC, August 4–6; 2013.
41. Mobarak AM, Rosenzweig M. Selling formal insurance to the informally insured. Economic growth center discussion paper no. 1007, New Haven: Yale University; 2012.
42. Dercon S, Hill RV, Clarke D, Outes-Leon I, Taffesse AS. Offering rainfall insurance to informal insurance groups: evidence from a field experiment in Ethiopia. J Dev Econ. 2014;106:132–43.
43. Vasilaky K, Osgood D, Martinez S, Stanimirova R. Informal networks within index insurance: randomizing distance in group insurance. New York: Columbia University; 2014.
44. McIntosh C, Povel F, Sadoulet E. Utility, risk, and demand for incomplete insurance: lab experiments with Guatemalan cooperatives. San Diego: University of California; 2015.
45. Omoyo NN, Wakhungu J, Oteng'i S. Effects of climate variability on maize yield in the arid and semi arid lands of lower eastern Kenya. Agric Food Sec. 2015;4(8):1–13.
46. FSD. Review of FSD's index based weather insurance initiatives. Nairobi: Financial Sector Deepening; 2013.
47. International Finance Corporation (IFC). Agriculture and climate risk enterprise (ACRE)—Kilimo Salama—Kenya. http://www.ifc.org/wps/wcm/connect/industry_ext_content/ifc_external_corporate_site/industries/financial+markets/retail+finance/insurance/agricultur e+and+climate+risk+enterprise. Accessed Nov 2015.
48. Greatrex H, Hansen JW, Garvin S, Diro R, Blakeley S. Scaling up index insurance for smallholder farmers: recent evidence and insights. CGIAR Research Program on Climate Change, Agriculture and Food Security (CCAFS), Report No. 14; 2015.

49. Kikulwe EM, Fischer E, Qaim M. Mobile money, smallholder farmers, and household welfare in Kenya. PLoS ONE. 2014;9(10):e109804.

50. Sina J, Jacobi P. Index-based weather insurance: international and Kenyan experiences. Nairobi: Adaptation to Climate Change and Insurance; 2012.

51. Ngetich KF, Mucheru-Muna M, Mugwe JN, Shisanya CA, Diels J, Mugendi DN. Length of growing season, rainfall temporal distribution, onset and cessation dates in the Kenyan highlands. Agric Forest Meteorol. 2014;188:24–32.

52. Adamowicz W, Boxall P, Williams M, Louviere J. Stated preference approaches for measuring passive use values: choice experiments and contingent valuation. Am J Agric Econ. 1998;80(1):64–75.

53. Anderson DR, Sweeney DJ, Williams TA, Martin RK. An introduction to management science: quantitative approaches to decision making. Kendallville: ICC Macmillan; 2008.

54. Louviere JJ, Hensher DA, Swait JD. Stated choice methods: analysis and applications. Cambridge: Cambridge University Press; 2000.

55. Hanley N, Mourato S, Wright RE. Choice modelling approaches: a superior alternative for environmental valuation. J Econ Surv. 2001;15(3):435–62.

56. Schipmann C, Qaim M. Supply chain differentiation, contract agriculture, and farmers' marketing preferences: the case of sweet pepper in Thailand. Food Policy. 2011;36(5):667–77.

57. Veettil PC, Speelman S, Frija A, Buysse J, Van Huylenbroeck G. Complementarity between water pricing, water rights and local water governance: a Bayesian analysis of choice behaviour of farmers in the Krishna river basin, India. Ecol Econ. 2011;70(10):1756–66.

58. Kouser S, Qaim M. Valuing financial, health, and environmental benefits of Bt cotton in Pakistan. Agric Econ. 2013;44(3):323–35.

59. Giné X, Yang D. Insurance, credit and technology adoption: field experimental evidence from Malawi. J Dev Econ. 2009;89(1):1–11.

60. Heimfarth LE, Musshoff O. Weather index-based insurances for farmers in the North China plain: an analysis of risk reduction potential and basis risk. Agric Finance Rev. 2011;71(2):218–39.

61. Bech M, Gyrd-Hansen D. Effects coding in discrete choice experiments. Health Econ. 2005;14(10):1079–83.

62. Ligon E, Thomas JP, Worrall T. Informal insurance arrangements with limited commitment: theory and evidence from village economies. Rev Econ Stud. 2002;69(1):209–44.

63. Hensher DA, Rose JM, Greene WH. Applied choice analysis: a primer. Cambridge: Cambridge University Press; 2005.

64. Kuhfeld WF. Marketing research methods in SAS: experimental design, choice, conjoint, and graphical techniques. Cary: SAS-Institute; 2010.

65. Train KE. Discrete choice methods with simulation. Cambridge: Cambridge University Press; 2003.

66. Hole AR. Estimating mixed logit models using maximum simulated likelihood. Stata J. 2007;7(3):388–401.

67. Hole AR, Kolstad JR. Mixed logit estimation of willingness to pay distributions: a comparison of models in preference and WTP space using data from a health-related choice experiment. Empir Econ. 2012;42(2):445–69.

68. Meijer E, Rouwendal J. Measuring welfare effects in models with random coefficients. J Appl Econ. 2006;21(2):227–44.

69. Dohmen T, Falk A, Huffman D, Sunde U, Schupp J, Wagner GG. Individual risk attitudes: measurement, determinants, and behavioural consequences. J Euro Econ Assoc. 2011;9(3):522–50.

70. Clarke DJ. A theory of rational demand for index insurance. Am Econ J Microecon. 2016;8(1):283–306.

71. Patt A, Suarez P, Hess U. How do small-holder farmers understand insurance, and how much do they want it? Evidence from Africa. Global Environ Change. 2010;20(1):153–61.

## 5

Climate change impacts and adaptation among smallholder farmers in Central America

Celia A. Harvey[1]*[ID], Milagro Saborio-Rodríguez[2,3], M. Ruth Martinez-Rodríguez[1], Barbara Viguera[2], Adina Chain-Guadarrama[4], Raffaele Vignola[2,5] and Francisco Alpizar[2]

Abstract

Background: Smallholder farmers are one of the most vulnerable groups to climate change, yet efforts to support farmer adaptation are hindered by the lack of information on how they are experiencing and responding to climate change. More information is needed on how different types of smallholder farmers vary in their perceptions and responses to climate change, and how to tailor adaptation programs to different smallholder farmer contexts. We surveyed 860 smallholder coffee and basic grain (maize/bean) farmers across six Central American landscapes to understand farmer perceptions of climate change and the impacts they are experiencing, how they are changing their agricultural systems in response to climate change, and their adaptation needs.

Results: Almost all (95%) of the surveyed smallholder farmers have observed climate change, and most are already experiencing impacts of rising temperatures, unpredictable rainfall and extreme weather events on crop yields, pest and disease incidence, income generation and, in some cases, food security. For example, 87% of maize farmers and 66% of coffee farmers reported negative impacts of climate change on crop production, and 32% of all smallholder farmers reported food insecurity following extreme weather events. Of the farmers perceiving changes in climate, 46% indicated that they had changed their farming practices in response to climate change, with the most common adaptation measure being the planting of trees. There was significant heterogeneity among farmers in the severity of climate change impacts, their responses to these impacts, and their adaptation needs. This heterogeneity likely reflects the wide diversity of socioeconomic and biophysical contexts across smallholder farms and landscapes.

Conclusions: Our study demonstrates that climate change is already having significant adverse impacts on smallholder coffee and basic grain farmers across the Central American region. There is an urgent need for governments, donors and practitioners to ramp up efforts to help smallholder farmers cope with existing climate impacts and build resiliency to future changes. Our results also highlight the importance of tailoring of climate adaptation policies and programs to the diverse socioeconomic conditions, biophysical contexts, and climatic stresses that smallholder farmers face.

Keywords: Adaptation strategies, Climate change, *Coffea arabica*, Ecosystem-based Adaptation, Smallholder farmers, *Zea mays*

*Correspondence: celiaharvey@stanfordalumni.org
[1] Conservation International, 2011 Crystal Drive Suite 50C, Arlington, VA 22202, USA
Full list of author information is available at the end of the article

Background

Climate change poses a significant threat to smallholder farmers and threatens to undermine global progress toward poverty alleviation, food security, and sustainable development [1, 2]. Globally, there are an estimated 475 million smallholder farmers cultivating less than 2 ha of land [3], many of whom are poor, experience food insecurity, and live in highly precarious conditions [4, 5]. Smallholder farmers are highly vulnerable to climate change because most depend on rain-fed agriculture, cultivate marginal areas, and lack access to technical or financial support that could help them invest in more climate-resilient agriculture [4, 6, 7].

While there is growing evidence of the vulnerability of smallholder farmers to climate change [5, 8] and increased interest in ensuring food security under climate change [1, 2], adaptation efforts are still hindered by the lack of information on how smallholder farmers are experiencing and responding to climate change. Policy makers, donors, and practitioners interested in developing policies, institutional responses, and strategies for smallholder farmer adaptation need detailed, context-specific information on what climate change impacts smallholder farmers are experiencing and whether (and how) they are adapting their management strategies to deal with these impacts [9, 10]. In addition, more information is needed on how smallholder farmer vulnerability and responses vary across different farming systems and socioeconomic conditions.

Understanding the impacts of climate change on smallholder farmers and developing appropriate adaptation strategies are critical issues in Central America, a region where small-scale agriculture is central to economic development, food security, and local livelihoods [11]. There are an estimated 2.3 million smallholder farmers in Central America [12], many of whom farm on steep lands with thin soils, are poor, and suffer seasonal food insecurity [13–15]. Two common smallholder farming systems in the region are basic grain (maize and beans) and small-scale coffee production. There are an estimated one million smallholder farmers growing maize and beans for subsistence and local consumption [16]. Many smallholder farmers also cultivate coffee, an export crop that is a significant contributor to agricultural GDP and accounts for employment of an estimated 4 million across the region [17]. Both maize and coffee production are of a significant cultural importance in the region [18, 19].

Smallholder coffee and basic grain farmers are highly vulnerable to climate change as their crops are sensitive to rising temperatures and changing rainfall patterns. Rising temperatures are known to negatively affect coffee growth, flowering, fruit set, and bean quality [20, 21].

Similarly, high temperatures and drought conditions have negative impacts on biomass production, flowering, and yields of maize and beans [16]. Since smallholder farmers in Central America depend entirely on rain-fed agriculture, they are vulnerable to extended droughts, irregular rainfall patterns, and extreme rain events [18, 22, 23] which can significantly reduce yields and exacerbate food insecurity and poverty. For example, a 3-year drought (2014–2016) in the dry Pacific region of Central America resulted in 1.6 million people becoming food insecure and 3.5 million requiring humanitarian assistance [24]. Hurricanes have had significant impacts on smallholder farmer livelihoods in recent years, with strong winds and torrential rainfall destroying coffee plantations and 'milpas' (the small fields where farmers cultivate basic grains), causing leaves, flowers, and coffee cherries to drop from coffee plantations, and resulting in a significant crop damage and even crop failure [25–27]. In addition, many farmers are routinely affected by hurricane damage to roads, bridges, and farm infrastructure [25, 28], which disrupts crop harvest, processing, and transportation. For example, in 2005 Hurricane Stan resulted in the loss of 20% of the coffee harvest (worth US 4 million) in the Pacific region of Guatemala alone [29].

The impacts of climate change on smallholder agriculture are likely to intensify in future years, as climate models project rising temperatures, more erratic rainfall, and a potential increase in the intensity and/or frequency of extreme weather events [30, 31]. Recent studies suggest that by 2025 climate change may reduce bean production in Central America by more than 20% and maize yields by as much as 15% in Honduras, El Salvador, and Nicaragua [16]. In addition to direct impacts on crop production, climate change will likely alter the areas suitable for smallholder production across the region. Crop suitability models suggest that 40% or more of the current coffee areas in Nicaragua, Costa Rica, and El Salvador will likely lose suitability for coffee production by 2050, as lower elevations become suboptimal for coffee production [32]. Changing climatic conditions may also indirectly affect crop production by altering the incidence and severity of pest and disease outbreaks. The recent coffee leaf rust (*Hemileia vastatrix*) outbreak in 2012 and 2013, which had major economic and social impacts across the region, is likely to have been caused, in part, by changing climatic conditions and, in part, by poor management resulting from high input prices and low coffee prices [33]. The coffee leaf rust outbreak decimated the region's coffee production, affecting 51.2% of the cultivated coffee area, causing the loss of > 264,000 jobs, and resulting in economic losses of 479.2 million USD [34].

Policy makers across Central America are increasingly aware of the urgency of helping smallholder farmers become more resilient to climate change [35–37]; however, they lack the necessary information on how farmers are being impacted by climate change and how they could be best supported. A recent survey of Central American policy makers reported that the lack of scientific information on climate change impacts on smallholder farmers is a major constraint to the development of agricultural adaptation policies [10]. While the scientific literature on climate change impacts on smallholder farmers has rapidly expanded in recent years [e.g., 18, 38–40], there is still little information on how climate change impacts and adaptation strategies vary across different smallholder farming systems and landscapes and the extent to which adaptation strategies need to be tailored to different smallholder contexts. There is also a lack of information on how smallholder maize and bean farmers in Central America are being affected by climate change (but see 40, 41 for information from Mexico), despite their importance for food security and poverty alleviation efforts.

To inform climate change adaptation planning for smallholder farmers, we surveyed smallholder farmers in 6 Central American landscapes to examine whether farmers perceive changes in climate, how they are being impacted by climate change, whether and how they are changing their agricultural systems to cope with or adapt to climate change impacts, and what adaptation support they require from government institutions. We also explored how climate change impacts and farmer responses varied across farming systems and landscapes. Our study provides policy-relevant information that is needed for developing robust and effective adaptation strategies for smallholder farmers across the region and mainstreaming smallholder farmer adaptation into climate change and sustainable development policies. It also provides important insights into the extent to which strategies for smallholder farmer adaptation need to be tailored for different socioeconomic and biophysical contexts.

Methods

Study landscapes

We explored perceptions of climate change, climate change impacts on crop production, and adaptation strategies with smallholder farmers located in 6 Central American landscapes (Turrialba and Los Santos in Costa Rica, Choluteca and Yoro in Honduras, and Chiquimula and Acatenango in Guatemala, Fig. 1) that are typical

Fig. 1 The six Central American study landscapes

of smallholder farmer landscapes in the region. The six landscapes were selected on the basis that they (a) were dominated by smallholder farming systems, (b) had coffee (*Coffee arabica*) and/or basic grain production (beans (*Phaseolus vulgaris*) and maize (*Zea mays*)) as the predominant agricultural land use, and (c) had farming communities with low adaptive capacity to climate change. We focused our study on coffee and basic grains as these the two most common types of smallholder systems in the region [39]. We characterized landscapes as having low adaptive capacity using expert mapping interviews, validation workshops, and expert online surveys, in which experts from the region characterized landscapes based on 20 variables (representing natural, human, social, physical, and financial capital) that contributed to farmer adaptive capacity (see [6] for details). The Turrialba and Los Santos landscapes are dominated by smallholder coffee production, Choluteca is dominated by basic grain production, while the remaining landscapes (Yoro, Chiquimula, and Acatenango) include a mix of coffee and basic grain production. Key characteristics of the landscapes, farms, and farmers surveyed can be found in Table 1.

Farmer surveys

In each of the six landscapes, we randomly selected smallholder farmers to be interviewed about climate change impacts, responses, and adaptation needs. The sampling method varied across countries due to differences in the availability of information on farmer populations in each country, but in all cases, the selection of farmers was random. In the Costa Rican landscapes, we selected farmers randomly from an existing list of coffee farms from the 2003–2006 coffee census [42]. In the Guatemalan and Honduran landscapes, we generated a sampling frame by using remote sensing imagery to detect household roofs and then randomly sampling households from this list of potential farms. In total, we sampled 860 randomly selected farmers (115–155 farmers per landscape). To ensure our sample size consisted of only smallholder farmers, we included only farmers who self-identified as smallholder farmers and whose farm area was within two standard deviations of the mean of the sampled population.

To document farmer perceptions of climate change, perceived climate change impacts, and adaptation strategies, we implemented a detailed household survey that collected information on farm characteristics, farmer and household socioeconomic characteristics, land use, farm management practices, farmer perceptions of climate change, climate change impacts, and farmer adaptation strategies. The survey was designed to allow us to establish causal links between farmer perceptions of climate change and perceptions of climate change impacts with the adaptation measures they had implemented. For example, we first asked farmers whether they had perceived changes in temperature over the last decade. If they responded positively, we then asked them about any perceived impact from the change in temperature on their crop production, and what farm management changes, if any, they had made to address these impacts. With this approach, we were able to distinguish adaptation decisions (i.e., decisions to adapt farm management practices in response to climate change) from the myriad of decisions taken by the farmer every day in response to other issues not related to climate change (such as fluctuations in coffee prices). The survey was piloted in the field prior to data collection and underwent an ethics review before implementation. Surveys were administered in the field by a team of enumerators who underwent formal training. For a subset of 300 randomly chosen farmers (50 per landscape), we also asked follow-up questions about what adaptation support they required.

All data were collected in handheld tablets using SurveyCTO software (www.surveycto.com), to minimize data entry errors. Surveys were conducted with the household head or family member in charge of the farm at the farmer's house or on the farm. Surveys took approximately 1 h to complete. All surveys were conducted between April and September 2014.

Data analysis

We used descriptive statistics to summarize the main trends in data relating to farmer perceptions of climate change, perceived impacts, and adaptation measures implemented in each landscape. We also used analysis of variance to compare means across the 6 landscapes and explore landscape-specific differences. In some cases, we analyzed data for coffee farmers, maize farmers, and bean farmers separately, to explore potential differences in perceptions, impacts, or adaptation strategies across farmers with different farming systems. In these instances, we classified 'coffee farmers' as all farmers who were growing and selling coffee ($n = 485$), 'maize farmers' as those growing maize on their land ($n = 490$), and 'bean farmers' as those who grow beans ($n = 383$). Of the 860 farmers, 129 farmers grew both coffee and basic grains on their farms. These farmers were included in both the coffee and basic grain categories in the analyses comparing different cropping systems. In the Turrialba and Los Santos landscapes, the number of farmers growing beans and maize present was insignificant ($n < 10$), so we opted not to report data for these two types of farmers in these landscapes. There were no coffee farmers in the Choluteca landscape, so no data on coffee farmers are

Table 1 Characteristics of landscapes, farmers, and farming systems where the impacts of climate change were studied

Characteristics	Turrialba, Costa Rica	Los Santos, Costa Rica	Acatenango, Guatemala	Chiquimula, Guatemala	Choluteca, Honduras	Yoro, Honduras	Total (all landscapes)	F and p values for comparisons across landscapes
Landscape characteristics								
Municipalities	Turrialba	Dota, Tarrazú, León Cortés	Acatenango, Alotenango, San Pedro Yepocapa	Quetzaltepeque, San Jacinto, San Juan Ermita	El Triunfo, Concepción de María	Yoro, Yorito, Victoria	—	—
Area (ha)	158,800	82,000	42,600	39,600	47,500	325,900	—	—
Holdridge life zone	Premontane Tropical Wet Forest, Premontane Rainforest	Lower Montane Wet Forest, Premontane Rainforest	Subtropical Wet Forest	Subtropical Moist Forest	Tropical Dry Forest	Subtropical Wet Forest Subtropical Moist forest	—	—
Mean annual temperature (°C)	19.1	17.5	19.5	22.2	27.0	22.2	—	—
Mean annual rainfall (mm)	2813	3371	2288	1210	1855	1389	—	—
Elevational range (masl)	41–3733	76–3368	471–3680	452–1779	4–775	73–2191*	—	—
Farmer and household characteristics								
n (all farmers surveyed)	144	151	149	115	155	146	860	—
# (and %) who were men	122 (84.7)	136 (90.1)	114 (76.5)	85 (73.9)	110 (71.0)	107 (73.3)	674 (78.4)	—
Mean (±SE) age of household head	58.2±1.2a	51.9±1.2b	48.2±1.2cd	46.0±1.4d	49.8±1.2bc	48.1±1.2cd	50.5±05	$F=11.8$ $p<0.0001$
Mean household size (±SE)	3.4±0.2b	3.7±0.2 b	5.0±0.2a	5.4±0.2a	5.4±0.2a	5.5±0.2a	4.7±0.03	$F=30.4$ $p<0.0001$
Mean number of years of farming experience (±SE)	29.0±1.4c	29.3±1.3c	29.7±1.5c	32.2±1.3bc	34.7±1.3b	39.6±1.1a	32.5±0.1	$F=11.8$ $p<0.0001$
# (and %) of farmers with at least one family member older than 18 years old that has migrated to another community	51 (35.4)	32 (21.2)	16 (10.7)	15 (13.0)	53 (34.2)	45 (30.8)	212 (24.7)	—

Table 1 (continued)

Characteristics	Turrialba, Costa Rica	Los Santos, Costa Rica	Acatenango, Guatemala	Chiquimula, Guatemala	Choluteca, Honduras	Yoro, Honduras	Total (all landscapes)	F and p values for comparisons across landscapes
Education levels								
# (and %) who have not completed primary school	54 (37.5)	33 (21.9)	103 (69.1)	79 (68.7)	102 (65.8)	99 (67.8)	470 (54.7)	–
% who only have a primary school education	68 (47.2)	83 (55)	28 (18.8)	27 (23.5)	45 (29)	33 (22.6)	284 (33.0)	–
# (and %) who have more than a primary education	22 (15.3)	35 (23.2)	18 (12.1)	9 (7.8)	8 (5.2)	14 (9.6)	106 (12.3)	–
Access to services								
# (and %) with access to electricity	139 (96.5)	149 (98.7)	138 (92.6)	98 (85.2)	68 (43.9)	55 (37.7)	647 (75.2)	–
# (and %) with access to tap water	137 (95.1)	140 (92.7)	139 (93.3)	92 (80.0)	62 (40.0)	115 (78.8)	685 (79.7)	–
# (and %) that own cell phone	126 (87.5)	143 (94.7)	114 (76.5)	95 (82.6)	115 (74.2)	92 (63.0)	685 (79.7)	–
# (and %) that have received at least on agronomic training in the past 2 years	18 (12.5)	68 (45.0)	27 (18.1)	17 (14.8)	23 (14.8)	33 (22.6)	186 (21.6)	–
# (and %) that have received a visit from an agronomist in past 2 years	32 (22.2)	83 (55.0)	22 (14.8)	5 (4.3)	20 (12.9)	29 (19.9)	191 (22.2)	–
Household living conditions and assets								
# (and %) of farmers living in a cement block and/or pre-fabricated house	102 (70.8)	127 (84.1)	97 (65.1)	53 (46.1)	26 (16.8)	21 (14.4)	426 (49.5)	–
# (and %) of farmers living in a house with dirt floor only	0	0	38 (25.5)	47 (40.9)	61 (39.4)	63 (43.2)	209 (24.3)	–
# (and %) that own a car or motorcycle	71 (49.31)	136 (90.07)	38 (25.5)	25 (21.74)	25 (16.13)	27 (18.49)	322 (37.44)	–

Table 1 (continued)

Characteristics	Turrialba, Costa Rica	Los Santos, Costa Rica	Acatenango, Guatemala	Chiquimula, Guatemala	Choluteca, Honduras	Yoro, Honduras	Total (all landscapes)	F and p values for comparisons across landscapes
# (and %) having enough food throughout the year	138 (95.8)	147 (97.4)	97 (65.1)	56 (48.7)	56 (36.1)	56 (38.6)	550 (64)	–
Farm characteristics								
Mean farm size (ha)	2.02±0.05b	3.20±0.05a	0.93±0.05c	1.12±0.06c	1.90±0.05b	2.16±0.05b	1.83±0.01	$F=32.22$ $p<0.0001$
# (and %) of farmers who own land	133 (92.4)	145 (96.0)	93 (62.4)	43 (37.4)	68 (43.9)	98 (67.1)	580 (67.4)	–
# (and %) of farmers who rent land	1 (0.7)	0	18 (12.1)	37 (32.2)	45 (29.0)	14 (9.6)	115 (13.4)	–
# (and %) of farmers who own and rent land	2 (1.4)	0	26 (17.4)	16 (13.9)	10 (6.5)	8 (5.5)	62 (7.2)	–
# (and %) with at least one piece of land under a different land tenure system (e.g., shared, communal or lent)	8 (5.6)	6 (4.0)	12 (8.1)	19 (16.5)	32 (20.6)	26 (17.8)	103 (12.0)	–
Coffee production								
# (and %) of farmers who grow coffee	144 (100)	151 (100)	112 (75.2)	26 (22.6)	–	81 (55.5)	514 (59.8)	–
# (and %) of coffee farmers who sell coffee	144 (100)	151 (100)	112 (100)	17 (65.4)	–	61 (75.3)	485 (94.4)	–
Mean area (±SE) under coffee production (ha)	1.13±0.05b	2.52±0.05a	0.75±0.05c	0.50±0.14c	–	1.47±0.07b	1.41±0.01	$F=32.1$ $p<0.0001$
Maize production								
# (and %) of farmers who grow maize	8 (5.6)	4 (2.7)	89 (59.7)	114 (99.1)	155 (100)	120 (82.2)	490 (57)	–
# (and %) of maize farmers who sell maize	3 (37.5)	2 (50.0)	20 (22.5)	23 (20.2)	51 (32.9)	23 (19.2)	122 (24.9)	–
Mean area (±SD) per farmer under maize production (ha)	–	–	0.48±0.04c	0.98±0.03 b	1.20±0.03a	0.88±0.03b	0.92±0.01	$F=25.35$ $p<0.0001$

Table 1 (continued)

Characteristics	Turrialba, Costa Rica	Los Santos, Costa Rica	Acatenango, Guatemala	Chiquimula, Guatemala	Choluteca, Honduras	Yoro, Honduras	Total (all landscapes)	F and p values for comparisons across landscapes
Bean production								
# (and %) who grow beans	10 (6.9)	3 (2.0)	64 (43.0)	108 (94.0)	98 (63.2)	100 (68.5)	383 (44.5)	–
# (and %) of bean farmers who sell beans	1 (10)	2 (66.7)	20 (31.3)	31 (28.7)	17 (17.4)	24 (24.0)	95 (24.8)	–
Mean area (± SE) per under bean production (ha)	–	–	0.26 ± 0.04c	0.45 ± 0.03 b	0.55 ± 0.03b	0.95 ± 0.03a	0.57 ± 0.01	$F = 34.23$ $p < 0.0001$
Mixed systems								
# (and %) who grow both maize and beans	5 (3.5)	3 (2.0)	60 (40.3)	108 (93.9)	98 (63.2)	95 (65.1)	369 (42.9)	–
# (and %) who grow both coffee and basic grains (maize and/or beans)	13 (9)	4 (2.6)	56 (37.6)	16 (13.9)	–	40 (27.4)	129 (15)	–

Different letters within the same row denote significant differences across landscapes, based on ANOVAs ($p < 0.05$)

reported for this landscape. All analyses were done using the statistical package InfoStat [43].

Results

Characteristics of smallholder farmers and farms

We interviewed 860 smallholder farmers, of whom 674 (78.4%) were men. The mean age of the farmers was 50.5 ($\pm$0.5 SE) years old, and they had an average of 32.5 ($\pm$0.1) years of farming experience. The farmers had a mean of 4.7 ± 0.03 family members, small farms (mean of 1.83 ± 0.01 ha), and low levels of education, with 54.7% lacking a primary school education (Table 1). Many lacked access to electricity, tap water, cell phones and transportation, and a subset reported being food insecure. Few reported having access to extension support or agricultural training events. However, there were significant differences across smallholder farmers in different landscapes. For example, Costa Rican farmers generally had higher education, higher food security, and greater access to services than farmers in other countries. Almost all farmers (>92%) in the Costa Rican landscapes owned the land they cultivated, while in the remaining countries, many farmers rented part of the land they cultivate or had alternative tenure arrangements.

Of the farmers interviewed, 59.8% had coffee plantations, with an average coffee area of 1.41 ha ($\pm$0.01). Of the farmers growing coffee, 94% sold their coffee production. Fifty-seven percent of the farmers grew maize, with an average area of 0.92 ha ($\pm$0.01). Most maize production was for home consumption, with less than a quarter of farmers selling their maize production. Forty-four percent of the farmers grew beans, with an average bean area of 0.57 ha ($\pm$0.01). Like maize production, most bean production was for home consumption and only 24.8% of farmers producing beans sold their produce. Of the 860 farmers, 15% of the farmers grew both coffee and basic grains (either maize and/or beans), and 42.9% grew both maize and beans.

Farmer perceptions of climate change and climate change impacts

Ninety-five percent of all farmers surveyed reported that they have perceived changes in local climate over the last decade. Of those who perceived the climate to be changing, 96.3% reported changes in temperature and 94.6% in rainfall (Table 2). The most commonly reported change among those farmers perceiving changes in overall temperature was temperature rise (mentioned by 96.1% of farmers). Among farmers perceiving changes in overall

Table 2 Percent of smallholder farmers reporting changes in climate over the last decade

% of farmers perceiving changes over the last decade	Turrialba, Costa Rica	Los Santos, Costa Rica	Acatenango, Guatemala	Chiquimula, Guatemala	Choluteca, Honduras	Yoro, Honduras	Total (across 6 landscapes)
Overall changes noted							
n (all farmers surveyed)	144	151	149	115	155	146	860
Overall change in climate	98.6	97.4	92.0	96.5	93.6	94.5	95.4
n (farmers perceiving change in climate)	142	147	137	111	145	138	820
Overall change in temperature	97.2	98.6	96.4	91.0	98.6	94.9	96.3
Overall change in rainfall	95.1	95.9	92.7	99.1	96.6	89.1	94.6
Specific perceived changes in temperature							
n (farmers perceiving changes in temperature)	138	145	132	101	143	131	790
The temperature has increased	97.1	96.6	96.2	94.1	95.8	96.2	96.1
Night and day are both warmer	2.9	0.7	21.2	15.8	21.0	6.9	11.1
Cool season and warm season are both warmer	4.4	8.3	8.3	8.9	4.9	8.4	7.1
Specific changes in rainfall noted							
n (farmers perceiving changes in rainfall)	135	141	127	110	140	123	776
Less rain falls in a year	66.7	66.7	46.5	85.5	90.0	56.1	68.6
When the rains will begin is unknown	26.7	19.2	15.8	13.6	29.3	30.9	22.8
Rainfall is concentrated in a shorter period	20.0	38.3	31.5	8.2	3.6	9.8	18.9
When it will rain is unknown	20.7	17.7	22.8	13.6	18.6	8.9	17.3
Rainy season begins later than usual	5.9	6.4	18.1	12.7	14.3	23.6	13.3
More rain falls in a year	4.4	1.4	27.6	7.3	2.9	5.7	8.0

The table shows only those changes that were reported by at least 5% of all farmers surveyed

rainfall the most common changes included lower annual rainfall (68.6%), and greater uncertainty in when the rains will begin (mentioned by 22.8%). Many farmers also reported changes in the seasonality of the rainy season. Farmer perceptions of how rainfall patterns had changed differed across landscapes (Table 2).

Most smallholder farmers reported that climate change has negatively impacted their crop production, with 87% of maize farmers, 78.4% of bean producers, and 66.4% of coffee farmers reporting reduced yields (Fig. 2a). In three of the four landscapes where both basic grains and coffee are grown, a greater proportion of farmers reported impacts on maize and bean production than

on coffee production. Smallholder farmers also attributed increases in pest and disease outbreaks to climate change, with 73.9% of coffee farmers, 78.4% of maize farmers, and 67.9% of bean farmers reporting climate-associated increases in crop pests and diseases over the past 10 years (Fig. 2b). However, the percent of farmers indicating climate-related disease and pest outbreaks differed across landscapes.

Effects of extreme weather events on agricultural production, food security, and income

Many farmers reported that their agricultural systems had been affected by extreme weather events during

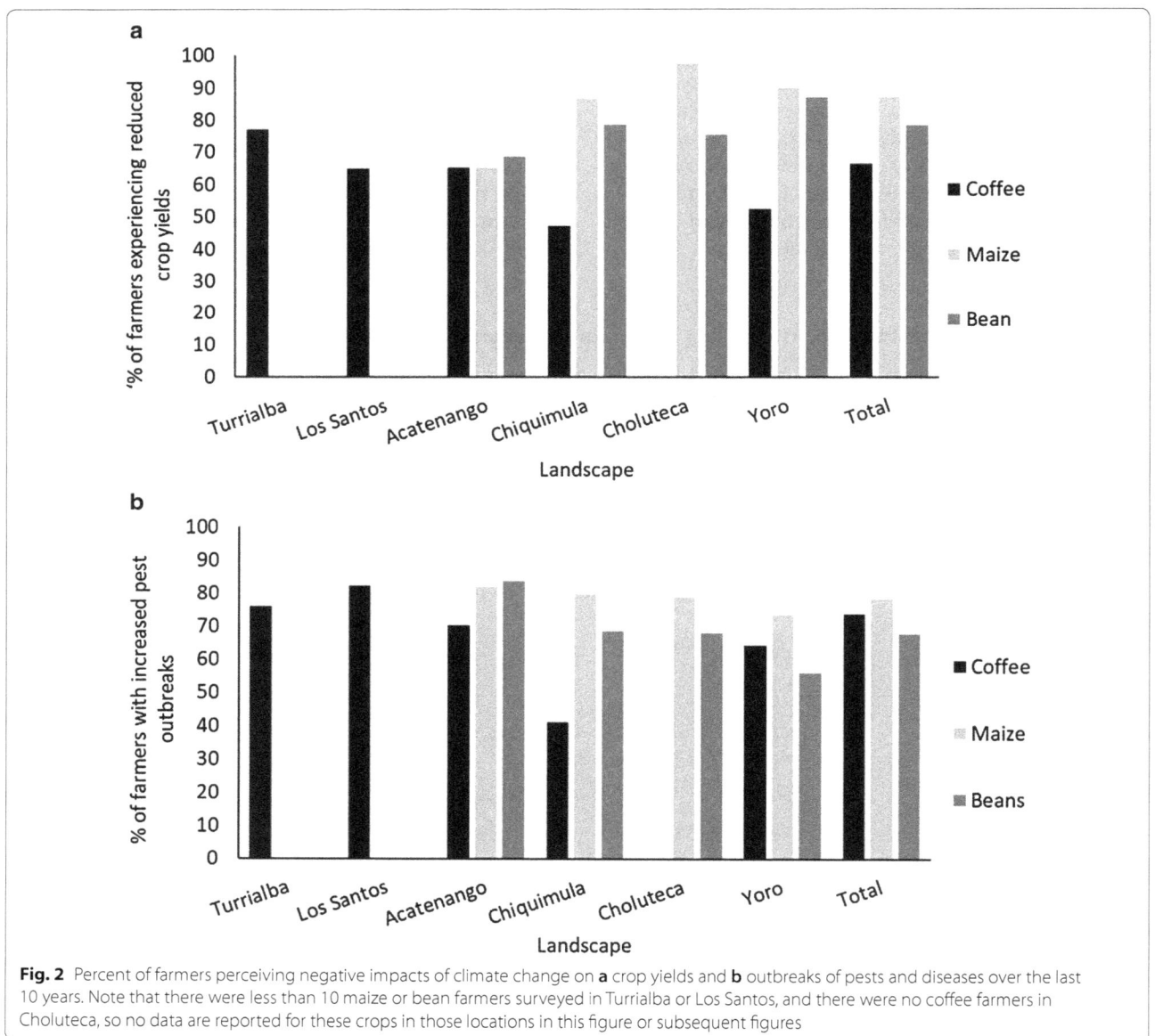

Fig. 2 Percent of farmers perceiving negative impacts of climate change on **a** crop yields and **b** outbreaks of pests and diseases over the last 10 years. Note that there were less than 10 maize or bean farmers surveyed in Turrialba or Los Santos, and there were no coffee farmers in Choluteca, so no data are reported for these crops in those locations in this figure or subsequent figures

the last decade (Fig. 3a): 57.3% reported having been impacted by droughts, 52.0% by heavy rainfall events, 32.1% by hurricanes, and 12.6% by floods. The percentages of farmers reporting these events and the main climatic stresses varied across landscapes. In Chiquimula, Choluteca, and Yoro, drought was the main climatic stress for farmers. In Acatenango, hurricanes and heavy rainfall were the main challenges, while in Turrialba and Los Santos, farmers were exposed to multiple climatic stresses.

Extreme weather events have had significant impacts on smallholder crop production: 94.6% of maize farmers, 78.8% of coffee farmers, and 77.5% of bean farmers reported that extreme weather events had negatively affected their crop yields over the last 10 years (Fig. 3b). Maize and bean farmers generally reported higher crop losses due to extreme weather events than did coffee farmers: 50.9% of all maize farmers reported losses of >50% of their crop due to extreme weather events, and an additional 37.6% reported losses of 25–50% of their crop. Bean production was also significantly impacted, with 44.8% of farmers growing beans reporting losses of more than half their crop due to extreme weather events. In contrast, only 16.4% of farmers growing coffee reported high (>50%) losses due to extreme weather events.

Extreme weather events have also led to increased food insecurity among smallholder farmers in some of the landscapes (Fig. 3c). Overall, 32.4% of smallholder farmers reported food insecurity following the most severe extreme weather events; however, there were pronounced differences across landscapes, with highest food insecurity in Choluteca (affecting 63% of farmers) and Yoro (affecting 44.5%). In contrast, in the Costa Rica landscapes, less than 7% of farmers indicated food shortages following extreme weather events. There were also differences in food insecurity levels across different types of farmers: 47% of both maize and bean farmers indicated they had experienced food shortages following extreme weather events, compared to only 15.1% of coffee farmers.

Across the 6 landscapes, 44.5% of farmers reported experiencing decreases in household income following extreme weather events (Fig. 3d). The percent of farmers experiencing income reductions following extreme weather events was highest in the Honduran landscapes (Choluteca and Yoro). A greater proportion of maize and bean farmers (56.7% in both cases) reported impacts on household income, compared to coffee (29.9%); maize and bean farmers also reported more significant income reductions than coffee farmers.

Adaptation strategies used by smallholder farmers

Of the 820 farmers perceiving changes in climate, 46.1% indicated that they had changed their farming practices in response to climate change (Fig. 4). The percent of farmers changing farming practices in response to climate change was highest in the Los Santos (78.2%) and lowest in Chiquimula (24.3%). There were also differences in the frequency of use of adaptation strategies across different types of farmers: 58.7% of coffee farmers had made changes in response to climate change, compared to only 35.5% of basic grain farmers. On average, farmers who had changed their management practices in response to climate change had implemented an average of 1.5 (± 0.12) adaptation practices.

Smallholder farmers had implemented a variety of adaptation practices, including agroforestry and restoration activities, the adoption of agroecological practices, the use of intensification, and the use of new crop varieties and technologies (Table 3). The most common practice used by both coffee and basic grain farmers was the planting of more trees on farms (reported by 48.1% of all coffee farmers who implemented adaptation measures and by 33.1% of all maize farmers who implemented adaptation measures). Among coffee farmers who had implemented adaptation practices, other common practices included applying using more pesticides, herbicides, and fungicides (17.2%), adopting soil and water conservation practices (15.3%), and using more fertilizers (14.8%). Among basic grain farmers who implemented adaptation measures, other specific adaptation practices included the adoption of soil and water conservation practices (13.4%) and changing agricultural calendars (15.7%).

(See figure on next page.)
Fig. 3 Percent of smallholder farmers who have experienced **a** an extreme weather event, **b** reductions in crop yields, **c** food insecurity or **d** reductions in household income due to extreme weather events. For figures **b** and **d**, the percentages reflect the farmer recollections of the percent of income or crop yield loss experienced following the extreme weather event that according to the farmer most affected its home, land, or parcel. Sample sizes vary across figures due to the sequencing of the survey: **a** the sample size corresponds to all 860 interviewed farmers, **b** sample size corresponds to total number or coffee ($n = 287$), maize ($n = 413$), and bean ($n = 318$) farmers indicating being affected by at least one extreme weather event and reporting general impacts on crop production; **c** sample size corresponds to coffee ($n = 398$), maize ($n = 457$), and bean ($n = 353$) farmers reporting being affected by at least on extreme event; and **d** sample size corresponds to total number or coffee ($n = 398$), maize ($n = 446$), and bean ($n = 342$) farmers indicating being affected by at least on extreme weather event and reporting general impacts on household income

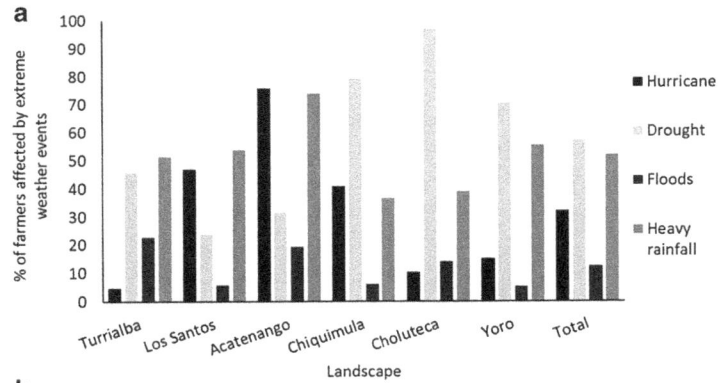

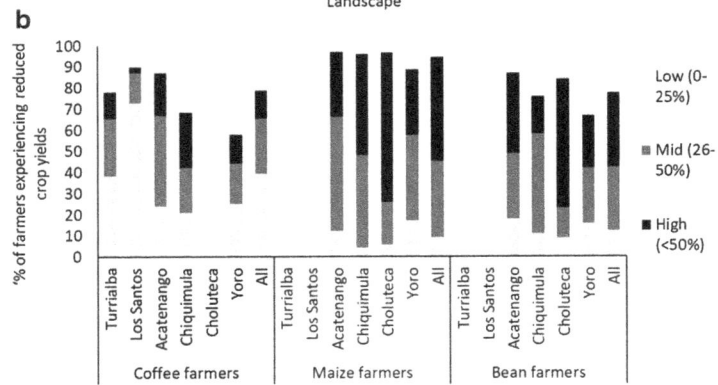

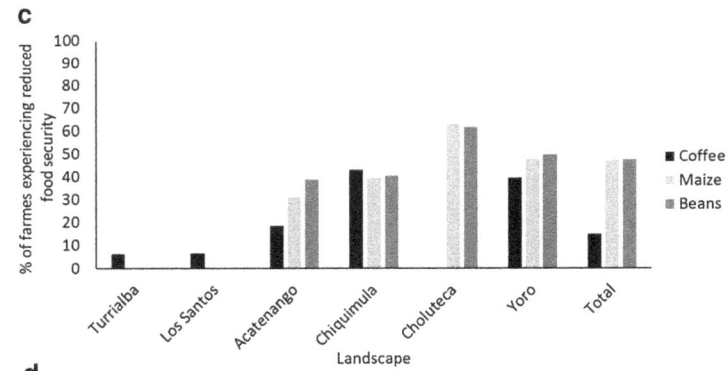

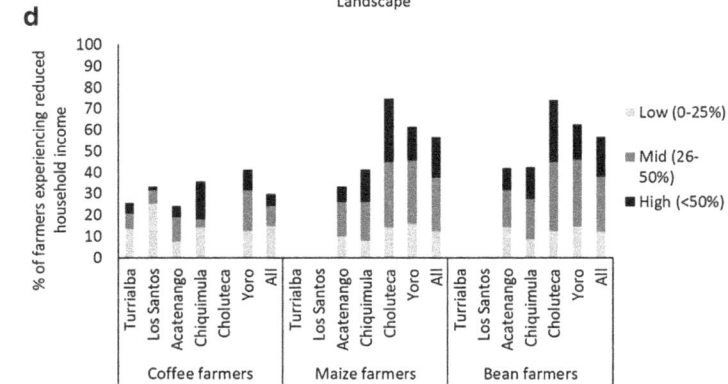

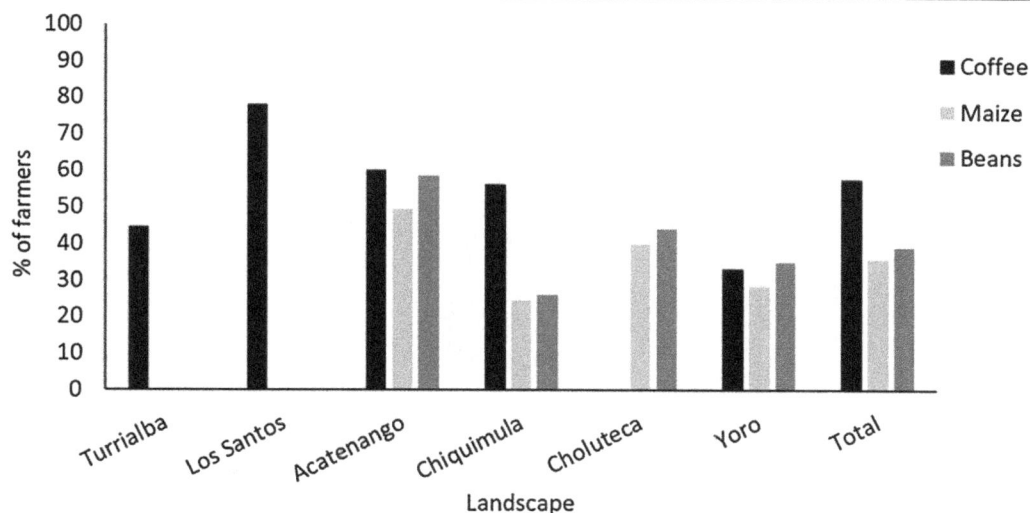

Fig. 4 Percent of smallholder farmers who have changed farming practices in response to climate change

Farmer adaptation needs

Smallholder farmers identified numerous ways in which governments, agricultural research centers, and other institutions could help them adapt to climate change (Additional file 1: Table S1). Among the subset of 188 coffee farmers who were asked about adaptation needs, the most commonly mentioned needs were the provision of fertilizers and agrochemicals (mentioned by 40.4%), technical support (40.4%), training (18%), improving coffee prices and marketing (17.0%), and providing access to finance (14.4%). However, there were landscape-level differences in the identified needs. For example, the most commonly mentioned need by coffee farmers in the Guatemalan and Honduran sites was the provision of agrochemicals, while in the Costa Rica sites, the highest demand was for technical support.

Among the subset of 169 basic grain farmers who were asked about adaptation needs, the most commonly mentioned need was the provision of agrochemical inputs (mentioned by 84.0%), followed by financial incentives (33.1%), and provision of improved varieties and seeds (30.2%). Other needs—such as technical support and training—were also mentioned, but with lower frequencies. Specific adaptation needs varied across landscapes (S1).

Discussion

Climate change perceptions, impacts, and responses of smallholder farmers

Our study indicates that climate change is already putting a significant pressure on smallholder coffee and basic grain farmers across Central America, and highlights the urgent need to build farmer resilience to sustain food security and maintain rural livelihoods under changing climatic conditions. Like studies of smallholder farmers in other developing regions [e.g., 44–47], we found that almost all of the Central American smallholder farmers surveyed had perceived changes in their local climate over the last decade, regardless of their farming system or the landscape in which they were located. The most commonly observed changes include rising temperatures, more variable rainfall, and changes in the onset and length of the rainy season. Farmers' perceptions of climate change generally mirrored historical climatic data for the region which show significant increases in mean temperatures and increases in maximum temperatures across most of Central America, but less clear trends in rainfall patterns [31, 48]. Most farmers (87.2%) also reported having been affected by at least one extreme weather event during the last decade, and many considered that the frequency and intensity of extreme weather events were increasing, as is also suggested by projections from climate models [30]. Smallholder farmers are keenly aware of changing climatic conditions because they plan their planting, management, and harvesting activities in response to seasonal rainfall patterns [15, 41, 49]. They also see visible impacts of extreme temperatures, droughts, or torrential rains on plant growth, flowering, coffee berry ripening, and pest and disease incidence [25, 26, 50].

Changing climatic conditions are already presenting a significant challenge to smallholder coffee and maize farmers in Central America. Across the region, most smallholder farmers attributed reductions in crop yields and changes in pest and disease outbreaks (such as coffee

Table 3 Number (and %) of smallholder farmers who have implemented adaptation measures in response to climate change

Adaptation measures		Coffee farmers						Maize/beans farmers				
		Turrialba	Los Santos	Acatenango	Chiquimula	Yoro	Total	Acatenango	Chiquimula	Choluteca	Yoro	Total
n (farmers who took an action to cope with changes in temperature and/or precipitation)		63	115	62	9	19	268	42	27	58	36	168
Agroforestry and restoration	Plant trees	16 (25.4)	73 (63.5)	26 (41.9)	3 (33.3)	11 (57.9)	129 (48.1)	11 (26.2)	7 (25.9)	25 (43.1)	11 (30.6)	56 (33.1)
	Restore degraded areas	1 (1.6)	1 (0.9)	1 (1.6)		2 (10.5)	5 (1.9)			7 (12.1)	3 (8.3)	11 (6.4)
	Stop cultivating certain areas	2 (3.2)		1 (1.6)			3 (1.1)	2 (4.8)		4 (6.9)	3 (8.3)	9 (5.2)
Agroecological practices	Introduce soil management and conservation practices	6 (9.5)	20 (17.4)	13 (21)		2 (10.5)	41 (15.3)	7 (16.7)		12 (20.7)	4 (11.1)	22 (13.4)
	Introduce water management and conservation practices	4 (6.3)	6 (5.2)	2 (3.2)		1 (5.3)	13 (4.8)	2 (4.8)		3 (5.2)		8 (4.7)
	Reduce use of pesticides, herbicides, fungicides	9 (14.3)	14 (12.2)	2 (3.2)	1 (11.1)	1 (5.3)	27 (10.1)	3 (7.1)	1 (3.7)	2 (3.5)		6 (3.5)
	Reduce fertilizer use	3 (4.8)	5 (4.3)	1 (1.6)			9 (3.4)	1 (2.4)		1 (1.7)		2 (1.2)
Intensification	Increase fertilizer	10 (15.9)	17 (14.8)	11 (17.7)		2 (10.5)	40 (14.9)	7 (16.7)	1 (3.7)	2 (3.5)	7 (19.4)	17 (9.9)
	Increased use of pesticides, herbicides, fungicides	10 (15.9)	32 (27.8)	4 (6.5)			46 (17.2)	4 (9.5)		1 (1.7)	3 (8.3)	9 (5.2)
	Cut trees	1 (1.6)	1 (0.9)	6 (9.7)			8 (3)	1 (2.4)				
	Increase the number of animals									2 (3.5)		2 (1.2)
New crops and technologies	Change agricultural calendar	1 (1.6)	9 (7.8)	3 (4.8)	1 (11.1)	1 (5.3)	15 5.6	5 (11.9)	3 (11.1)	7 (12.1)	12 (33.3)	26 (15.7)
	Introduce new crops	3 (4.8)	1 (0.9)	7 (11.3)	2 (22.2)	3 (15.8)	16 (6)	3 (7.1)	3 (11.11)	4 (6.9)	2 (5.6)	12 (7)
	Diversify production	2 (3.2)	1 (0.9)	1 (1.6)			4 (1.5)				1 (2.8)	1 (0.6)
	Crop irrigation	6 (9.5)	1 (0.9)	2 (3.2)		1 (5.3)	10 (3.7)	2 (4.8)	1 (3.7)		1 (2.8)	5 (2.9)
	Change variety	1 (1.6)	1 (0.9)	1 (1.6)		1 (5.3)	4 (1.5)	1 (2.4)		3 (5.2)	2 (5.6)	6 (3.5)
	Change general management practices	17 (27)	28 (24.4)	15 (24.2)	4 (44.4)	4 (21.1)	68 (25.4)	11 (26.2)	10 (37)	11 (19)	7 (19.4)	41 (24.4)

Table 3 (continued)

Adaptation measures	Coffee farmers						Maize/beans farmers				
	Turrialba	Los Santos	Acatenango	Chiquimula	Yoro	Total	Acatenango	Chiquimula	Choluteca	Yoro	Total
Other	10 (15.9)	7 (6.1)	14 (22.6)	3 (33.3)	–	34 (12.7)	10 (23.8)	7 (25.9)	6 (10.3)		22 (13.4)
Mean number ($\pm$ SE) of specific changes in farming practices made by farmers to cope with changes in temperature and/or precipitation	1.62 ± 0.11 a	1.89 ± 0.10 a	1.77 ± 0.13 a	–	1.53 ± 0.18 A	1.73 ± 0.13	1.67 ± 0.13 a	1.22 ± 0.08 b	1.55 ± 0.11 ab	1.56 ± 0.11 ab	1.50 ± 0.12

Note that there were less than 10 maize or bean farmers surveyed in Turrialba or Los Santos, and there were no coffee farmers in Choluteca, so no data are reported for these crops in those locations

Different letters within the same row denote significant differences among coffee farmers (or maize/bean farmers) across landscapes, based on ANOVAs ($p < 0.05$)

leaf rust, fall armyworm, and others) to rising temperatures and changing precipitation patterns. They also reported detrimental impacts of extreme weather events on crop yields, pest and disease incidence, household income, and, in some cases, household food security. Although the perceived impacts varied across households and landscapes, the magnitude of potential climate change impacts on smallholder farmer was significant: of the smallholder farmers who were affected by an extreme weather in the last decade, 32.4% reported being food insecure following the extreme weather event and 27.5% reported losing more than half their household income. Our study provides novel information on how climate change is affecting smallholder basic grain farmers across Central America, a group that is critical for regional food security, and strengthens the evidence based on coffee farmers are being impacted across Mesoamerica [e.g., 18, 28, 38, 39]. Collectively, these results highlight the potential for climate change to have significant economic and social impacts across the region, unless action is taken to help smallholder farmers cope with and adapt to these changes.

Interestingly, although most coffee and basic grain smallholder farmers perceived climate change to be occurring and reported significant impacts on their farms and livelihoods, less than half of them had actively changed their farm management practices to minimize impacts or adapt to these changes. Even among those who did change their farm management practices in response to climate change, most had adopted only two or two practices (mean of 1.5 practices). This gap between the perceptions of climate change and implementation of adaptation measures has been noted in studies of smallholder farmers elsewhere [e.g., 18, 51, 52] and is thought to reflect their low adaptive capacity. In our study, the limited use of adaptation strategies probably reflects the fact that farmers have small plots of land, limited capital and labor, low education, and little access to finance or technical support and are therefore constrained in their ability to invest in their farms and adopt management practices which could enhance farm resiliency. In addition, some smallholder farmers (particularly maize and bean farmers in Honduras and Guatemala) have insecure land tenure and either rent the land that that they cultivate, use communal land, or share land with other farmers. Our parallel field survey of farm practices used by these smallholder farmers [53] found that smallholder farmers with insecure land tenure were less likely to have implemented adaptation strategies than those farmers who owned their land, because they were unwilling to make long-term investments in practices that yield long-term benefits. Insecure land tenure, limited capital, low education, and lack of access to financial and technical

support have also been identified as key constraints to adaptation elsewhere [e.g., 6, 16, 54, 55]. The fact that farmers identified the need for government support in providing agricultural inputs, technical support, training, and access to finance further corroborates that these factors serve as constraints to farmer adaptation.

Among those smallholder farmers who implemented adaptation measures, the most common actions included planting of trees, adopting soil conservation practices, increasing fertilizer or agrochemical use, and introducing new crops. Some of these practices are considered 'Ecosystem-based Adaptation' (i.e., adaptation practices that are based on the conservation, restoration, and sustainable management of biodiversity and ecosystem services [56]) and are known to improve the long-term resiliency of smallholder farming systems [57]. For example, the incorporation of trees as shade is known to buffer extreme temperatures within fields [21, 58], mitigate the impacts of extreme weather events [59, 60], and ensure the provision of ecosystem services [61, 62] and can therefore help enhance adaptation of farming systems. Farmers in Chiapas, Mexico, for example, planted more shade trees in their coffee plots as a response to Hurricane Stan [26]. Similarly, the planting of fruit trees has been reported as a means of reducing food insecurity of farmers under changing climatic conditions [49, 63]. However, other practices being used by smallholder farmers in our study landscapes, such as the increased use of fertilizers and agrochemicals to ensure yields under adverse climatic conditions, are resource-intensive solutions that are short-term fixes that are unlikely to contribute to climate resilience and could even be counterproductive to long-term adaptation efforts [15].

Differences in climate change impacts and responses across smallholder farming systems and landscapes

While almost all smallholder farmers perceived climate change to be happening and reported climate change impacts, farmers varied in the impacts they experienced, their use of adaptation strategies, and their adaptation needs. In general, maize and bean farmers appeared to be slightly more affected by impacts of climate change than coffee farmers, with a larger percentage of basic farmers reporting impacts of climate change on their crop production than coffee farmers. Smallholder basic grain farmers also appeared to suffer greater impacts of extreme weather events, with a greater proportion of basic grain farmers suffering food insecurity (47 vs. 15.1% of coffee farmers), and experiencing a loss in household income (56.7% vs. 29.9%) following extreme events. In addition, the magnitude of crop loss and income losses following extreme weather events were higher among smallholder farmers. The higher vulnerability of basic grain farmers probably reflects the fact that these farmers tend to cultivate slightly smaller plots of land, are directly dependent on their plot for food security, are less likely to own the land they cultivate, and are often much poorer than their coffee counterparts [41, 64]. It also reflects the differential vulnerability of the agroecosystems: in the landscapes we studied, coffee is traditionally planted in agroforestry systems [53], which are more likely to endure climate change impacts than non-agroforestry systems such as basic grains [65, 66]. Basic grain farmers are also less likely to have access to technical support than coffee farmers: only 13.1% of basic grain farmers had been visited by agronomists in the last 2 years, compared to 31.3% of coffee farmers. Whereas coffee farmers are also highly vulnerable to climate change and face many of the same constraints to adaptation, they appear to be slightly better off, as they obtain annual income from coffee production, are more integrated into markets, and are more likely to have access to some technical advice or support to cope with climate change impacts [67].

There was also heterogeneity in climate change impacts and farmer responses across the six landscapes studied. Smallholder farmers in different landscapes were exposed to different types of climatic stresses, with some farmers being most affected by droughts and others struggling with the impacts of excessive rainfall or hurricanes. There were also landscape-level differences in the percent of farmers reporting different climate change impacts and the severity of these impacts. One clear difference was the fact that smallholder farmers in Costa Rica were much less likely to be food insecure following extreme weather events than farmers in Honduras or Guatemala. Costa Rican farmers were generally better off than farmers in the surrounding countries, with higher education levels, better homes, and better access to services (communication, electricity, water), and it is likely that these characteristics improve farmers' ability to cope with climatic stresses. Another key difference was that a much higher proportion of farmers in Los Santos, Costa Rica, had adopted adaptation measures in response to climate change than in other landscapes. We suspect this is due to the presence of strong coffee cooperatives in the region (e.g., CoopeTarrazu, CoopeDota) which provide farmers with technical support and advice. While more research is needed to understand the factors that influence the use of adaptation measures by smallholder farmers, the differences among smallholder farmers cultivating different crops and living in different geographic settings point to the very specific nature of climate change impacts, the varying adaptive capacities of smallholder farmers, and the need for locally tailored adaptation solutions.

Conclusions and policy implications

Our study has several key implications for policy makers, donors, and practitioners interested in enhancing smallholder farmer resilience to climate change in the Central American region. First, our study suggests that there is an urgent need to ramp up efforts to help smallholder farmers cope with existing changes and adapt to future climatic conditions. Climate change is already having significant adverse impacts on smallholder coffee and basic grain farmers across the region and could undermine national and regional efforts to alleviate poverty, achieve food security, and enhance economic development. Facilitating smallholder farmer adaptation to climate change will require a combination of policy, technical, and research solutions, including the development of adaptation policies and programs targeted at smallholder farmers, the creation of incentives, credits, and other financing mechanisms to support farmer adaptation efforts, the development of research to identify the most effective adaptation options, and the strengthening of extension services to provide technical support to farmers on how to enhance their resilience to climate change [10, 52, 53], among other activities. As Central American countries develop their national climate adaptation strategies and plans for achieving the UN Sustainable Development Goals, the adaptation needs of smallholder farmers merit special attention.

Second, there is a need to identify adaptation strategies that are accessible for smallholder farmers and fit their agroecological and socioeconomic contexts. A wide range of adaptation options have been proposed for helping farmers adapt to climate change. These include planting new crop varieties that are heat tolerant, drought resistant, or less susceptible to pests and diseases [38, 39], increasing fertilizer and pesticide use [52], improving water management through irrigation and water harvesting [39, 47], changing farm management practices such as changes in planting dates or crop rotations [46], adopting soil conservation practices such as live barriers, cover crops, and terracing [64, 68], diversifying crop production and household income sources [49, 62, 69], and restoring degraded areas and risk-prone sites [38], among others. All of these options merit inclusion in adaptation strategies. However, many of the technological adaptation strategies (such as the planting of new varieties, establishment of irrigation systems, or an increased use of fertilizers and agrochemicals) are resource-intensive and are often beyond the reach of smallholder farmers who have limited capital, family labor, and access to credit or finance [51]. Ecosystem-based Adaptation measures, such as adding trees to coffee systems to buffer the impacts of extreme weather events or diversifying crop production to reduce the risk of crop losses, may be more accessible to smallholder farmers as they are based on the management of existing resources [53, 57]. Adaptation plans should therefore support a diverse menu of adaptation practices that farmers can select from and modify based on their contexts, needs, and experiences. Farmers should also be encouraged to develop adaptation strategies that combine multiple adaptation practices that together provide long-term resilience, rather than adopting individual practices which, on their own, may provide more limited adaptation benefits.

Finally, given the heterogeneity of different smallholder farmer contexts and experiences with climate change, it will be important to develop adaptation strategies that are flexible and can be tailored to specific farming contexts and climatic stresses. Adaptation strategies need to consider the diversity of farming systems, socioeconomic conditions (e.g., poverty, land tenure, food insecurity), and climatic stresses that smallholder farmers face. Emphasis should be placed on building farm resilience to both climate change and other stressors [9], and ensuring farmers can both cope with existing changes and adapt to future conditions. For smallholder farmers who are living on the edge, any efforts to increase their resiliency to climate change must begin by addressing the underlying poverty and food insecurity they face, securing access to electricity, running water and other key services, improving crop productivity and income generation, and securing land tenure [18], as these stresses make them highly vulnerable to climate change and, if unaddressed, will undermine adaptation efforts [46]. Access to disaster relief following extreme weather events, such as droughts or hurricanes, is also critical for ensuring smallholder farmers do not get trapped in poverty [67]. Once the basic needs of smallholder farmers are met and their vulnerability is reduced, emphasis can be placed on building capacity, developing knowledge networks to exchange experiences, providing technical assistance, and facilitating access to credit and finance to implement adaptation measures that enhance long-term resiliency. The design and implementation of adaptation strategies that build the resilience of smallholder farmers to climate change will be challenging, but is necessary if the region is to achieve its goals of alleviating poverty, achieving food security, and enhancing economic development.

Authors' contributions
CH, FA, MSR, MRMR, and RV generated the research idea, designed the study, and developed the household survey instrument. MS, FA, and BV oversaw field data collection and organized the database. AC and CH led data analysis and interpretation. CH and AC led the writing of the manuscript. MSR, MRMR, RV, BV and FA contributed to the writing of the manuscript.

Author details
[1] Conservation International, 2011 Crystal Drive Suite 500, Arlington, VA 22202, USA. [2] Tropical Agriculture and Higher Education Center (CATIE), Apdo 7170, Turrialba, Costa Rica. [3] University of Costa Rica, San Pedro de Montes de Oca 11501, Costa Rica. [4] Turrialba, Costa Rica. [5] Present Address: Wageningen University, Hollandseweg 1, 6706 KN Wageningen, The Netherlands.

Acknowledgements
We are grateful to the 860 farmers for participating in the survey; Federico Castillo, Juan Manuel Medina, and Jorge Albizúrez for support with field work; Lucia Contreras and Tabare Capitan for data entry and management; Sergio Vílchez-Mendoza for help with data analysis; Kellee Koenig for preparing Fig. 1; Nishina Nambiar for accessing scientific literature, and Vlasova Gonzalez for administrative support. This work was conducted as part of the CASCADE project ('Ecosystem-based Adaptation for Smallholder Subsistence and Coffee Farming Communities in Central America'), which was funded by the International Climate Initiative (IKI) of the German Federal Ministry for the Environment, Nature Conservation, Building and Nuclear Safety (BMUB). BMUB supports this initiative on the basis of a decision adopted by the German Bundestag.

Competing interests
The authors declare that they have no competing interests.

Funding
All data collection, data analysis, and write-up of the study were funded by the International Climate Initiative (IKI) of the German Federal Ministry for the Environment, Nature Conservation, Building and Nuclear Safety (BMUB). The funder had no role in the design of the study, collection, analysis, and interpretation of data, or writing of the manuscript.

References
1. Vermeulen SJ, Aggarwal PK, Ainslie A, Angelone C, Campbell BM, Challinor AJ, Hansen JW, Ingram JSI, Jarvis A, Kristjanson P, Lau C. Options for support to agriculture and food security under climate change. Environ Sci Policy. 2012. https://doi.org/10.1016/j.envsci.2011.09.003.
2. Lipper L, Thornton P, Campbell BM, Baedeker T, Braimoh A, Bwalya M, Caron P, Cattaneo A, Garrity D, Henry K, Hottle R. Climate-smart agriculture for food security. Nat Clim Change. 2014. https://doi.org/10.1038/nclimate2437.
3. Lowder SK, Skoet J, Raney T. The number, size, and distribution of farms, smallholder farms, and family farms worldwide. World Dev. 2016. https://doi.org/10.1016/j.worlddev.2015.10.041.
4. Morton JF. The impact of climate change on smallholder and subsistence agriculture. Proc Natl Acad Sci. 2007. https://doi.org/10.1073/pnas.0701855104.
5. Cohn AS, Newton P, Gil JD, Kuhl L, Samberg L, Ricciardi V, Manly JR, Northrop S. Smallholder agriculture and climate change. Annu Revi Environ Resour. 2017. https://doi.org/10.1146/annurev-environ-102016-060946.
6. Holland MB, Shamer SZ, Imbach P, Zamora JC, Medellín C, Leguía E, Donatti CI, Martínez-Rodríguez MR, Harvey CA. Mapping agriculture and adaptive capacity: applying expert knowledge at the landscape scale. Clim Change. 2017. https://doi.org/10.1007/s10584-016-1810-2.
7. Donatti CI, Harvey CA, Martinez-Rodriguez MR, Vignola R, Rodriguez CM. Vulnerability of smallholder farmers to climate change in Central America and Mexico: current knowledge and research gaps. Clim Dev. 2018. https://doi.org/10.1080/17565529.2018.1442796.
8. Harvey CA, Rakotobe ZL, Rao NS, Dave R, Razafimahatratra H, Rabarijohn RH, Rajaofara H, MacKinnon JL. Extreme vulnerability of smallholder farmers to agricultural risks and climate change in Madagascar. Philos Trans R Soc. 2014. https://doi.org/10.1098/rstb.2013.0089.
9. Castellanos EJ, Tucker C, Eakin H, Morales H, Barrera JF, Diaz R. Assessing the adaptation strategies of farmers facing multiple stressors: lessons from the coffee and global changes project in Mesoamerica. Environ Sci Policy. 2013. https://doi.org/10.1016/j.envsci.2012.07.003.

10. Donatti CI, Harvey CA, Martinez-Rodriguez MR, Vignola R, Rodriguez CM. What information do policy makers need to develop climate adaptation plans for smallholder farmers? The case of Central America and Mexico. Clim Change. 2017. https://doi.org/10.1007/s1058 4-016-1787-x.

11. Hannah L, Donatti CI, Harvey CA, Alfaro E, Rodriguez DA, Bouroncle C, Castellanos E, Diaz F, Fung E, Hidalgo HG, Imbach P, Landrum J, Solano AL. Regional modeling of climate change influence on ecosystems and smallholder agriculture in Central America. Clim Change. 2017. https://doi.org/10.1007/s10584-016-1867-y.

12. PRESANCA, FAO. Centroamérica en Cifras. Datos de Seguridad Alimentaria Nutricional y Agricultura Familiar. 2011. http://www.fao.org/fileadmin/user_upload/AGRO_Noticias/docs/CentroAm%C3%A9ricaEnCifras.pdf. Accessed 10 Oct 2017.

13. Hellin J, Schrader K. The case against direct incentives and the search for alternative approaches to better land management in Central America. Agric Ecosyst Environ. 2003. https://doi.org/10.1016/S0167 -8809(03)00149-X.

14. Morris KS, Méndez VE, Olson MB. 'Los meses flacos': seasonal food insecurity in a Salvadoran organic coffee cooperative. J Peasant Stud. 2013. https://doi.org/10.1080/03066150.2013.777708.

15. Bacon CM, Sundstrom WA, Stewart IT, Beezer D. Vulnerability to cumulative hazards: coping with the coffee leaf rust outbreak, drought, and food insecurity in Nicaragua. World Dev. 2017. https://doi.org/10.1016/j.world dev.2016.12.025.

16. Eitzinger A, Läderach P, Sonder K, Schmidt A, Sian G, Beebe S, Rodríguez B, Fisher M, Hicks P, Navarrete-Firas C, Nowak A. Tortillas on the roaster: Central America's maize–bean systems and the changing climate. CIAT Policy Brief No. 6. Cali: Centro Internacional de Agricultura Tropical (CIAT); 2012.

17. CEPAL (Comisión Económica Para América Latina y el Caribe). El impacto de la caída de los precios de café en el 2001. 2002. http://www.fondo minkachorlavi.org/cafe/docs/cepal2002.pdf. Accessed 1 Oct 2016.

18. Tucker CM, Eakin H, Castellanos EJ. Perceptions of risk and adaptation: coffee producers, market shocks, and extreme weather in Central America and Mexico. Glob Environ Change. 2010. https://doi.org/10.1016/j.gloenvcha.2009.07.006.

19. Altieri MA, Funes-Monzot F, Petersen P. Agroecologically efficient agricultural systems for smallholder farmers: contributions to food sovereignty. Agron Sustain Dev. 2012. https://doi.org/10.1007/s13593-011-0065-6.

20. Gay C, Estrada F, Conde C, Eakin H, Villers L. Potential impacts of climate change on agriculture: a case study of coffee production in Veracruz, Mexico. Clim Change. 2006. https://doi.org/10.1007/s1058 4-006-9066-xcc.

21. Lin BB. Agroforestry management as an adaptive strategy against potential microclimate extremes in coffee agriculture. Agric For Meteorol. 2007. https://doi.org/10.1016/j.agrformet.2006.12.009.

22. Conde C, Liverman D, Flores M, Ferrer R, Araújo R, Betancourt E, Villarreal G, Gay C. Vulnerability of rainfed maize crops in Mexico to climate change. Clim Res. 1997;9:17–23.

23. Rahn E, Läderach P, Baca M, Cressy C, Schroth G, Malin D, Van Rikxoort H, Shriver J. Climate change adaptation, mitigation and livelihood benefits in coffee production: where are the synergies? Mitig Adapt Strat Glob Change. 2014. https://doi.org/10.1007/s11027-013-9467-x.

24. FAO (Food and Agriculture Organization. Dry corridor-situation report June 2016. 2016. http://www.fao.org/emergencies/resources/docum ents/resources-detail/en/c/422097. Accessed 16 Jun 2017.

25. Philpott SM, Lin BB, Jha S, Brines SJ. A multi scale assessment of hurricane impacts on agricultural landscapes based on land use and topographic features. Agric Ecosyst Environ. 2008. https://doi.org/10.1016/j.agee.2008.04.016.

26. Cruz-Bello GM, Eakin H, Morales H, Barrera JF. Linking multi-temporal analysis and community consultation to evaluate the response to the impact of Hurricane Stan in coffee areas of Chiapas, Mexico. Nat Hazards. 2011. https://doi.org/10.1007/s11069-010-9652-0.

27. Eakin H, Benessaiah K, Barrera JF, Cruz-Bello GM, Morales H. Livelihoods and landscapes at the threshold of change: disaster and resilience in a Chiapas coffee community. Reg Environ Change. 2012. https://doi.org/10.1007/s10113-011-0263-4.

29. Haggar J, Schepp K. Coffee and climate change. Desk study: impacts of climate change in four pilot countries of the coffee and climate initiative. 2011. http://www.coffeeandclimate.org/tl_files/Themes/CoffeeAndC limate/Country%20profiles/C711_Coffee%20and%20Climate%20Cha nge_synthesis%20report_final.pdf. Accessed 10 Sept 2017.

30. Magrin G, Gay García C, Cruz Choque D, Giménez JC, Moreno AR, Nagy GJ, Nobre C, Villamizar A. Latin America. In: Parry ML, Canziani OF, Palutikof JP, van der Linden PJ, Hanson CE, editors. Climate change 2007: impacts, adaptation and vulnerability, contribution of working group II to the fourth assessment report of the intergovernmental panel on climate change. Cambridge: Cambridge University Press; 2007. p. 581–615.

31. Imbach P, Beardsley M, Bouroncle C, Medellín C, Läderach P, Hidalgo H, Alfaro E, Van Etten J, Allan R, Hemming D, Stone R, Hannah L, Donatti CI. Climate change, ecosystems and smallholder agriculture: an introduction to the special issue. Clim Change. 2017. https://doi.org/10.1007/s1058 4-017-1920-5.

32. Läderach P, Ramirez-Villegas J, Navarro-Racines C, Zelaya C, Martinez-Valle A, Jarvis A. Climate change adaptation of coffee production in space and time. Clim Change. 2017. https://doi.org/10.1007/s10584-016-1788-9.

33. Avelino J, Cristancho M, Georgiou S, Imbach P, Aguilar L, Benemann G, Läderach P, Anquetil F, Hruska A, Morales C. The coffee rust crises in Colombia and Central America (2008–2013): impacts, plausible causes and proposed solutions. Food Secur. 2015. https://doi.org/10.1007/s1257 1-015-0446-9.

34. ICO (International Coffee Organization). Report on the outbreak of coffee leaf rust in Central America and action plan to combat the pest. 2013. http://www.ico.org/documents/cy2012-13/ed-2157e-report-clr.pdf. Accessed 01 Sept 2017.

35. MARN. Política Nacional de Cambio Climático. 2009. http://www.marn. gob.gt/Multimedios/56.pdf. Accessed 01 Jun 2017.

36. DNCC (Dirección Nacional de Cambio Climático). Plan de Acción de la Estrategia Nacional de Cambio Climático. 2015. http://cambioclimatico cr.com/2012-05-22-19-42-05/estrategia-nacional-de-cambio-climatico. Accessed 01 Oct 2017.

37. Mesa de Trabajo en Cambio Climático de la Secretaría de Agricultura y Ganadería. Estrategia Nacional de Adaptación al Cambio Climático para el sector Agroalimentario de Honduras (2015–2025), versión ejecutiva. Tegucigalpa: Programa de Adaptación al Cambio Climático en el sector forestal (CLIFOR); 2015.

38. Schroth G, Läderach P, Dempewolf J, Philpott S, Haggar J, Eakin H, Castillejos T, Moreno JG, Pinto LS, Hernandez R, Eitzinger A. Towards a climate change adaptation strategy for coffee communities and ecosystems in the Sierra Madre de Chiapas, Mexico. Mitig Adapt Strat Glob Change. 2009. https://doi.org/10.1007/s11027-009-9186-5.

39. Baca M, Läderach P, Haggar J, Schroth G, Ovalle O. An integrated framework for assessing vulnerability to climate change and developing adaptation strategies for coffee growing families in Mesoamerica. PLoS ONE. 2015. https://doi.org/10.1371/journal.pone.0088463.

40. Eakin H, Appendini K, Sweeney S, Perales H. Correlates of maize land and livelihood change among maize farming households in Mexico. World Dev. 2015. https://doi.org/10.1016/j.worlddev.2014.12.012.

41. Eakin H. Smallholder maize production and climatic risk: a case study from Mexico. Clim Change. 2000. https://doi.org/10.1023/A:1005628631 627.

42. INEC (Instituto Nacional de Estadística y Censos). Censo Cafetalero: Turrialba y Coto Brus 2003, Valle Central y Valle Central Occidental 2004, y Pérez Zeledón Tarrazú y Zona Norte 2006. Principales resultados. 1st ed. San José: INEC; 2007.

43. Di Rienzo JD, Casanoves F, Balzarini MG, Gonzalez L, Tablada M, Robledo CW. InfoStat. Córdoba: Universidad Nacional de Córdoba; 2016.

44. Maddison DJ. The perception of and adaptation to climate change in Africa. Policy Research Working Paper 4308. Washington: World Bank; 2007.

45. Deressa TT, Hassan RM, Ringler C. Perception of and adaptation to climate change by farmers in the Nile basin of Ethiopia. J Agric Sci. 2011. https://doi.org/10.1017/S0021859610000687.

46. Esham M, Garforth C. Agricultural adaptation to climate change: insights from a farming community in Sri Lanka. Mitig Adapt Strat Glob Change. 2013. https://doi.org/10.1007/s11027-012-9374-6.

28. Ruiz Meza LE. Adaptive capacity of small-scale coffee farmers to climate change impacts in the Soconusco region of Chiapas, Mexico. Clim Dev. 2015. https://doi.org/10.1080/17565529.2014.900472.

47. Chengappa PG, Devika CM, Rudragouda CS. Climate variability and mitigation: perceptions and strategies adopted by traditional coffee growers in India. Clim Dev. 2017. https://doi.org/10.1080/17565529.2017.1318740.

48. Hidalgo HG, Alfaro EJ, Quesada-Montano B. Observed (1970-1999) climate variability in Central America using a high-resolution meteorological dataset with implication to climate change studies. Clim Change. 2017. https://doi.org/10.1007/s10584-016-1786-y.

49. Bacon CM, Sundstrom WA, Gómez MEF, Méndez VE, Santos R, Glottis B, Dougherty I. Explaining the 'hungry farmer paradox': smallholders and fair-trade cooperatives navigate seasonality and change in Nicaragua's corn and coffee markets. Glob Environ Change. 2014. https://doi.org/10.1016/j.gloenvcha.2014.02.005.

50. Jaramillo J, Muchugu E, Vega FE, Davis A, Borgemeister C, Chabi-Olaye A. The influence and implications of climate change on coffee berry borer (*Hypothenemus hampei*) and coffee production in East Africa. PLoS ONE. 2011. https://doi.org/10.1371/journal.pone.0024528.

51. Bryan E, Ringler C, Okoba B, Roncoli C, Silvestri S, Herrero M. Adapting agriculture to climate change in Kenya: household strategies and determinants. J Environ Manag. 2013. https://doi.org/10.1016/j.jenvman.2012.10.036.

52. Burnham M, Ma Z. Linking smallholder farmer climate change adaptation decisions to development. Clim Dev. 2016. https://doi.org/10.1080/17565529.2015.1067180.

53. Harvey CA, Martínez-Rodríguez MR, Cárdenas JM, Avelino J, Rapidel B, Vignola R, Donatti CI. The use of Ecosystem-based Adaptation practices by smallholder farmers in Central America. Agric Ecosyst Environ. 2017. https://doi.org/10.1016/j.agee.2017.04.018.

54. Adger WN, Huq S, Brown K, Conway D, Hulme M. Adaptation to climate change in the developing world. Prog Dev Stud. 2003. https://doi.org/10.1191/1464993403ps060oa.

55. Muttarak R, Lutz W. Is education a key to reducing vulnerability to natural disasters and hence unavoidable climate change? Ecol Soc. 2014. https://doi.org/10.5751/ES-06476-190142.

56. CBD (Convention on Biological Diversity). Connecting biodiversity and climate change mitigation and adaptation: Report of the second ad hoc technical expert group on biodiversity and climate change. CBD Technical Series No. 41. Montreal: Convention on Biological Diversity; 2009.

57. Vignola R, Harvey CA, Bautista-Solis P, Avelino J, Rapidel B, Donatti C, Martínez MR. Ecosystem-based adaptation for smallholder farmers: defi-

nitions, opportunities and constraints. Agric Ecosyst Environ. 2015. https://doi.org/10.1016/j.agee.2015.05.013.

58. Hellin J, William LA, Cherrett I. The Quezungual system: an indigenous agroforestry system from western Honduras. Agrofor Syst. 1999;46:228–37.

59. Holt-Giménez E. Measuring farmers' agroecological resistance after Hurricane Mitch in Nicaragua: a case study in participatory, sustainable land management impact monitoring. Agric Ecosyst Environ. 2002. https://doi.org/10.1016/S0167-8809(02)00006-3.

60. Lasco RD, Delfino RJP, Catacutan DC, Simelton ES, Wilson DM. Climate risk adaptation by smallholder farmers: the roles of trees and agroforestry. Curr Opin Environ Sustain. 2014. https://doi.org/10.1016/j.cosust.2013.11.013.

61. Verchot LV, Van Noordwijk M, Kandji S, Tomich T, Ong C, Albrecht A, Mackensen J, Bantilan C, Anupama KV, Palm C. Climate change: linking adaptation and mitigation through agroforestry. Mitig Adapt Strateg Glob Change. 2007. https://doi.org/10.1007/s11027-007-9105-6.

62. Cerda R, Alline C, Gary C, Tixier P, Harvey CA, Krolczyk L, Mathieu C, Clément E, Aubertite JN, Avelino J. Effects of shade, altitude and management on multiple ecosystem services in coffee agroecosystems. Eur J Agron. 2016. https://doi.org/10.1016/j.eja.2016.09.019.

63. Caswell M, Méndez VE, Bacon CM. Food security and smallholder coffee production: current issues and future directions. ARLG Policy Brief# 1. Burlington: University of Vermont; 2012.

64. Hellin J, Ridaura- López S. Soil and water conservation on Central American hillsides: if more technologies are the answer, what is the question? AIMS Agric Food. 2016. https://doi.org/10.3934/agrfood.2016.2.194.

65. Lin BB. The role of agroforestry in reducing water loss through soil evaporation and crop transpiration in coffee agroecosystems. Agric For Meteorol. 2010. https://doi.org/10.1016/j.agrformet.2009.11.010.

66. Altieri MA, Nicholls CI, Henao A, Lana MA. Agroecology and the design of climate change-resilient farming systems. Agron Sustain Dev. 2015. https://doi.org/10.1007/s13593-015-0285-2.

67. Bacon CM. Confronting the coffee crisis: can fair trade, organic, and specialty coffees reduce small-scale farmer vulnerability in Northern Nicaragua? World Dev. 2005. https://doi.org/10.1016/j.worlddev.2004.10.002.

68. Frank E, Eakin H, López-Carr D. Social identity, perception and motivation in adaptation to climate risk in the coffee sector of Chiapas, Mexico. Glob Environ Change. 2011. https://doi.org/10.1016/j.gloenvcha.2010.11.001.

69. Lin BB. Resilience in agriculture through crop diversification: adaptive management for environmental change. Bioscience. 2011. https://doi.org/10.1525/bio.2011.61.3.4.

The determinants of crop yields in Uganda: what is the role of climatic and non-climatic factors?

Terence Epule Epule[1*], James D. Ford[1,2], Shuaib Lwasa[3], Benon Nabaasa[3] and Ambrose Buyinza[3]

Abstract

Background: It is widely accepted that crop yields will be affected by climate change. However, the role played by climate in affecting crop yields vis-a-vis non-climatic stresses, is often unclear, limiting decision choices around efforts to promote increased production in light of multiple stresses.

Results: This study quantifies the role of climatic and non-climatic factors affecting multiple crop yields in Uganda, utilizing a systematic approach which involves the use of a two-stage multiple linear regression to identify and characterize the most important drivers of crop yield, examine the location of the key drivers, identify the socio-economic implications of the drivers and identify policy options to enhance agricultural production. We find that non-climatic drivers of crop yields such as forest area dynamics ($p = 0.012$), wood fuel ($p = 0.032$) and usage of tractors (0.041) are more important determinants of crop yields than climatic drivers such as precipitation, temperature and CO_2 emissions from forest clearance. Climatic drivers are found to multiply existing risks facing production, the significance of which is determined by variability and inadequate distribution of precipitation over the crop growing seasons.

Conclusion: The significance and validity of these results is observed in an f-statistic of 50 for the final optimized model when compared to the initial model with an f-statistic of 19.3. Research and agricultural policies have to be streamlined to include not only the climatic elements but also the non-climatic drivers of global, regional and national agricultural systems.

Keywords: Crop yields, Climatic drivers, Non-climatic drivers, Uganda

Background

Increasing global population, changes in consumption patterns and dietary needs and a rising demand for green energies has triggered a global need for increased food production. It is estimated that one in seven people lack access to food or are faced by malnutrition caused by poverty and rising food prices. Rising food prices caused by market speculations, expansion of crop cultivation to generate bioenergy and climate shocks are rendering the global food security problem even worse [1–3]. Even if these problems of access to food are solved, production would need to double to meet up with projected demand due to dietary changes and population growth [3–10]. Global food production will have to increase by ~ 70% to meet global food needs this century [1, 2, 11–13]. The task of feeding more people globally is even more daunting given global, regional and national patterns of climate and land use change [3, 7, 8, 10, 11, 14, 15].

Globally in general and in sub-Saharan Africa (SSA) in particular, agriculture is important in meeting dietary needs and underpinning economic growth. In Africa for example, about 50% of the gross domestic product (GDP) of most countries is from agriculture [16]. In most SSA countries, huge differentials exist between actual crop yields and projected yields mainly because production is often at subsistence levels [17]. The crop yield deficits recorded in most SSA countries are accentuated by climatic and non-climatic factors, including limited access

*Correspondence: terence.epule@mail.mcgill.ca
[1] Department of Geography, McGill University, 805 Sherbrooke St. W., Burnside Hall 416, Montreal, QC H3A 0B9, Canada
Full list of author information is available at the end of the article

to sufficient farm inputs such as tractors, unsustainable methods of cultivation, limited use of agroecology-related inputs and a significant climate variability [18, 19].

In Uganda, agriculture contributes about 20% to the GDP, 48% to export earnings, and employs about ~ 73% of the population [20]. In addition, more than 4 million Ugandan households depend on small-scale farming for their sustenance [20]. The pertinence of agriculture in Uganda is further seen as poverty reduction is tied to improvements in the agricultural sector [21, 22]. Agricultural systems in Uganda are highly sensitive to both land use and climatic conditions. However, it remains unclear which of the land use and climate-related drivers are most important in determining agricultural production. It has been argued, for instance, that the significance of changes in precipitation in affecting Ugandan crop yields is a function of variability and inadequate distribution of precipitation throughout the cropping seasons as opposed to the precipitation change *per se* [23]. Mubiru and Banda [24] added that due to frequent delays in precipitation during the March–May cropping season by close to 30 days, much of the precipitation that comes later (mid-April) is often not available for crops. It has also been argued that low agricultural productivity can be attributed to other non-climatic drivers such as limited use of external inputs, nutrient mining, soil erosion, deforestation, slash and burn cultivation *inter alia* as well as climate-related drivers such as precipitation, temperature and GHG emissions [24]. While multiple drivers affect crop yields in Uganda, the relative importance of different drivers is not well understood, constraining efforts to direct attention to the main impediments to maintaining and increasing crop yield in light of multiple stresses.

This study develops a systematic approach to identify and characterize the drivers of crop yields in Uganda. Identified key drivers are then used to determine specific policies that can enhance crop yields in light of multiple stresses. The work contributes to scholarship seeking to tease out the role of climate and non-climatic factors in affecting specific outcomes, with the approach developed having application to multiple contexts. To our knowledge, the study is the first to examine at a national level in Uganda the role of multiple factors in affecting crop yield.

Methods
Study area
Uganda is a SSA country located in East Africa (Fig. 1), and according to the 2013 census, Uganda has a population of ~ 36 million people [24]. Located within a humid equatorial region, prevailing winds and water bodies occasion differences in precipitation patterns. Mean

annual precipitation ranges from 800 to 1500 mm. Precipitation is bimodal in the south (March–May and September–November) and unimodal in the north (April–October) [25, 26]. Though located close to the equator, Uganda's climate is diverse due to the country's unique biophysical characteristics influenced by large rivers, water bodies and mountain ranges to the east and west [25]. Variations in sea surface temperatures in the distant tropical pacific and Indian Oceans strongly influence the timing of annual precipitation in Uganda [25]. Northern Uganda experiences less precipitation than the south and is more susceptible to droughts [15, 25].

Theoretical approach and data acquisition
This study seeks to identify and characterize the relative contributions and importance of *climatic and non-climatic variables* (independent variables) in affecting crop yields (dependent variable) in Uganda using a systematic modelling approach. The study is based on national scale data of 10 non-climatic variables, 3 climatic variables and 1 dependent variable. The data points for each variable span a period of 53 years (1961–2014).

The *dependent* variable whose predictors this study attempts to determine is *crop yield*. Crop yield data were culled from the Food and Agricultural Organization's department of Statistics-FAOSTAT [27]. A total of 31 food and cash crops (Additional file 1: Table S1) were aggregated to obtain the crop yield data per year in hectograms/hectare (Hg/Ha). However, the yield data were converted to tons/hectare (t/Ha) for standardization purposes. The rational of using several food and cash crops (31) as a representation of crop yields was based on:

- The availability of complete time series data for all the 31 food and cash crops over the period 1961 and 2014.
- The crops present a true picture of the crop yield scenarios of Uganda as they represent both arable and permanent crops.
- The crops in question are crops whose production tallies are proportionate with the recognized crop growing seasons in Uganda for crops that grow all year round.
- All the crop data were in the same units of measurement (Hg/Ha), as such comparisons and converting into (t/Ha) were easy to perform.
- All the crops with available data were included to avoid the miss-representation that is often associated with using national scale growing season precipitation and temperature data over a single crop.

Climatic drivers were identified based on the availability of data and on the importance of climatic drivers in

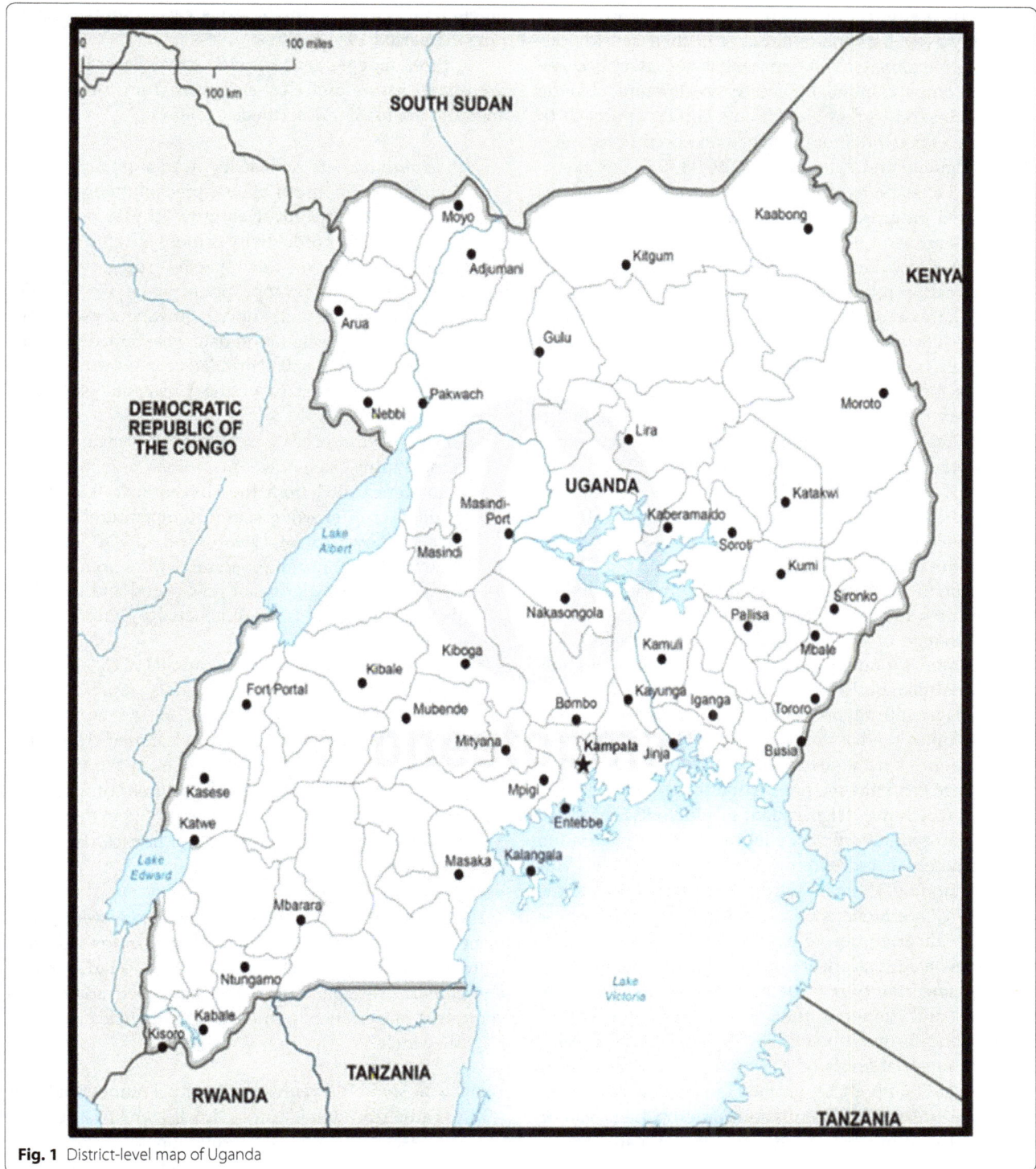

Fig. 1 District-level map of Uganda

the growth of crops [28]. Adequate precipitation during the crop growing season as well as average annual precipitation and temperature enhances crop growth [29]. Since most of the 31 crops included in this study grow all year round, the mean annual precipitation and gridded temperature spanning the period 1961–2014 were used to fit the models. The use of crop growing season and mean annual precipitation and gridded temperature helps to capture the net effect of climate on all the crops used in this study as well as the differences in their sensitivities to changes in climate. This approach is necessary because different crops respond differently to different climatic stresses. Several studies have used a similar approach to verify the effects of climate change on several crops at a global, regional and country scale [9, 93–95]. For example, Lobell et al. [9] examined the effects of changes in climate based on growing season and annual precipitation and temperature at a global scale for the following crops: maize, wheat, soybeans and rice. Also, Sarker et al. [93] used regression models to examine the effects of climate change on three major rice crops: Aus, Aman and Boro in Bangladesh using mean growing season and mean annual temperature and total growing season and annual precipitation. Furthermore, a study by Lobell and Field [94] comes very close to this current study from a methodological perspective. The study developed statistical models of yield response to climate change at a global scale. The crops included rice, maize, soybeans, barley and sorghum. The study used gridded mean monthly/annual temperature and rainfall data for the period 1961–2002 to verify the effects of climate change on yields.

Atmospheric CO_2 on the other hand may trigger higher temperatures and reduce precipitation or generally cause climate change in SSA and reduce crop yields [28]. An example here is found in a study by [96] which shows how interannual changes in CO_2 emissions in Ethiopia have reduced agricultural productivity for major crops and livestock up to the baseline period of 2030. On the positive side, it has been argued that rising CO_2 content in the atmosphere can enhance production if there is aerial fertilization which may trigger plant productivity, enhanced water management efficiency and reduced transpiration and ultimately increased crop productivity [28, 29]. This notwithstanding, it is generally difficult to investigate the effects of CO_2 emissions on agricultural systems because of many uncertainties, quantitative estimates of CO_2 emissions on agriculture are often of low confidence [97]. A major uncertainty is to be able to project changes in emissions and climate in the context of crop yields [97]. Ideally, long-term estimates of CO_2 emissions covering the pre-industrial levels of 275 ppm and 2 °C increased warming at 550 ppm CO_2 provide longer timescales over which the effects of CO_2 emissions on crop yields can be computed with much certainty. The non-climatic drivers were selected based on the availability of data and on the argument that the outcome of crop yields is not only a function of climatic drivers but also due to land tenure systems [30–32]. Since it is still unclear which of these drivers determine crop yields the most, an experiment to verify this hypothesis was established in this study. Data

on climatic and non-climatic drivers were sourced spanning the period 1961 to 2014.

The three independent *climatic drivers* are: *precipitation, temperature* and *CO_2 emissions* from forest clearance and are further described as follows:

- *Precipitation* here represents mean annual precipitation (mm). The mean annual precipitation was used because the crops involved grow all year round and therefore their productivity is more affected by mean annual precipitation data. Specific crop growing season data are more appropriate when we consider crops like maize with shorter growing seasons. The mean annual precipitation data were culled from the collaborative 0.5° × 0.5° gridded crop growing season precipitation database of the University of Oxford and UNDP [33, 34].

- The *temperature* data represent the mean annual temperatures (°C). Like precipitation, temperature data were culled from the collaborative 0.5° × 0.5° gridded crop growing season temperature database of the University of Oxford and UNDP [33, 34]. Again, mean annual temperatures were used because the crop involved grow all year round and therefore their productivity is more affected by mean annual temperature data.

- The third climatic variable, annual *CO_2 emissions* from forest clearance in Uganda, represents the stock of annual CO_2 emitted to the atmosphere due to deforestation or forest clearance (Houghton 1991); these data were collected from the Food and Agricultural Organization's department of Statistics-FAOSTAT [27]. For a summary of the theoretical linkages between the climatic and non-climatic drivers, see Fig. 2.

The 10 independent *non-climatic* variables were culled from the Food and Agricultural Organization's department of Statistics-FAOSTAT [27]. These 10 independent *non-climatic* variables have been described as being of very great importance in impacting crop yields in Africa [27]; they include:

- *Cattle stock* This represents the total number of herds of cattle reared in Uganda. Usually, when cattle stock increases, this imposes reductions in crop yields because herds require land for grazing and in some cases herds of cattle affect yields by taking up land (for range land purposes), eating up crops and potentially occasioning farmer-grazier conflicts [35, 36].

- *Wood fuel* This is the amount of wood extracted from the forest and used for energy purposes. Its unit of

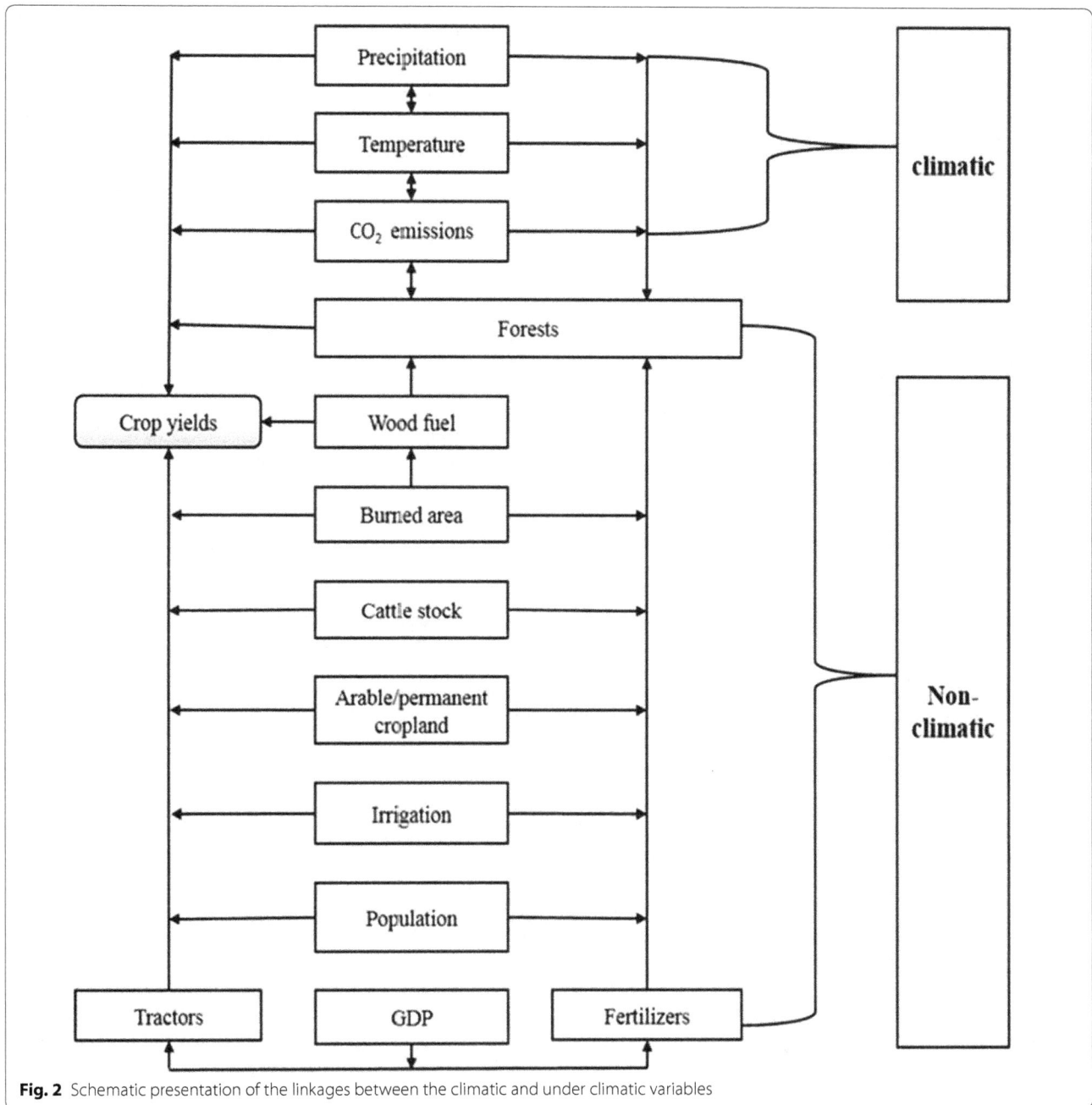

Fig. 2 Schematic presentation of the linkages between the climatic and under climatic variables

measurement is (m^2). This reflects the total forest area lost due to wood fuel extraction. This variable often reduces crop yields when it entails destruction of forests associated with reduced forest carbon stock, reduced soil organic carbon (SOC) and nitrogen (SON) which are vital to crops [35–37].

- *Arable and permanent cropland* This is the amount of land under temporal and permanent crops expressed in thousand hectares (K/Ha). The more this variable increases, the higher the yields in the short term and the lower the yields per Ha if agriculture is not mechanized in the long term. However, without mechanization of agriculture, this may not increase yields per Ha. Increasing crop yields in most of Africa is based on farmland expansion [35, 36]. However, with the prospects of getting more land becoming less feasible, relying on this option to increase crop yields is becoming a challenge [35–37].

- *The total area equipped for irrigation* This is expressed in thousand hectares (K/Ha) and represents the farmlands that have access to irrigation facilities. In regions that are faced with problems of recurrent droughts as is the case of Uganda, irrigation facilities serve as safety nets that do sustain crop yields during periods of dryness.
- *Forest area* This represents the amount of available forest expressed in thousand hectares (K/Ha). This variable often reduces crop yields when destruction of forests is associated with reduce forest carbon stock, reduced SOC and SON which are vital to crops [38].
- *Population* This is the number of people who live in the country over time. This study used total population because population reflects total food needs of a country. Crop yields are influenced by both farming and non-farming populations. Normally, when the population growth is high, the food needs of the country are likely to be high, and this creates food security challenges if population growth fails to be matched by increased food production capacity [39]. Population here is therefore used to serve as a proxy for total food needs in the absence of total food needs data.
- *Gross domestic product (GDP) per capita* This is expressed in millions and is an indicator used to gauge the health of the economy per head. It is the dollar value of all goods and services produced in Uganda over a specific period of time. A higher GDP has the potential to increase crop yields as most farmers will be able to access farm inputs.
- *Burned area* This is expressed in hectares, and it is the total area of forest or grassland that is burnt and transformed into farmland. Burning is a common phenomenon in SSA, and it is used to create new farmlands and to provide ash that serves as a temporal source of nutrients to the soil. Normally, the more the burning, the more farmland is available, the more likely the increase in yields in the short term due to more land and ash from the burnt forests and grasslands [35].
- *Total fertilizers used* This is expressed in tons. Fertilizers constitute a very important input into the agricultural system. When farmers have adequate access to fertilizers, their crop yields are likely to increase. However, in most SSA countries including Uganda, access to inorganic fertilizers is often limited by purchasing power because most of the farmers are poor. Organic fertilizers which are often free and agroecological and more sustainable have not been sufficiently valorized to levels at which they can sustain yields without inorganic fertilizers [40–42]. In this study, total fertilizers used entail both organic and inorganic fertilizers. Organic fertilizers include manure and compost, while inorganic fertilizers include chemical fertilizers such as nitrogen, potassium, phosphorus and magnesium fertilizers inter alia. In FAOSTAT (database from which data were culled), by simply selecting the options total organic and inorganic fertilizers, an excel sheet of the data for the years selected is obtained.
- *Agricultural tractors* This represents the total number of agricultural tractors used. Tractors represent a very important input in any agricultural system, and the larger the number of tractors typically results in higher yields due to a faster and more efficient production process [43]. However, due to poverty, the majority of farmers in SSA who operate on predominantly small scale are unable to have access to such heavy capital inputs [43–45].

Analysis

We draw upon the systematic approach (SA) of Muller et al. [46] which is composed of a 4-step process for data analysis (Fig. 3).

All 14 independent and dependent variables are subjected to detrending which removes all linear models of the actual time series by dividing the expected time series by the actual time series data. Detrending removes the effects of increased technology, shows annual variations and reduces the effects of inconsistent errors in reporting. For the equation used in determining the expected time series of all the 14 variables, see Eq. 1.

The first step of the systematic approach involves the use of a multiple linear regression approach (MLR) to identify the drivers of crop yields from among climatic and non-climatic variables. MLR provides the feedbacks between the dependent and independent variables through the t and p values which helps us to detect the level of importance of a given variable, and has been used in comparable contexts to this study, e.g. [47–50]. For the equations used in computing the initial model (IM) and the final optimized model (FOM) based on MLR see Eqs. 2 and 3 below.

$$\text{EXP}_y = ax + b \tag{1}$$

where EXP_y is the expected maize yield, χ is the year, a is the linear trend, b is the intercept when $\text{EXP}_y = ax$.

$$
\begin{aligned}
Y_{\text{CYIM}} = \alpha_0 &+ \alpha_1 X_{\text{PPT}} + \alpha_2 X_{\text{TEMP}} + \alpha_3 X_{\text{CO}_2} \\
&+ \alpha_4 X_{\text{WF}} + \alpha_5 X_{\text{CS}} + \alpha_6 X_{\text{APC}} + \alpha_7 X_{\text{IRRI}} \\
&+ \alpha_8 X_{\text{FA}} + \alpha_9 X_{\text{POP}} + \alpha_{10} X_{\text{GDP}} \\
&+ \alpha_{11} X_{\text{BA}} + \alpha_{11} X_{\text{FERT}} + \alpha_{11} X_{\text{AGT}}
\end{aligned}
\tag{2}
$$

$$Y_{\text{CYFOM}} = \alpha_0 + \alpha_1 X_{\text{FA}} + \alpha_2 X_{\text{WF}} + \alpha_3 X_{\text{AGT}} \tag{3}$$

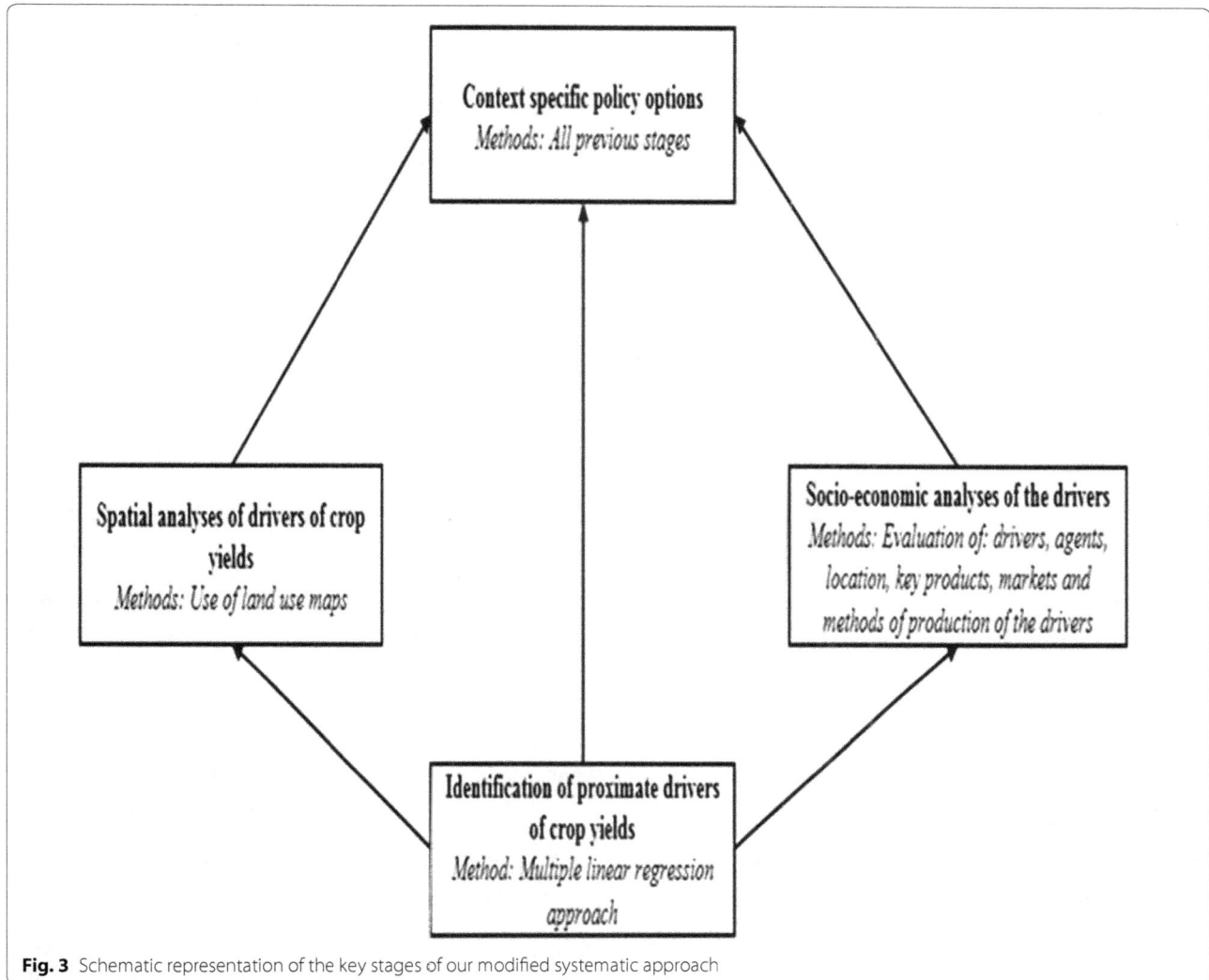

Fig. 3 Schematic representation of the key stages of our modified systematic approach

where Y_{CYIM} represents crop yields of the IM (dependent variable), Y_{CYFOM} represents crop yields of the FOM (dependent variable), α_0 is the regression intercept, $\alpha_1 X_{PPT}$ is the partial regression coefficient and mean annual precipitation, $\alpha_2 X_{TEMP}$ is the partial regression coefficient and mean annual temperature, $\alpha_3 X_{CO_2}$ is the partial regression coefficient and annual CO_2 emissions from forest clearance, $\alpha_4 X_{WF}$ is the partial regression coefficient and wood fuel, $\alpha_5 X_{CS}$ is the partial regression coefficient and cattle stock, $\alpha_6 X_{APC}$ is the partial regression coefficient and arable and permanent cropland, $\alpha_7 X_{IRRI}$ is the partial regression coefficient and area equipped for irrigation, $\alpha_8 X_{FA}$ is the partial regression coefficient and forest area, $\alpha_9 X_{POP}$ is the partial regression coefficient and population, $\alpha_{10} X_{GDP}$ is the partial regression coefficient and GDP, $\alpha_{10} X_{BA}$ is the partial regression coefficient and burned area, $\alpha_{11} X_{FERT}$ is the partial regression coefficient and total fertilizers used,

and $\alpha_{11} X_{AGT}$ is the partial regression coefficient and agricultural tractors.

The two-step regression is used because the initial regression model (IM) has weaknesses such as collinearity and inconsistent p and t values such that a second-stage regression analysis described herein as the final optimized model (FOM) is required to enhance the model and reduce the effects of the weaknesses of the IM on the FOM.

To optimize the IM, we sequentially removed all the ten other variables that did not make the list of the three most important variables and rerun the model. This process was repeated ten times until only the three most important variables remained (forest area, wood fuel and agricultural tractors). A model comprised of these three variables is then fitted to produce the FOM. To verify the importance of some of these drivers and to guide the analyses, the following hypotheses were established:

the null hypothesis (H_0) is that there is no difference in the impacts that all the climatic and non-climatic drivers/predictors have on crop yields. In other words, this means that both climatic and non-climatic drivers are of equal importance in terms of their relationship with crop yields. The alternate hypothesis (H_1) states that some drivers/predictors of crop yields are more important than others. The model with the higher f-statistic is usually the more significant one.

Second, we examine the socio-economic aspects of the key drivers of crop yields in Uganda. The first part of this involves an examination of the population or organization involved with the drivers and the characteristics of the population involved in the operationalization of the drivers called agents. The second aspect of the socio-economic analysis involves the identification of the location or actual zones in the country where the drivers are dominant, a task which is similar to the spatial analysis. The third phase of the socio-economic analysis encompasses an identification of key products associated with any given driver. This is followed by an analysis of the methods of production of the products associated with the identified drivers and the markets where the products are sold (Table 3). The socio-economic analysis enables us to be able to understand in detail some of the more intricate issues behind the drivers that are often only specific to a given region.

Third, a detailed spatial analysis of the key drivers is carried out by identifying the various zones in the country where the different drivers are dominant. This enhances our understanding of where or answers the question, in what region or part of the country is a particular driver dominant. This phase is very similar to the location analysis carried out in the previous section.

The final phase of the approach involves a synthesis of stages one, two and three from which recommendations are made on possible context-specific policies that can be used to enhance crop yields in Uganda based on the drivers. With these, we proposed policies that are context specific and only in relation to the most important drivers of crop yields.

Results

Non-climatic factors are the main drivers of crop yields in Uganda

From the MLR simulations, the five most significant predictors/drivers of crop yields in Uganda do not encompass any climatic variable, and they are: forest area (t value: -3.56), wood fuel (t value: -3.44), GDP (t value: 2.82), tractors (t value: 2.92) and population (t value: -1.55) (Table 1). This model has an f-statistic of 19.3 which indicates that the model is significant. This initial model (IM) is flawed however, by: (1) inconsistencies between the t and p values for most of the predictors and (2) extensive multicollinearity between some independent variables such as wood fuel, forest area and population.

Given these flaws, the IM was optimized by sequentially removing in a step-by-step fashion 10 variables, with the model rerun until only the three most significant predictors remained. The final optimized model (FOM) has the following non-climatic predictors in order of importance: forest area dynamics (t value: -11.11; p value: 0.012(1.20%); R: -0.5), wood fuel (t value: -9.40; p value: 0.032(3.16%); R: 0.3) and tractors used (t value: 8.46; p value: 0.041(4.09%); R: 0.2). The correlation coefficients obtained for the three most important predictors are also consistent: forest area dynamics has the

Table 1 Outputs of the initial regression model based on all the 13 independent variables

Independent variable	Unstandardized coefficient	Standard error	Standardized coefficient	t value	p value
Cattle stock	− 3.49	3.26	− 0.28	− 1.04	0.304
Wood fuel/logging concessions	− 1	2.91	− 2.59	− 3.44	0.001
CO_2 emissions from forest clearance	0.002	0.002	0.19	0.75	0.452
Arable and permanent crop land	− 0.01	0.01	− 0.43	− 0.64	0.522
Total area equipped for irrigation	2.209	4.37	0.2	0.50	0.616
Forest	− 0.22	0.062	− 5.68	− 3.56	0.0009
Population	− 0.013	0.008	− 3.55	− 1.55	0.127
GDP per capita	− 0.14	0.049	− 0.72	2.82	0.007
Burned area	0.00011	0.00014	0.55	0.77	0.441
Total fertilizers used	− 0.00028	0.0013	− 0.03	− 0.21	0.831
Number of agricultural tractors	0.051	0.018	2.48	2.92	0.005
Mean growing season precipitation	− 0.203	0.191	− 0.08	− 1.06	0.292
Mean growing season temperature	− 9.22	7.92	− 0.16	− 1.16	0.251

$R = 0.92$; adjusted $R^2 = 0.86$; f-statistics $= 19.3$

highest correlation coefficient of about ~ − 50%, wood fuel record ~ 30% and tractors ~ 20% (Additional file 2: Table S2a and Fig. S2a). The FOM is more significant than the IM with a higher f-statistic of 50. The t and p values of this model are also consistent as the most important predictor of crop yields (forest area) has the highest t value and the lowest p value (Table 2). With these results, H_1 hypothesis that states that some predictors of crop yields are more important than others in Uganda is valid, while H_0 is rejected.

While climatic predictors are found to be of limited statistical importance when compared to the non-climatic predictors, a model of how the 3 climatic variables perform among themselves shows that temperature is the most important climatic predictor of crop yields in Uganda followed by precipitation and CO_2 emissions from forest clearance (Additional file 2: Table S2b); this may vary for individual crops as in the case of maize in which for example precipitation is more important. Mean annual temperature is about 22 °C, while mean annual precipitation ranges between 800 and 1500 mm with wide variation between the north and the south. This can be explained by the fact that there is a large dependence on irrigation for agricultural purposes due to unreliable precipitation, a situation that is even more daunting northwards. According to statistics, irrigation for agricultural purposes is the second most important water withdrawal component in Uganda. The greatest water user in Uganda in 2008 was 'municipalities' which withdrew about 328 million m^3 (51%) of water, while agriculture was second and it used 259 million m^3 (41%), while industries withdrew 50 million m^3 [89]. Uganda has a huge irrigation potential with most of the potential still unexploited; the 2011 Irrigation Master Plan classified a total of 567,000 Ha (5070 Km^2) of potential irrigation land [89]. Some examples of huge irrigation projects exist around Lake Kyoga basin, the Western Region, the Albert Nile Valley and the Jinja districts on Lake Victoria in the south-east of the country [89].

In addition to the variables considered in this analysis, there are other socio-economic and demographic variables that can affect food production. For example, with a population of ~ 36 million in 2013, Uganda has a population growth rate of 3.26%, a dependency rate of 108%, average annual income of 303, 700 Uganda Shillings (UGX) and a literacy rate of 69.6% [90]. Unfortunately, due to the absence of long-term time series and established standardized reliable data on these variables at national scale, it is not possible to use them in our computations.

Agents, location, products, methods of production and markets of key drivers of crop yields in Uganda

In this section, an analysis of the agents, location, products, method of production and markets of the main drivers of crop yields in Uganda is performed. This will help identify and further understand the elements that shape crop production in Uganda.

Forest area

Based on our analysis, forest area is the most important determinant of crop yields in Uganda. The p value of 0.012 or 1.2% shows that the chances of having observations that will differ from the current results are lowest among all the results as there is only a 1.2% chance of having a different observation. The low probabilities for changes depict generally a more reliable model. When the rate of decline in forest area is high, crop yields rise in the short term because of the availability of more land for cultivation, but in the long term reduced SOC and SON will go a long way to contribute to reduced crop yields due to reduced forest area. The farms will also become exposed to erosion. Due to high demand for more land from agriculture, an amount of 88,150 Ha or 1.86% of forest was lost annually during the period 1990–2010 [51]. Uganda experienced a total forest area growth between 51,000 and 2,988,000 Ha of planted trees during the last 2 and a half decade. About 462 tons of carbon is stored in 1 Ha of forest [51]. The agents here include the population of Uganda (~ 36 m) and foreign companies. In terms of location or spatial analysis, most of the forest in Uganda are found in three key areas which are the eastern shoulders of the East African Rift valley, the northern shores of Lake Victoria and isolated montane forests in the north and east. However, pockets of forest can also be found in the north due to an ongoing reforestation drive. Forests are mainly in three key areas of

Table 2 Outputs of the optimized regression model based on the 3 most significant independent variables

Independent variable	Unstandardized coefficient	Standard error	Standardized coefficient	t value	p value	Rank of t value
Wood fuel/logging concessions	− 2.01	2.14	− 5.23	− 9.4	0.032 (3.16%)	2
Forests	− 0.13	0.011	− 3.43	− 11.11	0.012 (1.2%)	1
Number of agricultural tractors	0.05	0.006	2.79	8.46	0.041 (4.09%)	3

$R = 0.86$; adjusted $R^2 = 0.75$; f-statistics $= 50$

the south-west and east which are: the west including the eastern shoulders of the rift valley, the northern shores of Lake Victoria and around isolated montane forests in the north and east [52]. The key products are: wood fuel, pulp, timber and lumber. The methods of exploitation are either small scale, large scale, manual or mechanized, while the products are either sold nationally or internationally within East Africa and Europe (Table 3).

Uganda's forest is declining at a significant rate (between 1990 and 2010, a total of about 88,150 Ha or 1.86% of Ugandan forest were lost yearly) [51]. At this rate, the landscape will experience not only increased soil erosion, reduced SOC and SON but also an increased amount of atmospheric carbon. At the above rate of deforestation, a total of 40,725,300 tons of carbon stock (Ha/year) will be lost to the atmosphere (business as usual BAU assuming 462 tons of carbon is lost per Ha). Several scenarios starting with a 75% reduction in deforestation (22,037.5 Ha/year) show lower mean carbon emissions and higher SOC and SON. However, as we move to other scenarios (50, and 25%), the amount of forest lost increases and the carbon emissions also increase, and SOC and SON tend to reduce with increased deforestation (Fig. 4). In general, it can be said that the lower the rate of deforestation, the lower the amount of carbon emissions and the higher the SOC and SON.

Wood fuel

In most African countries including Uganda, wood fuel is the most important source of energy and most of the people use it for domestic and small-scale industrial purposes. The p value of 0.032 or 3.16% shows that there is a 3.16% chance of having observations that will differ from the current. The low probabilities for changes depict generally a reliable model. Charcoal, a derivative of wood fuel, is mostly used in urban areas. Due to persistent energy crisis in Uganda, the demand and consumption of wood fuel in Uganda is estimated at a rate of 3%/year [53–56].

In Uganda, wood fuel collection is practised by about 70–80% of the local population and over 90% of energy used for domestic activities is obtained from wood fuel in the form of firewood and charcoal [54, 55, 57–60]. In terms of location or spatial analysis, the main areas of gathering are within the rural areas and urban outskirts in the south-west, east and around isolated forest patches in the north and north-east. In terms of products, firewood, charcoal, wood pellets, wood ash, production of bricks and tiles, are the most common products of wood fuel [56]. The methods of extraction are either small or large-scale, manual or mechanized with the use of chain saws. The products are sold and used locally and nationally (Table 3). The north and north-east of the country are essentially grasslands, but one can find pockets of forest due to current reforestation efforts in this fragile ecoregion. Increased gathering of wood fuel increases deforestation and atmospheric CO_2 and thereby reduces SOC and SON while also exposing the soil to erosion and consequently reducing crop yields [61].

Tractors

Mechanization plays an important role in enhancing crop yields globally [41]. Here, a p value of 0.041 or about 4.095% depicts a 4.095% chance of having an observation that is different from the current. The low probabilities for changes depict generally a reliable model. In Uganda, they constitute the third most important non-climatic determinant of crop yields. In terms of agents, in 2014, Uganda had more than 4000 agricultural tractors. As concerns location or spatial analysis, tractors are used all over the country. They are used mainly to harvest, plant, till the soil and to transport crops. The tractors used in Uganda are essentially manufactured by foreign companies (Table 3). The significance of tractors as major elements of agricultural intensification and yield increase is seen as they enhance the entire production process by making tillage, sowing, harvesting and transportation relatively easier than human labour and thereby enhancing crop yields.

Table 3 Socio-economic analysis of the main non-climatic predictors of crop yields in Uganda

	Forest area	Wood fuel	Tractors
Agents	About 36 million people local and national population, foreign companies	70–80% of the local population	More than 4000 in 2014
Location	The eastern shoulders of the rift valley, northern shores of Lake Victoria, isolated montane forests in the north and east	Forest zones of the south-west, east of Uganda and patches in the north and north-east	All over the country
Products	Wood fuel, pulp, timber, lumber	Firewood, charcoal, wood pellets and wood ash, bricks and tiles	Used to: harvest, plant, till the soil and transport crops
Production	Small scale, large scale, manual and mechanized	Large and small scale, manual, mechanized	By foreign companies
Markets	Internally, East African region, internationally	Locally and nationally	All over the country

The determinants of crop yields in Uganda: what is the role of climatic and non-climatic...

81

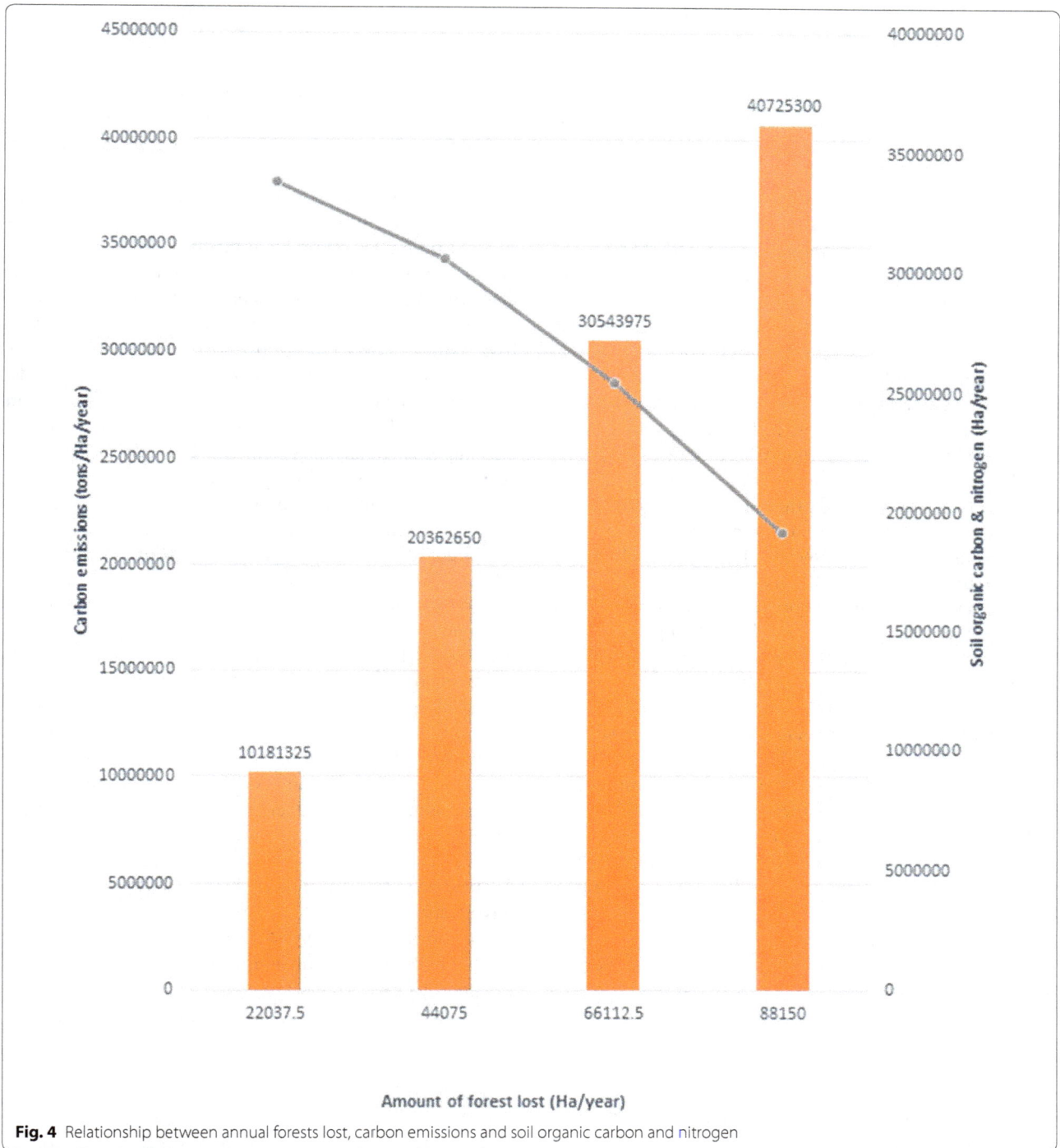

Fig. 4 Relationship between annual forests lost, carbon emissions and soil organic carbon and nitrogen

Discussion

Forest area, wood fuel and tractors have varying effects on crop yields in SSA

A common way of increasing crop yields in SSA is through farmland expansion by deforestation; in the short term this increases crop yields since more land is brought under cultivation [62–65]. With the current trends of declining forest due to deforestation for further expansion of farmlands and wood fuel gathering (Additional file 2: Fig. S2a), farms will be exposed to soil erosion and decline in crop yields; this has been illustrated in the ability of forest ecosystems to reduce soil erosion and to maintain and supply SOC and SON. In the longer term, the inputs from the leaves of plants can increase soil aeration and enhance SOC and SON and consequently increase crop yields [61].

The use of tractors has the potential of increasing crop yields in Uganda as in other parts of the world (Additional file 2: Fig. S2a). Increased mechanization would lead to increased food production, improved land use, enhanced rural prosperity and greater exports and less reliance on imports [16, 45]. In 1798, Malthus advocated the principle of preventive checks in his book, 'An Essay on the principle of population'. Today, these preventive checks may include investments that would prevent famine and subsequent population decline. Examples of such checks in today's world would include machines and fertilizers that would enhance productivity [63]. In the rest of Africa, it has been argued that mechanization of agriculture is either facing stagnation or retrogression [16, 66]. The causes of this are the absence of effective strategies and policies to enhance investments in tractors and other farm inputs such as irrigation equipments and absence of coordination between governments and private sector [16, 45]. On the negative side, the expansion in the use of tractors recorded in Uganda may be responsible for some of the deforestation in Uganda as reflected by a rapidly declining forest area; this is linked to the relative ease with which mechanized agriculture can enhance forest clearance.

Non-climatic factors are the main drivers of crop yields in most of sub-Saharan Africa

It is expected that declining crop yields in most of SSA are strongly linked to declining precipitation. However, SSA seems to have experienced a relative increase in precipitation from the 1990s [67, 68]. The evidence that precipitation has increased in SSA heralds the argument that non-climatic factors can better explain the yield trends observed in most of SSA. This view point seems to be tenable since the SSA is a zone of sharp seasonal contrast with fluctuations in precipitation at interannual and decadal scales. This makes SSA to be considered further as a region of climatic variability. As such, the observed increase in precipitation does not come with lots of doubt. In fact, the key issue here is that the increased precipitation is often not properly distributed within the growing season and within crop production areas [68].

SSA experienced declining precipitation between the 1950s and early 1980s. However, from the 1990s, trends of an increase in precipitation in most parts of the Sahel and SSA have been recorded [69]. While projections indicate that by 2020, food shortages would have reduced in most developing countries, the Sahel and SSA will still be facing problems with food shortages due to inappropriate land uses such as deforestation, over-cultivation,

cattle stock rearing, rapid population growth and inadequate precipitation during the crop growing seasons [70]. Even if crop yields are projected to rise in SSA, it is possible that due to the envisaged increase in population the growth in production will be worthless and the amount of malnourished children will keep on a pessimistic turn towards 2020 [70, 71].

The question is, are there any other studies that support the argument that the influence of climatic drivers such as precipitation on crop yields is declining? In the affirmative, several studies have argued in support of this assertion [72–74]. In fact, it has been discussed that agricultural production has declined in Africa due mainly to land use and cover change [62, 75]. Land cover changes through deforestation have repercussions on crop yields, and this can be linked to over-grazing and population pressure on dry land [76]. Stephenne and Lambin [77] also support the above statement when they argue that land cover changes through anthropogenic deforestation, cattle rearing and population growth are responsible for grain declines in the Sahel. It has also been argued that due to an increase in precipitation in the West African Sahel, the role of precipitation as a primary and sole driver of environmental change is now questionable because land use changes are increasingly becoming more important [89]. In India, Nepal and Bangladesh, it has been observed that in terms of changes in farming practices (proxy for adaptation) a majority of farmers responded to market-related drivers than climatic stressors [91]. Market-related variables such as yield and better market opportunities and resource-related variables such as declining fertility and labour have been stronger drivers of change compared to climatic drivers [92].

Adaptation options to enhance crop yields based on the drivers of crop yields

Based on the three most important drivers of crop yields identified, a number of adaptation options can be designed to meet the twin challenges of increasing crop yields and maintaining a sustainable environment (see Fig. 5).

Forest area adaptations

Forest area loss is the most important determinant of crop yields in Uganda. To moderate this, the establishment of quotas could be used to restrict logging for commercial and agricultural purposes. This involves establishing a limit beyond which farmers and commercial logging companies are not expected to exceed. The quotas could be in hectares or number of trees that each farmer or company can cut within a year. For this to be successful, the current inventory of forest stock available must be known and all monitoring and reporting

Fig. 5 Specific non-climatic adaptations to enhance crop yields in Uganda

systems strengthened, including tackling persistent challenges with corruption. To be effective, monitoring must be shared between officials and the local population to establish ownership and improve compliance. To handle cases of non-compliance, heavy and well-documented and defined financial penalties should be levied. Non-compliance can also be handled through the suspension of licences in the case of commercial logging companies [78]. One way of enhancing compliance could be by enforcing mechanisms like the Reduction of Emissions from Deforestation and Forest Degradation (REDD+) in which farmers and companies that use less of their

deforestation quotas get compensated. In the forest zones of the south-west and east of Uganda where most of the people depend on deforestation either to expand agricultural land or for energy, livelihoods can be diversified through the introduction of small-scale businesses, harnessing of local resources, such as medicinal herbs/plants, gathering of fruits, mushrooms, hunting and bee keeping. In addition, the pressure on the forests could be further reduced through investments in renewable energies such as solar, wind and geothermal energies [81]. For these to be successful, governments have to create safety nets through commitments to invest in these sectors as it is the case with biofuels in Brazil [79, 80]. According to Angelson [81], most East African governments are beginning to invest in renewable energies.

Wood fuel adaptations

To be able to restrict the over-exploitation of forests for wood fuel, strict enforcement of quotas in which limits of exploitation are set needs to be observed [78]. Exploiters that exceed their quotas will be liable to pay penalties for non-compliance. Those that are dependent on wood fuel as a source of energy and income will require a diversification of livelihoods through renewable energies such as solar, wind and geothermal and the enhancement of access to new livelihood options such as small-scale businesses, harnessing of local resources, such as medicinal herbs/plants, gathering of fruits, mushrooms, hunting and bee keeping [78, 82–84]. Again, just as in the case of forests, the government needs to establish the required safety nets to make sure that the populations can have access to these alternatives.

Tractors/mechanization adaptations

In Uganda, the surge in the use of tractors has positive effects on crop yields. Further approaches have to be identified towards making this accessible to the people. In a country in which most agriculture is in the hands of small-scale farmers, the question that comes up is, how can these farmers can gain access if not ownership of such heavy capital equipments? The importance of agricultural equipment on crop yields has historical origins.

In most of SSA, humans still constitute the principal source of power and farm labour, cultivating about ~ 10% of the total area with tractors and 65% of the total area with draught animals [85–88]. In Asia, ~ 30% of the land is cultivated by hand, 30% by draught animals and ~ 40% by tractors. In North Africa and the Middle East, ~ 20% of the land is cultivated by hand and another 20% by draught animals, while tractors cultivate about ~ 60% of the land (Clarke and Bishop 2005; Sims et al. 2007). Latin America and the Caribbean also have a well-established record of use of tractors [85–88]. Within SSA, there is a variation in the use of mechanization with manual power being dominant in central Africa (~ 85%), draught animals being dominant in East Africa (~ 32%) and tractors being dominant in southern Africa (~ 25%).

More recent use of tractors in Africa began in the 1940s. They were first used in commercial white-owned farms, but they spread through tractor hire schemes for small-scale farmers through aid programmes, donor countries and tractor manufacturers before governments came in to support. So by the 1960, several tractorization schemes had been established. At independence of most African countries in 1961, the number of tractors in SSA (172,000 units) exceeded that of Asia (120,000 units). Ten

years later in 1971, the number of tractors in Asia had exceeded those in Africa fivefold to 600,000 units and to 6 million units by 2000. During the same period, in SSA, the number of tractors increased slowly and peaked to 275,000 units in 1990 and declined to 221,000 units in 2000. The increase in the use of tractors in Asia illustrates the effects of the green revolution which triggered increased demand for farm power [85–88].

For such small-scale holders in Uganda, small tractors could be made available at various community extension agricultural centres. The farmers that need such tractors could ask the extension services that will either train the farmers on how to use the equipments or an official might come to the farmers' farms to perform the task for a fee. Farmers could also form cooperatives through which they can come together and purchase tractors. For the larger-scale farmers, the governments could still help them get the tractors through low interest loans that they will be expected to pay over a given period of time. This should be based on a careful evaluation of the farmers' assets. Care should be taken to distinguish between large-scale and small-scale farmers to avoid forcing small-scale farmers who are often poor from trying to own tractors they might never be able to pay for and whose profitability is low owing to the amount of land under cultivation.

At independence, most governments in SSA promoted the use of tractors in an effort to meet up with increased demand for food and cash crops. This was achieved through government run tractor hire schemes. Here commercial banks provided soft loans at low interest rates, while farmer groups and cooperatives were encouraged to purchase tractors. This resulted in an increase in the use of tractors at independence. However, this system soon started failing due to poor economic performance, weak infrastructure and poor management. Under the government tractor hire scheme, the areas cultivated per hectare per machine were small and the cost to run the machines was fixed [86–88] as such much was spent and little gains were made; the schemes were bound to fail. Therefore, if the tractorization programme in Uganda has to be successful, the area under cultivation has to be taken into consideration to avoid these weaknesses as well as avoid a system that lacks the basic infrastructure to support machines. A scheme that is entirely subsidy driven, poorly maintained, expensive to repair machines and difficult in obtaining spare parts should be avoided.

Conclusions

This paper examines the drivers of crop yields in Uganda, identifying non-climatic factors—primarily forest area, wood fuel, and tractors—to be more important than climatic factors in determining crop yields. These results are consistent with what currently obtains in most of SSA as

The determinants of crop yields in Uganda: what is the role of climatic and non-climatic...

85

most of the literature argues that precipitation in Africa is currently increasing but not properly distributed across the growing seasons. As a result, precipitation decline during crop growing seasons has only come to make a bad situation worse due to land use dynamics such as forest area loss, wood fuel dynamics and agricultural inputs such as tractors have emerged to be more significant. The systematic approach is a context-specific approach that argues that it is wrong to up-lift policies that have worked elsewhere and apply them in other areas; like most climate smart agricultural approaches, the right approach is to base policies and adaptations on the actual drivers of the problem that is under investigation.

In terms of specific policies, forest area dynamics and fuel wood gathering could be controlled through: establishment of quotas, renewable energies, penalties and enforcement, and diversification of livelihoods. In terms of tractors, large-scale farmers could get low interest loans to purchase tractors, while small-scale farmers could benefit by forming cooperatives that can buy the tractors or extension services could be put in place by the government to enhance access to these services. It is however suggested that, since this study is based on large-scale national-level data for the whole of Uganda, further experiments should be carried out touching on specific districts based on the perceptions of farmers. Also, the response of individual crops to these variables should be verified including the possible future outcomes. Data on the influence of market demand on food production from specific sites across Uganda and other SSA countries could be obtained from population perceptions based on household surveys. Studies such as Bhatta et al. [91, 92] have used this approach. The latter is necessary because for Uganda and other SSA countries, there exist very limited long-term reliable time series data on market demand. It is for this reason that for this national scale study, time series population trends have been used as a proxy for national-level market demand or food needs in Uganda. This is valid as in any country; population numbers exert pressure on food production and thereby control demand.

Additional files

Additional file 1. Table S1. Summary of 31 food and cash crops used in this study and all other data in excel

Additional file 2. Table S2a. Correlation matrix and summary of variables used in this study and their sources. **Table S2b**. Outputs of the optimized regression model based on the 3 most sign ficant climatic variables. **Figure S2a**. Scatter plots of (a): forest area and yield; (b): wood fuel and yield; (c) agricultural tractors and yield.

Authors' contributions

TEE collected the data, performed the analysis and wrote the paper. JF is the postdoctoral advisor cf TEE, he had extensive discussions with TEE on the design and elaboration of the paper, and he also participated in writing and editing the paper and supervised the entire study. SL had discussions with TEE. BN and AB were field assistants during Dr. Epule's field visit to Karamoja, north-east Uganda, and thus assisted in conducting interviews and culling other data. All authors read and approved the final manuscript.

Author details

Department of Geography, McGill University, 805 Sherbrooke St. W., Burnside Hall 416, Montreal, QC H3A 0B9, Canada. [2] Priestley International Centre for Climate, University of Leeds, Leeds, UK. [3] Department of Geography, Makerere University, P.O. Box 7062, Kampala, Uganda.

Acknowledgements

We wish to thank the authors whose works we consulted and the University of Oxford, the UNDP and FAO for providing the data used in this study.

Competing interests

The authors declare no conflict of interests.

Funding

This work was supported by a Grant from the Social Science and Humanities Research Council of Canada Grant Number 756-2016-0003.

References

1. International Assessment of Agricultural Knowledge (IAASTD). Agriculture at a crossroads. Global report Chs 1, 4. Island Press. 2009. http://www.agassessment.org/reports/IAASTD/EN/Agriculture at a Crossroads Global Report (English).
2. The Royal Society. Reaping the benefits: science and the sustainable intensification of global agriculture 1–10, 47–50 (The Royal Society). 2009. http://royalsociety.org/.
3. Foley JA, Ramankutty N, Brauman KA, Cassidy E, Greber J, Johnston M, Mueller ND, O'Connell C, Ray DK, West PC, Balzer C, Bennett EM, Carpenter SR, Hill J, Monfredo C, Polasky S, Rockstrom J, Sheehan J, Siebert S, Tilman D, Zaks DPM. Solutions for a cultivated planet. Nature. 2011;478:337–478. https://doi.org/10.1038/nature10452.
4. Kearney J. Food consumption trends and drivers. Philos Trans R Soc B. 2010;365:2793–807.
5. Cirera X, Masset E. Income distribution trends and future food demand. Philos Trans R Soc B. 2010;365:2821–34.
6. Tester M, Langridge P. Breeding technologies to increase crop production in a changing world. Science. 2010;327:818–22.
7. Sanchez PA. Tripling crop yields in tropical Africa. Nat Geosci. 2010;3:299–300.
8. DeFries RS, Rudel T, Uriarte M, Hansen M. Deforestation driven by urban population growth and agricultural trade in the twenty-first century. Nat Geosci. 2010;3:178–81.
9. Lobell DB, Schlenker W, Costa-Roberts J. Climate trends and global crop production since 1980. Science. 2011;333(6042):616–20.
10. Naylor R. Expanding the boundaries of agricultural development. Food Secur. 2011;3:233–51.
11. Schmidhuber J, Tubiello FN. Global food security under climate change. PNAS. 2007;104:19703–8.
12. Food and Agriculture Organization of the United Nations (FAO). The State of Food Insecurity in the World: Economic crises—Impacts and lessons learned 8–12, FAO, Rome, Italy. 2009. http://www.fao.org/3/a-i0876e.pdf.
13. Thurow R, Kilman S. Enough: why the world's poorest starve in an age of plenty Chs 2, 4, 12. New York: Perseus Books; 2009.
14. Battisti DS, Nay or RL. Historical warnings of future food insecurity with unprecedented seasonal heat. Science. 2009;2009(323):240–4.

15. Moss RH, Edmonds JA, Hibbard KA, Manning MR, Rose SK, van Vuuren DP, Carter TR, Emori S, Kainum M, Kram T, Meehl GA, Mitchell JFB, Nakicenovic N, Riahi K, Smith SJ, Stouffer RJ, Thomson AM, Weyant JP, Wilbanks TJ. The next generation of scenarios for climate change research and assessment. Nature. 2010;463:747–56. https://doi.org/10.1038/nature08823.

16. Food and Agricultural Organization of the United Nations Statistics Division (FAO) and United Nations Industrial Development Organization (UNIDO). Agricultural mechanization in Africa…Time for action. Planning investment for enhanced agricultural productivity report of an expert group meeting in January 2008, Vienna Austria. FAO, Rome. 2008. http://www.unido.org/fileadmin/user_media/Publications/Pub_free/agricultural_mechanization_in_Africa.pdf.

17. Lobell DB, Cassman KG, Field CB. Crop yield gaps: their importance, magnitudes and causes. Ann Rev Environ Resour. 2009;34:179–204.

18. Negin J, Remans R, Karuti S, Fanzo JC. Integrating a boarder notion of food security and gender empowerment into the African green revolution. Food Secur. 2009;1:351–60.

19. Denning G, Kabambe P, Sanchez P, Malik A, Flor R, Harawa R, Nkhoma P, Zambia C, Banda Magombo C. Inputs subsidies to improve smallholder maize productivity in Malawi: towards an African green revolution. PLoS Biol. 2009;7:e1000023.

20. Kaizzi K. Application of the GYGA approach to Uganda. Uganda. 2014. http://www.yieldgap.org/gygamaps/excel/GygaUganda.xlsx.

21. Poate CD. A review of methods for measuring crop production from small-holder producers. Exp Agric. 1988;24:1–14.

22. IFAD. Enabling poor rural people to overcome poverty in Uganda: rural poverty in Uganda. Rome, Italy. 2012. www.ifad.org.

23. Mubiru DN, Kyazze FB, Radeny M, Zziwa A, Lwasa J, Kinyangi J. Climatic trends, risk perceptions and coping strategies of smallholder farmers in rural Uganda. CCAFS Working Paper no. 121. CGIAR Research Program on Climate Change, Agriculture and Food Security (CCAFS). Copenhagen, Denmark. 2015. www.ccafs.cgiar.org.

24. Mubiru J, Banda EJKB. Monthly average daily global irradiation map for Uganda. A location in the equatorial region. Renew Energy. 2012;41:412–5.

25. Farley C, Farmer A. Uganda climate change vulnerability report. USAID. 2013. http://community.eldis.org/.5b9bfce3/ARCC-Uganda%20VA-Report.pdf.

26. Government of Uganda, Ministry of Water and Environment. 2008 Inception Report: climate change vulnerability assessment, adaptation strategy and action plan for the water resources sector in Uganda. One World Sustainable Investments. Directorate of Water Resource Management. 2008.

27. Food and Agricultural Organization of the United Nations Statistics Division (FAO). FAOSTAT. 2016. http://faostat3.fao.org/download/Q/QC/E.

28. Epule TE, Peng C, Lepage L, Chen Z. The causes, effects and challenges of Sahelian droughts: a critical review. Reg Envron Change. 2014;14:145–56. https://doi.org/10.1007/s10113-013-0473-z.

29. Prince SD, Brown De Coulston E, Kravitz LL. Evidence from rain-use efficiencies does not indicate extensive Sahelian desertification. Glob Change Biol. 1998;4:359–74.

30. Maynard K, Royer JF, Chauvin F. Impact of greenhouse warming on the West African summer monsoon. Clim Dyn. 2002;19:499–514.

31. Rotstayn LD, Lohmann U. Tropical rainfall trends and the indirect aerosol affect. J Clim. 2002;14:2103–16.

32. Knutson TR, Delworth TL, Dixon KW, Held IM, Lu J, Ramaswamy V, Schwarzkopf MD, Stenchikov G, Stouffer RJ. Assessment of twenty-century regional surface temperature trends using the GFDL CM2 coupled models. J Clim. 2006;19:1624–51. https://doi.org/10.1175/JCLI3709.1.

33. McSweeney C, New M, Lizcano G. UNDP climate change country profiles: Uganda. 2010. http://country-profiles.geog.ox.ac.uk/. Accessed 15 Mar 2017.

34. McSweeney C, New M, Lizcano G, Lu X. The UNDP climate change country profiles improving the accessibility of observed and projected climate information for studies of climate change in developing countries. Bull Am Meteorol Soc. 2010;91:157–66.

35. Epule TE, Changhui P, Laurent L, Zhi C. Forest loss triggers in Cameroon: a quantitative assessment using multiple linear regression approach. J Geogr Geol. 2011;3(1):30–40.

36. Epule TE, Changhui P, Laurent L, Zhi C, Nguh BS. The environmental quadruple: forest area, rainfall, CO_2 emissions and arable production interactions in Cameroon. Br J Environ Clim Change. 2012;2(1):12–27.

37. Houghton RA. Tropical deforestation and atmospheric carbon dioxide. Clim Change. 1991;19:99–118.

38. Mertens B, Lambin E. Spatial modeling of deforestation in southern Cameroon: spatial disaggregation of diverse deforestation processes. Appl Geogr. 1997;17(2):143–62.

39. Carr D, Suter L, Barbieri A. Population dynamics and tropical deforestation: state of the debate and conceptual challenges. Popul Environ. 2005;27:90–113.

40. Matson PA, Parton WJ, Power AG, Swift MJ. Agricultural intensification and ecosystem properties. Science. 1997;277:504–9.

41. Lindell L, Astrom M, Oberg T. Land-use versus natural controls on soil fertility in the Subandean Amazon, Peru. Sci Total Environ. 2010;408(4):965–75.

42. Epule TE, Bryant CR, Akkari C, Daouda O. Can organic fertilizers set the pace for a greener arable agricultural revolution in Africa? Analysis, synthesis and way forward. Land Use Policy. 2015;47(1):179–87. https://doi.org/10.1016/j.landusepol.2015.01.033.

43. Matson PA, Naylor R, Ortiz-Monasterio I. Integration of environmental, agronomic, and economic aspects of fertilizer management. Science. 1998;280:112–4.

44. Yohanna JK, Fulani AU, Akaama W. Survey of mechanization problems of small scale (peasant) Farmers in the middle belt of Nigeria. J Agric Sci. 2011;3(2):262–6.

45. Yohanna JK. Farm machinery utilization for sustainable agricultural production in Nasarawa State of Nigeria. Int J Food Agric Res. 2007;1&2:193–9.

46. Muller R, Pistorius T, Rohde S, Gerold G, Pacheco P. Policy options to reduce deforestation based on a systematic approach of drivers and agents in lowland Bolivia. Land use Policy. 2013;30:895–907. https://doi.org/10.1016/j.landusepol.2012.06.019.

47. Hector A, Schmid B, Beierkuhnlein C, Caldeira C, Diemer M, Dimitrakopoulos PG, Finn JA, Freitas H, Giller PS, Good J, Harris R, Hogberg P, Huss-Danell K, Joshi J, Jumpponen A, Korner C, Leadley PW, Minns A, Mulder PH, Donovan GO, Otway SJ, Pereira JS, Prinz A, Read DJ, Scherer-Lorenzen M, Schulze ED, Siamantziouras ASD, Spehn EM, Terry AC, Troumbis AY, Woodward FI, Yachi S, Lawton JH. Plant diversity and productivity experiments in European grassland. Science. 1999;286(5442):1123–7. https://doi.org/10.1126/science.286.5442.1123.

48. Neumann K, Verburg PH, Stehfest E, Muller C. The yield gap of global grain production: a spatial analysis. Agric Syst. 2010;103:316–26.

49. Stanhill G, Cohen S. Global dimming: a review of the evidence for a widespread and significant reduction in global radiation with discussion of its probable causes and possible agricultural consequences. Agric For Meteorol. 2010;107:225–78.

50. Calderini D, Slafer G. Changes in yield and yield stability in wheat during the 20th century. Field Crops Res. 1998;57:335–47.

51. Food and Agricultural Organization of the United Nations Statistics Division (FAO). Forest Resource Assessment. FAO Forestry paper no. 163. Rome: FAO; 2010. http://www.fao.org.

52. Hamilton AC. Distribution patterns of forest trees in Uganda and their historical significance. Vegetation. 1974;29:21–35.

53. Wood GB, Wiant HV Jr. editors. Modern methods of estimating tree and log volume. In: Proceedings of IUFRO conference, Morgantown, West Virginia, 14–16 June 1993. West Virginia University Publication Services. 1993.

54. Kohlin G, Amacher G. Welfare implications of community forest plantations in developing countries: the Orissa Social Forestry Project. Am J Agric Econ. 2006;87(4):855–69.

55. Arnold JEM, Köhlin G, Persson R. Woodfuels, livelihoods, and policy interventions: changing perspectives. World Dev. 2006;34(3):596–611.

56. Godfrey AJ, Denis K, Daniel W, Akais OC. Household firewood consumption and its dynamics in Kalisizo sub-county, central Uganda. Ethnobot Leaflets. 2010;14:841–55.

57. DFID. Energy for the poor: underpinning the millennium development goals. London: Department for International Development; 2002. http://www.dfid.gov.uk/Documents/publications/energyforthepoor.pdf.

The determinants of crop yields in Uganda: what is the role of climatic and non-climatic...

87

58. World Bank. Report of the AFTEG/AFTRS joint seminar on household energy and woodland management. Washington: World Bank; 2002.

59. Tabuti JRS, Dhilliona SS, Lye KA. Firewood use in Bulamogi County, Uganda: species selection, harvesting and consumption patterns. Biomass Bioenergy. 2003;25:581–96.

60. Fisher M. Household welfare and forest dependence in Southern Malawi. Environ Dev Econ. 2004;9:135–54.

61. Dan X, Deng Q, Li M, Wang W, Zhang Q, Cheng X. Reforestation of *Pinus massoniana* alters soil organic carbon and nitrogen dynamics in eroded soil in south China. Ecol Eng. 2013;52:154–60.

62. Geist HJ, Lambin EF. Proximate causes and underlying driving forces of tropical deforestation. Bio Sci. 2002;52:143–50.

63. Rosegrant MW, Cline SA. Global food security: challenges and policies. Science. 2003;302:1917–9.

64. Zhao S, Peng C, Jiang D, Lei X, Zhou X. Land use change in Asia and ecological consequences. Ecol Res. 2006. https://doi.org/10.1007/s11284-006-0048-2.

65. Zak RM, Cabido DC, Diaz S. What drives accelerated land cover change in central Argentina? Synergistic consequences of climatic, socioeconomic, and technological Factors. Environ. Manag. 2008;42:181–9.

66. Kepner RA, Bainer R, Barger EL. Principles of farm machinery. 3rd ed. Westport: AVI Publishing Company, Inc.; 1978.

67. Eklundh L, Olsson L. Vegetation index trends for the African Sahel 1982–1999. Geophy Res Lett. 2003;30(8):1430. https://doi.org/10.1029/2002GL016772.

68. Hulme M. Climatic perspective on Sahelian desiccation: 1973–1998. Global Environ Change. 2001;11:19–29.

69. Wang G, Eltahir AB. Role of vegetation dynamics in enhancing the low-frequency variability of the Sahel rainfall. Water Resour Res. 2000;36(4):1013–21.

70. International Fund for Agricultural Development (IFAD). The Challenge of ending rural poverty. Rural Poverty Report. 2001. http://www.ifad.org.

71. Tucker CJ, Nicholson SE. Variations in the size of the Sahara desert from 1980 to 1997. Ambio. 1999;2009(28):587–91.

72. Herrmann SM, Anyamba A, Tucker CJ. Recent trends in vegetation dynamics in the African Sahel and their relationship to climate. Glob Environ Change. 2005;15:394–404.

73. Olsson L, Eklundh L, Ardo J. The recent greening of the Sahel-trends, patterns and potential causes. J Arid Environ. 2005;63(3):556–66.

74. Anyamba A, Tucker CJ. Analysis of Sahelian vegetation dynamics using NOAA-AVHRR NDVI data from 1981–2003. J Arid Environ. 2005;63:595–614.

75. Ewert F, Rounsevell MDA, Reginster I, Metzger MJ, Leemans R. Future scenarios of European agricultural land use. Estimating the changes in crop productivity. Agric Ecosyst Environ. 2005;107:101–16.

76. Lambin E, Rounsevell MDA, Geist HJ. Are agricultural land-use models able to predict changes in land-use intensity? Agric Ecosyst Environ. 2000;82:321–31.

77. Stephenne N, Lambin EF. A dynamic simulation model of land-use changes in Sudano-Sahelian countries of Africa (SALU). Agric Ecosyst Environ. 2001;85:145–61.

78. Tollefson J. Paying to save the rainforests. Nature. 2009;460:936–7.

79. Nepstad D, Soares-Filho B, Merry F, Lima A, Moutinho P, Carter J, Bowman M, Cattaneo A, Rodrigues H, Schwartzman S, McGrath DG, Sticker CM, Lubowski R, Piris-Cabezas P, Rivero S, Alencar A, Almeida O, Stella O. The end of deforestation in the Brazilian Amazon. Science. 2009;326(5958):1350–1.

80. Gunilla E, Olsson A, Outtara S. Opportunities and challenges to capturing the multiple potential benefits of REDD+ in a traditional Savana-Woodland Region in West Africa. Ambio. 2013;42:309–19.

81. Angelson A. Realising REDD+ : national strategy and policy options. Centre for International Forestry Research, Bogor, Indonesia. 2009. http://www.cifor.org/publications/pdf_files/Books/BAngelsen0902.pdf.

82. Mwape C, Gumbo D. Communities' reorganization for REDD+ implementation in Zambia. In: Pathways for implementing REDD+. UNEP Perspectives Series2010. 2010. http://cd4cdm.org/Publications/PathwaysImplementingREDDplus.pdf.

83. Naughton-Treves L, Day C. Lessons about land tenure, forest governance and REDD+: case studies from Africa, Asia and Latin America. Land Tenure Center, UW-Madison. 2012. https://www.nelson.wisc.edu/ltc/docs/Lessons-about-Land-Tenure-Forest-Governance-and-REDD.pdf.

84. Brockhaus M, Obidzinski K, Dermawan A, Laumonier Y, Luttrell C. An overview of forest and land allocation policies in Indonesia: is the current frame-work sufficient to meet the needs of REDD+? For Policy Econ. 2012;18:30–7.

85. Winrock AK. Assessment of animal agriculture in sub-Saharan Africa. Little Rock: Winrock International; 1992.

86. Clarke L, Bishop C. Farm power-present and future availability in developing countries. Agricultural Engineering International: the CIGR Journal of Scientific Research and Development. Invited Overview Paper. Presented at the Special Session on Agricultural Engineering and International Development in the Third Millennium. ASAE Annual International Meeting/CIGR World Congress, 30 July 2002, Chicago, IL. USA. vol. IV. 2002. https://ecommons.cornell.edu/bitstream/handle/1813/121/Clarke%20and%20Bishop%2018aOct2002.pdf?sequence=17.

87. Sims BG, Kienzie ., Cuevas R, Wall G. Addressing the challenges facing agricultural mechanization input supply and farm product processing. In: Proceedings of an FAO workshop held at the CIGR world congress on agricultural engineering Bonn, Germany, 5–6 Sept 2006. FAO, Rome, Italy. 2007. ftp://ftp.fao.org/docrep/fao/010/a1249e/a1249e.pdf.

88. Yohanna JK. An appraisal of farm power and equipment operation and management in Nasarawa State of Nigeria. J Eng Sci Technol. 2006;1(1):58–61.

89. FAO. AQUASTAT. FAO. 2015. http://www.fao.org/nr/water/aquastat/countries_regions/UGA/.

90. Uganda Bureau of Statistics, UBOS. Uganda national household survey findings 2009/2010. 2010. http://www.ubos.org/UNHS0910/chapter2_introduction.html Accessed 7 March 2017.

91. Bhatta GD, Aggarwal PK, Shrivastava AK, Sproule L. Is rainfall gradient a factor of livelihood diversification? Empirical evidence from around climatic hotspots in Indo-Gangetic plains. Environ Dev Sustain. 2016;18:1657–78.

92. Bhatta GD, Aggarwal PK, Kristjanson P, Shrivastava AK. Climate and non-climatic factors influencing changing agricultural practices across different rainfall regimes in South Asia. Curr Sci. 2016;110:7.

93. Sarker MA, Alam K, Gow J. Exploring the relationship between climate change and rice yield in Bangladesh: an analysis of time series data. Agric Syst. 2012;112:11–6.

94. Lobell DB, Field CB. Global climate crop yield relationships and the impacts of recent warming. Eviron Res Lett. 2007;2(1–7):014002.

95. Schlenker W, Lobell DB. Robust negative impacts of climate change on African agriculture. Envion Res Lett. 2010;5(1–8):014010.

96. Mulatu DW, Eshete ZS, Gatiso RG. The impact of CO2 emissions on agricultural productivity and household welfare in Ethiopia. *Environ. For Dev.* Discussion paper Series; 2016. http://www.efdinitiative.org/sites/default/files/publications/efd-dp-16-08.pdf.

97. Jaggard K, Qi A, Ober ES. Possible change to arable crop yields by 2050. Philos Trans R Soc B. 2010;365:2835–51.

Agricultural extension and its effects on farm productivity and income: insight from Northern Ghana

Gideon Danso-Abbeam[1]* ⓘ, Dennis Sedem Ehiakpor[1] and Robert Aidoo[2]

Abstract

Background: In agricultural-dependent economies, extension programmes have been the main conduit for disseminating information on farm technologies, support rural adult learning and assist farmers in developing their farm technical and managerial skills. It is expected that extension programmes will help increase farm productivity, farm revenue, reduce poverty and minimize food insecurity. In this study, we estimate the effects of extension services on farm productivity and income with particular reference to agricultural extension services delivered by Association of Church-based Development NGOs (ACDEP).

Methods: The study used cross-sectional data collected from 200 farm households from two districts in the Northern region of Ghana. The robustness of the estimates was tested by the use of regression on *covariates*, regression on *propensity scores* and *Heckman treatment effect* model.

Results: The study found positive economic gains from participating in the ACDEP agricultural extension programmes. Apart from the primary variable of interest (ACDEP agricultural extension programme), socio-economic, institutional and farm-specific variables were estimated to significantly affect farmers' farm income depending on the estimation technique used.

Conclusions: The study has reaffirmed the critical role of extension programmes in enhancing farm productivity and household income. It is, therefore, recommended that agricultural extension service delivery should be boosted through timely recruitment, periodic training of agents and provision of adequate logistics.

Keywords: Agricultural extension, ACDEP, Heckman treatment effect, Productivity, Regression on propensity scores

Background

The millennium development goals (MDGs) of reducing hunger and to promote food security are rooted in increasing agricultural productivity, especially from the crop sector. This is because agriculture is considered as the engine of growth in many developing economies, particularly in sub-Saharan Africa (SSA). The policy direction of growth and poverty reduction strategy (GPRS II) formulated by Ghana was to achieve accelerated and sustainable growth and poverty reduction through agricultural productivity. Although poverty had declined by 7.7% over the years (2005–2013), about 25% of Ghana's population is still poor while under a tenth of the population are living in extreme poverty [1]. This suggests that the role of agriculture has not been sufficient to elevate many people above the poverty line, especially the rural folks who contribute immensely to agricultural production in Ghana. Ghana's agricultural sector, which employs about 42% of the workforce, is dominated by smallholder farmers (about 90% farming on less than 2 hectares of land) who are using traditional production methods and farm inputs [2]. Asfaw et al. [3] argue that achieving productivity growth in the agricultural sector can only be successful through the development and dissemination of improved agricultural technologies to these smallholder

*Correspondence: dansoabbeam@uds.edu.gh
[1] Department of Agricultural and Resource Economics, University for Development Studies, Tamale, Ghana
Full list of author information is available at the end of the article

farmers in the rural areas. Rural farmers farming on small hectares of land can be attributed to conditions such as lack of adequate credit, lack of access to product market, lack of adequate extension contacts, among others. Among these constraints, inadequate extension services have been identified as one of the main limiting factors to the growth of the agricultural sector and rural community development at large [3]. With recent threats of climate change and the rapid advancement in technology, more farmers require capital investment in agriculture and human capacity development to at least continue to make their living out of farming. Thus, the role of agricultural extension today goes beyond the transfer of technology and improvement in productivity, but also, it includes improvement in farmers' managerial and technical skills through training, facilitation and coaching, among others.

Incognizant of these problems, the government of Ghana through the Ministry of Food and Agriculture (MoFA) over the years has invested so much in building the capacities of smallholder farmers through agricultural extension programmes. Other stakeholders in the agricultural sector such as international and local funding agencies, non-governmental organizations and financial institutions have and continue to make investments in the delivery of extension services to farmers. One of the prominent NGOs that have been implementing extension programmes in Northern Ghana is the Association of Church-based Development NGOs (ACDEP). ACDEP is a network of over 40, mostly, but not only church sponsored development NGOs in Northern Ghana with the vision of improving the economic well-being of its actors, particularly farmers through improved farm technology and access to input and output markets. The ACDEP NGOs are also engaged in other fields such as primary health care, HIV/AIDS, water and sanitation, rural enterprise development for women and value chains. One of the key components of ACDEP programmes in the field of agriculture is the delivery of agricultural extension services.

The ACDEP agricultural extension programmes include capacity building in good agricultural practices (GAPs), creating linkages among the value chain actors (input dealers, farmers, wholesalers and retailers) and other value addition techniques.[1] Thus, wider dissemination of information regarding farmer skill development, the use of improved farm technologies, general farm management practices and easy access to input and output markets have been the fundamental principles

underlying delivery of ACDEP agricultural extension services. All these are geared towards improvement in productivity, reduction in poverty and enhancement in food security.[2] Given the scale of investment from ACDEP, the value for money regarding an increase in farm income is an important policy question. In this study, we hypothesized that participation in ACDEP agricultural extension programmes positively affects the welfare of the participating farm households through improvement in farm productivity and income. Many studies had dealt with issues relating to improving agricultural technologies such as improved crop varieties, adoption of fertilizer, etc. in Ghana [4–8]. Also, some other studies had focused on the impact of government extension programmes concerning the use of technology, adoption rates, farm productivity and efficiency, and farm output levels [9]. These studies had provided excellent information on factors shaping adoption and adoption intensities of farm technologies. However, rigorous studies on impact evaluation of agricultural extension services delivered by non-governmental organizations such as ACDEP on smallholders' farm productivity and income remain very rare in the Ghanaian agricultural literature. Thus, there are many gaps with regard to what is known about the effects of extension services on the productivity and income of farmers—the body of empirical evidence does not match the scale of implementation, particularly in Ghana's agricultural sector. Hence, the aim of the study is to estimate the effects of agricultural extension services with particular reference to ACDEP on the farm productivity and income of smallholder farmers in the ACDEP operational areas in the Northern region of Ghana.

The role of extension services in agriculture

Agricultural extension programmes have been one of the main conduits of addressing rural poverty and food insecurity. This is because, it has the means to transfer technology, support rural adult learning, assist farmers in problem-solving and getting farmers actively involved in the agricultural knowledge and information system [10]. Extension is defined by FAO [11] as; "systems that should facilitate the access of farmers, their organizations and other market actors to knowledge, information and technologies; facilitate their interaction with partners in research, education, agribusiness, and other relevant institutions; and assist them to develop their own technical, organizational and management skills and practices". By this definition, an extension is deemed as a primary

---

[1] The study used "ACDEP extension programmes" and "ACDEP agricultural extension programmes" interchangeably.

[2] Note: Information in this paragraph was sourced from ACDEP website (http://acdep.org/s.te/index.php/home-7/14-acdep-home/24-about-acdep) and its secretariat office in Tamale, Ghana.

tool for making agriculture, its related activities as well as other economic activities more effective and efficient to meet the needs of the people. It is, therefore, regarded as a policy tool for promoting the safety and quality of agricultural products. Agricultural extension is aimed primarily at improving the knowledge of farmers for rural development; as such, it has been recognized as a critical component for technology transfer. Thus, agricultural extension is a major component to facilitate development since it plays a starring role in agricultural and rural development efforts [12].

Bonye et al. [12] argued that extension provides a source of information on new technologies for farming communities which when adopted can improve production, incomes and standards of living. Extension service providers make an innovation known to farm households, act as a catalyst to speed up adoption rate and also control change and attempt to prevent some individuals in the system from discontinuing the diffusion process [13]. In reaching farmers, extension officers demonstrate a technology to farmers but with much concentration on early adopters since the laggards would learn later from the early adopting farmers. Through extension services, farmers' problems are identified for further investigation and policy direction. Swanson [14] argued that extension service goes beyond technology transfer to general community development through human and social capital development, improving skills and knowledge for production and processing, facilitating access to markets and trade, organizing farmers and producer groups, and working with farmers towards sustainable natural resource management. Where market failures such as limited access to credit and non-competitive market structures that provide a disincentive to farmers to produce exist, extension services tend to provide solutions.

Methods
The study area, data source and description of variables
The study analyses the effects of ACDEP extension programmes on farm productivity and income using a sample obtained from the farming communities in the Tolon and Kumbugu districts of the Northern region of Ghana. The two districts are predominantly rural communities with the majority being smallholder farm households. Tolon district has a population of 72,990 (males constitute 49.8% while females constitute 50.2%) of which 92.7% are engaged in agriculture, while Kumbugu has a population of 39,341 (equally distributed between males and females, i.e. 50%) of which 95.4% of the households are engaged in agriculture [15]. Crop farming is the main agricultural activities in the two districts with about 98% engaged in it [15]. The districts are characterized by a single rainy season, which starts in late April with

little rainfall, rising to its peak in July–August and declining sharply after that and coming to a complete halt in October–November. The dry season starts from November to March with day temperatures ranging from 33 to 39 °C, while mean night temperature ranges from 20 to 26 °C. The mean annual rainfall ranges between 950 and 1200 mm.

The data were obtained mainly from primary sources, through the use of structured questionnaires. The study followed a multi-stage random sampling technique in selecting the two districts and communities from the Northern region and farm households from each community. In the first stage, the two districts were randomly selected from a number of ACDEP operational districts in the Northern region of Ghana. In the second stage, a random sampling was used to select four ACDEP operational communities from each district. Four non-ACDEP operational communities were selected from Tolon district, while three were selected from Kumbugu district. Thus, the survey covered fifteen communities, eight from Tolon and seven from Kumbugu. In the final stage, 10–15 maize farm households were randomly selected from each operational community, while 10–12 maize farm households were selected from each of the non-operational communities. The total sample size of the study is 200 maize farm households consisting of 110 farm households who had participated in ACDEP extension programmes and 90 maize farm households who did not take part in the programme. Key informants such as some heads of departments of ACDEP office in Tamale, Ghana and community leaders were approached to discuss challenges and opportunities relating to the programme and how the programme can aid in increasing productivity and farm income.

Table 1 shows the descriptive statistics of the sampled farm households. Two main categories of variables are described here: the dependent variables (outcome variables) and the explanatory variables.

Dependent variables
The dependent variables consist of maize farm productivity (yields), maize farm income per hectare, total household income and household income per capita. Farm productivity is defined as the total output of maize in kilograms per hectare.[3] The average yield of maize of the participating farm households is 1811 kg per hectare, while that of the non-participants is 1511 kg per hectare. Thus, there is no statistical difference between yields of participants and non-participants of the ACDEP agricultural extension programme. However, difference could

---

[3] In this study, farm productivity and farm yields are used interchangeably.

Table 1 Descriptive statistics of the sampled farm households

Description of variables	Participants		Non-participants	
	Mean	SE	Mean	SE
Dependent variables				
Maize productivity/yield (kg/ha)	1811.27	179.49	1511.24	122.87
Maize farm income (total sales in GH¢ per ha)	1177.33[c]	116.67	982.3	79.86
Total household income (GH¢)	13,710.85[a]	183.43	8239.94	1158.60
Total household income per capita (GH¢)	2222.01[a]	313.36	788.04	90.04
Explanatory variables				
Socio-economic characteristics				
Gender of the household head (dummy; 1 = male, 0 otherwise)	0.69[a]	–	0.46	–
Age of the household head (years)	40.65	11.11	41.69	10.44
Household size (count)	6.17[a]	0.545	10.45	5.86
Number of years in formal education (years)	6.35	0.58	5.73	0.59
Number of years in crop farming	19.89	11.57	20.72	12.06
Plot characteristics				
Number of maize plots (count)	1.25[b]	0.66	1.47	0.611
Farm size (farmland allocated to maize cultivation in ha)	2.14[a]	2.03	1.3	0.75
Institutional factors				
Access to credit	0.28	–	0.07	–
Farmer-based organization	0.66		0.73	
Distance to local market (kilometres)	14.13[c]	10.22	16.08	12.12
Distance to regional market (kilometres)	24.32	11.34	23.23	14.56

[a,b,c] denote 1, 5 and 10% levels of significance, respectively

partly be ascribed to the over-concentration of the programme on linking farmers to the output market with the adoption of farm technology lagging.

The income from maize (measured as the total sales per hectare) shows a difference of 20% in favour of the participants which is significant at 10% level of significance. One of the key components of the ACDEP agricultural extension programme is to link farmers to the output market, and so farmers who participated in the programme might have had their products sold at the right time compared with their counterparts who did not. As part of the programme, some of the participants were engaged in contract farming where their farm product was pre-negotiated. Hence, might be responsible for the significant difference in the farm income between the participants and non-participants. The total household income is the summation of the revenues from maize and other crops, revenues from the sales of livestock and income from non-farm economic activities such as wages, salaries and other self-employed businesses earned by members of the household (e.g. household head, spouse and other economically active members). The total household income per capita defines the total income of the households relative to their size. The total household income of the ACDEP participating households is GH¢ 13,710, while that of the non-participants

is GH¢8239 showing a significant difference between the two groups.[4] Similarly, the total income per capita is GH¢2222 and GH¢788 for participants and non-participants, respectively. The large difference in household income is partly coming from the gap in the farm income, and probably the treated group were more engaged in other farming activities such as livestock rearing due to the knowledge they might have gained through the programme. The difference in household income per capita may emanate from the fact that non-participating households have a larger family size compared to the participating households. However, these descriptive statistics are limited regarding their implications for causality, as they fail to quantify and account for selectivity biases that may emanate from participation in the extension programmes.

Explanatory variables

Consistent with pieces of literature, the study hypothesized that participation in the ACDEP extension programme, as well as the determinants of farmers' farm income, can be explained by socio-economic, farm-specific and institutional factors [9, 15–17]. Some of the

---

[4] Average exchange in 2016: GH¢ = US$ 1.

socio-economic factors include gender, age, household size, the number of years in formal education and the number of years in crop farming (experience). The farm-specific factors included in the empirical models of this study are the hectare of agricultural land allocated to maize cultivation (farm size) and the number of maize farm plots. The institutional and policy variables include membership of a farmer-based organization (FBO), distance to the local district market and distance to the main regional city market as well as access to agricultural credit.

Econometric technique

The study uses a combination of three econometric techniques to assess the effects of the ACDEP agricultural extension services on farm productivity and income of the smallholder maize farmers. First, we use a simple regression referred to as *regression on covariates*. Second, a *regression on propensity scores* was used to account for selectivity bias in estimating the effects of the extension services. In the third stage, we complement the results with the *Heckman treatment effect model* to test the robustness of the results.

We start by considering that the outcome variables (in our case maize farm yields, farm income per capita, total household income and total household income per capita) are a linear function. This linear function consists of a vector of households, farm-specific and institutional factors (X_i). The simplest approach to assessing the effect of the ACDEP agricultural extension programme is to include in the outcome equation ACDEP agricultural extension participation variable (AE_i) denoting one (1) if farmer participates in the ACDEP programme and zero (0) if otherwise, and then apply ordinary least squares (OLS) estimation technique. This can be specified as:

$$Y_i = \beta + \alpha_i AE_i + \delta_i X_i + \varepsilon_i \qquad (1)$$

where Y_i is the outcome variable, AE_i is a dummy variable for ACDEP extension programme participation, X_i represents other explanatory variables, α_i and δ_i are parameters to be estimated. The causal effect of AE_i on the outcome variables can be measured by estimating the parameter α_i. However, estimating Eq. (1) with OLS might lead to biased estimates because it assumes that AE_i is random and exogenously given, while the selection of ACDEP participants is non-random and AE_i variable is potentially endogenous [16, 18]. The non-random sample selection problem arises from self-selection where the farmers themselves decide whether or not to participate in the extension programme, probably due to differences in resource endowments. Thus, the selection bias emanates from the fact that treated individuals may be systematically different from the non-treated for

reasons other than the treatment status. The endogeneity problem may arise becuase the programme may target farmers with specific characteristics (smallholder farmers, commercial farmers, poor farmers or relatively wealthy farmers). These may result in biased estimates of the coefficient of AE_i which measures the effects of extension programme participation on the outcome variables. Selection bias could also arise from the selection on observable or unobservable. Selection on observables could be controlled by including some set of variables in the model. However, selection on unobservables is usually difficult to control by adding variables. This is because unobserved variables such as farmers' managerial ability, motivation, among others are not observed, hence, difficult to capture. Excluding these unobserved variables gives biased estimates of α_i in Eq. (1). The study addresses the problem of selectivity bias in three ways.

The first is to include a set of observable covariates to account for potential selection bias due to selection on observables as applied in [13]. These variables include distance to local district market, distance to regional city market and farm size. We can then re-specify Eq. (1) as:

$$Y_i = \beta + \alpha_i AE_i + \chi_i Z_i + \delta_i X_i + \varepsilon_i \qquad (2)$$

where Z_i is a vector of variables to control for the selection bias and χ_i is the parameter to be estimated. As stated earlier, this is called *regression on covariates*.

The second approach is to estimate the propensity score or conditional probability to participate in the ACDEP agricultural extension programme and use it as an additional control variable in the regression model.[5] This is referred to as *regression on propensity scores* as used in Alemu et al. [13] and Asres et al. [19]. The propensity score as a control variable in the regression model reduces the potential biases created by selection on observable characteristics [17]. The new regression model can be specified as:

$$Y_i = \beta + \alpha_i AE + \varphi_i PS_i + \chi_i Z_i + \delta_i X_i + \varepsilon_i \qquad (3)$$

where $PS_i = p(AE_i = \frac{1}{X})$ and ϕ_i is the estimate of the propensity score PS_i. Other variables are defined earlier.

The third approach to deal with sample selection bias is the Heckman treatment effect model. The Heckman treatment effect model is one of the most widely used procedure to account for sample selection bias and offers a mean/way of correcting for biases that may arise from unobservable factors, and thus results in unbiased and consistent estimates [16]. The Heckman treatment effect model is an extension of the Heckman two-stage model.

---

[5] Note: A probit model was used to generate the propensity scores used in Eq. (3).

The only difference is that the dependent variable in the selection equation becomes one of the explanatory variables in the outcome equation of the former but not in the latter model. The principle behind the treatment effect model is to estimate the selection equation (usually a probit model) and use the predicted values of the dependent variable as a selection control factor called the inverse mills ratio (IMR). The IMR is then used as an additional regressor in the outcome equation to correct sample selection and free other explanatory variables from any biases. In this way, the true effects of participation in the ACDEP agricultural extension programme on outcome variables are measured [20]. Thus, in the treatment effect model, the treatment condition enters the outcome equations as an explanatory variable to measure the true effects on the outcome variables [18].

The model can be specified in two steps. The selection equation which is usually a probit is given as:

$$AE_i = \beta + \delta_i X_i + \varepsilon_i \tag{4}$$

where AE_i is a latent endogenous variable (participation in ACDEP agricultural extension programmes), X_i is a set of exogenous variables determining the selection of farm households into the extension programme, δ_i is a parameter to be estimated and ε_i is the error term. The substantive equation can be specified as:

$$Y_i = \beta + \delta X_i + \alpha_i AE_i + \varepsilon_i \tag{5}$$

where α_i measures the effect of ACDEP extension services on the outcome variables. To correct for self-selection biases in the substantive Eq. (5), an IMR denoted by the symbol λ was generated and added as an additional explanatory variable. The formulation process of IMR is given as:

$$\lambda = \frac{\phi(-\delta_i X)}{1 - \Phi(\delta_i X)} \tag{6}$$

where ϕ and Φ are normal probability density function and cumulative density function, respectively, of the standard normal distribution. Adding the IMR to Eq. (5) translates into Eq. (7) as:

$$Y_i = \beta + \delta_i X_i + \alpha_i AE + \gamma_i \lambda_i + \mu_i \tag{7}$$

where γ_i is an estimate of the IMR (λ_i) and μ_i is a two-sided error term with $N(0, \sigma_v^2)$. The rest are as defined earlier. A significant coefficient of the IMR implies that there is self-selection problem, while a non-significant coefficient indicates the absence of sample selection. Ignoring the addition of the IMR will render the results from Eq. (5) as biased [21]. Thus, the inclusion of the selectivity term makes the coefficient α_i (measuring the effects of the treatment variable on the outcome

Table 2 Determinants of participation in ACDEP agricultural extension programme

Variable	Coefficient	SE
Age of the household head	0.112[b]	0.050
Household size	0.014	0.018
Number of years in formal education	−0.002	0.017
Number of years in crop farming (experience)	0.016[c]	0.009
Access to agricultural credit	0.669[b]	0.283
Membership of farmer-based organization (FBO)	0.309[b]	0.149
Farm size allocated to maize cultivation in hectares	0.423[a]	0.107
Number of maize farm plots	0.043	0.180
Constant	−1.259	0.373
LR Chi^2 (8)	26.47	
Prob > Chi^2	0.000	
Wald Chi^2 (8)	28.37	
Prob > Chi^2	0.000	
Sensitivity (% correctly classified among participants)	69.37%	
Specificity (% correctly classified among non-participants)	69.66%	
Total correctly classified (%)	69.50%	

[a, b,c] denote significance levels at 1, 5 and 10%, respectively

variables) unbiased, albeit it is inefficient as the disturbance term (μ_i) is heteroscedastic [16]. The problem of heteroscedasticity can be corrected by the use of bootstrap standard errors or re-sampling. However, the STATA software package used in generating the estimates automatically adjusts for that bias in the standard errors [22].

Empirical results and discussions
Determinants of ACDEP agricultural extension programme participation

The results from the probit model for the participation in ACDEP agricultural extension programme are presented in Table 2. The model fits the data reasonably well as indicated by the Wald test that all the coefficients are jointly equal to zero is rejected [Chi^2 (8) = 28.37; $p = 0.000$].

The probit model also correctly classified 69.37% of the maize farm households among the participants and 69.66% among the non-participants with a total accurate prediction rate of 69.50% for the entire sample.

From the table, the probability of participating in the ACDEP extension programmes is significantly influenced by the age of the household head, the number of years in crop farming, access to agricultural credit, membership of a farmer-based organization and the size of plots allocated to maize production (farm size). The positive and significant influence of age on the probability of participation in the ACDEP agricultural extension programme is against the notion that older farmers are

Table 3 Effects of ACDEP extension programmes on yield, farm and household income

Outcome variable	Regression on covariates		Regression on propensity score		Heckman treatment effects	
	Coefficient	SE	Coefficient	SE	Coefficient	SE
Maize yield (kg/ha)	0.011	0.039	0.012	0.040	0.113[a]	0.018
Maize farm income (total sales in GH¢/ha)	0.135	0.120	0.113[c]	0.063	1.113[a]	0.192
Total household income (GH¢)	0.361[a]	0.038	0.233[a]	0.043	0.853[a]	0.153
Total household income per capita	0.347[a]	0.043	0.216[a]	0.048	1.104[a]	0.194

[a,b,c] denote 1, 5 and 10% significance levels, respectively. SE denotes standard errors

usually reluctant to accept new information and ideas as reported by Asres et al. [19] and Genius et al. [23]. However, the result is consistent with the studies by Tiwari et al. [24] and Mendola [25]. Farmers with longer years in farming business have a higher likelihood of participating in the ACDEP programmes to optimize their farm productivity and income. Hence, the positive and significant effect of this variable is expected.

Similarly, access to agricultural credit will encourage farmers to participate in the extension programme to get more information that may help to maximize their yield to repay the credit on time. Group membership such as FBO enhances farmer-to-farmer extension services where knowledge and ideas on farm business and other off-farm activities are transferred from one farmer to the other. Thus, farmers who are members of FBOs are likely to get sufficient awareness and knowledge on farm technologies and, hence, are sensitized to join extension programme for more information on their farm business. Similarly, participants in the ACDEP agricultural extension programme tend to have larger farm sizes than their non-participants counterparts as supported by the descriptive statistics in Table 1. Usually, members of such extension programmes are encouraged to consider their farm as a business entity rather than a cultural way of life and are, therefore, poised to achieve higher output through expansion and productivity. Gebreegziabher [26] reported a positive effect of plot size on the probability of participating in an extension programme in Ethiopia.

Effects of ACDEP extension programme on yield, farm income and household income

The econometric results of the effects of the extension programme on the four outcome variables (maize productivity, farm income per hectare, total household income and total household income per capita) are reported in Table 3. These outcome variables are already defined in Table 1. The following paragraphs discuss the effects of ACDEP agricultural extension programmes on each of the performance indicators across the three models.

Table 3 indicates that ACDEP agricultural extension programmes had positive effects on the productivity and farm income of the households in the study area. The results from different estimation approaches are quantitatively similar in terms of the direction, indicating the robustness of the results to changes in the estimation techniques. The estimated results show that the ACDEP extension programme had no significant effects on maize farm productivity when regression on covariates and regression on propensity scores were used. However, in the Heckman treatment effect model, the effect is positive and significant. Since all the variables are specified in logarithmic terms, it suggests that participating in the ACDEP extension programme has increased farm productivity by 11.3% points. The result from the Heckman model is at variance with the study of Feder et al. [27] who found no contribution of extension programmes to crop productivity. However, it is in line with other previous studies [28–31] reporting positive effects of extension programmes on crop farm productivity. The estimated effects for maize farm income per hectare vary from 0.113 to 1.13 depending on the estimation procedure. These coefficients imply that the extension programme has led to an increase in farm income by 11.3 to 111.3 percentage points. The effect is smaller in a regression on propensity score than the Heckman treatment effect model as indicated in Table 3. The wide divergence in the magnitude of the effects may partly be attributed to the difference in the unobserved heterogeneity among the maize farm households. The regression on propensity score minimizes selection biases based on observed covariates, while Heckman treatment effects correct for selection bias arising from unobserved factors, hence, the two estimation techniques are most likely to produce different estimates. Moreover, given the market-oriented nature of the extension programmes nowadays, the large effect on farm income is not surprising. Thus, the focus is more on linking farmers to the market where products are purchased at the right time, rather than technology adoption.

For Heckman treatment effects, regression on propensity score and regression on covariate models,

Table 4 Regression on covariates, propensity score and Heckman treatment effects (maize farm income per capita as dependent variable)

Variable	Regression on covariates		Regression on propensity scores		Heckman treatment effect	
	Coefficient	SE	Coefficient	SE	Coefficient	SE
ACDEP extension	0.135	0.120	0.113[c]	0.063	1.113[a]	0.192
ACDEP *pscore*			3.663[a]	0.749		
Distance to local market	− 0.142[a]	0.029	− 0.136[a]	0.036		
Distance to regional market	− 0.037[b]	0.014	− 0.019	0.088		
Gender of household head	0.165	0.138	0.136	0.131	0.209	0.121
Formal education	0.009	0.011	0.024[b]	0.007	0.014[a]	0.012
Household size	0.091[a]	0.011	− 0.117[a]	0.011	0.097[a]	0.012
Farming experience	0.002	0.006	0.021[a]	0.007	− 0.008	0.006
Access to credit	0.052	0.159	0.801[a]	0.232	0.179[b]	0.018
Membership of FBO	0.138	0.137	0.030	0.122	0.309[a]	0.049
Number of maize farm plots	0.106	0.093	0.319[b]	0.125	− 0.025	0.107
Farm size (maize)	0.345[a]	0.043	0.217[a]	0.048	0.312[a]	0.048
Constant	6.228[a]	0.206	5.596a	0.235	6.040	0.234
Adjusted R-squared	0.535		0.585			
Lambda					0.632	0.107[a]
LR Chi^2(1) = 13.43					0.000*	
Breusch–Pagan/Cook–Weisberg test						
Chi^2(1)		0.380	0.240			
Prob > Chi^2		0.538	0.622			
VIF (mean)		1.390	2.65			
Observations	200		200		200	

*denotes significance level of LR Chi^2 and it is at 1% level

[a,b,c] denote significance levels at 1, 5 and 10%, respectively

extension programmes lead to 85.3%, 23.3% and 36.1% increase in total household income, respectively. Gebrehiwot [9] observed that extension programmes lead to 7% and 10% points in farm household income when estimated with regression on covariates and propensity score matching using stratification technique, respectively. Similarly, Asres et al. [19] reported 6% and 18% increase in household income through participation in extension programme when estimated with OLS and Heckman treatment effects, respectively. The results also corroborate with that of [28] and [32]. The effects of ACDEP agricultural extension programmes on per capita income differ across the three models regarding the magnitudes of the effects. The extension programmes improve per capita income by 34.7% and 21.6% for regression on covariates and regression on the propensity scores, respectively. It increases by 110.4% in the Heckman treatment effect model. The greater impact of the ACDEP programme on the per capita income could emanate from the large effect of farm income coupled with the smaller household size of the participants compared with that of the non-participants as indicated in the descriptive statistics.

Apart from participation in extension programme, some other factors have been estimated to affect farm households' income. Table 4 presents the full results of the regression on covariates, regression on propensity scores and the second stage of the Heckman treatment model. The two regressions (regression on covariates and regression on propensity scores) are free from multicollinearity as indicated by the variance inflation factor (VIF) mean values of 1.39 and 2.65 for regression on covariates and regression on propensity scores, respectively. The VIF for each independent variable was less than the critical value of 10 indicating non-existence of multicollinearity [33]. The results are also free from heteroscedasticity as indicated by the small values of the Chi^2 generated by the Breusch–Pagan/Cook–Weisberg heteroscedasticity test.

The results indicate the significance of household locations and their effects on their farm income. The distance to market centres, both local and regional markets, is found to negatively affect their farm income. These findings may be attributed to the fact that ACDEP programmes are targeted at the remote districts and communities where poverty is pervasive. The other side is that farmers sometimes find it difficult in transporting their produce to the urban market centres as they are being constrained by finance and other factors such as road networks.

Other socio-economic factors such as the educational attainment (measured as the number of years in formal education), household size and farming experience (measured as the number of years in crop farming) significantly affect farm income with different estimation techniques. Similarly, institutional variables such as access to agricultural credit and social capital variable (FBO membership) are estimated to have positive and significant effects on farm income. The number of farm plots owned is positive and significant in the regression on propensity score but not significant in the regression on covariates and Heckman treatment effect model. Furthermore, the size of the plot allocated to maize production (farm size) had a positive and significant influence on farm income across the three estimation techniques. This underscores the importance of farm size to increasing farm income among smallholder rural farm households.

Conclusions

The study has assessed the effects of ACDEP agricultural extension programme on the productivity and income of farm households using primary data from two districts in the Northern region, Ghana. Since the agricultural extension programmes require a substantial amount of investment, understanding its effects on the beneficiaries (farmers) is very important. After controlling for selectivity bias, we found out that participation in the ACDEP agricultural extension programme improves welfare through an increase in farmers' income. However, the effect levels are different depending on the empirical estimation procedure adopted. The central government and development partners should commit more human, financial and logistical resources to agricultural extension delivery in the country to boost agricultural productivity, farm incomes and total household income. Also, access to agricultural credit and formation of farmer groups such as farmer-based organizations should be promoted for agricultural extension service delivery to realize its full impact.

Abbreviations

ACDEP: Association of Church-based Development NGOs; FAO: Food and Agriculture Organization; GAP: good agricultural practices; GPRS II: growth and poverty reduction strategy II; MDG: millennium development goals; OLS: ordinary least squares; SSA: sub-Saharan Africa.

Authors' contribution

GD-A designed the study and analysed the data. DSE collected the data and wrote the manuscript. RA revised the manuscript and conducted some additional analysis. All authors read and approved the final manuscript.

Author details

[1] Department of Agricultural and Resource Economics, University for Development Studies, Tamale, Ghana. [2] Department of Agricultural Economics, Agribusiness and Extension, Kwame Nkrumah University of Science and Technology, Kumasi, Ghana.

Acknowledgements

The authors are very grateful to the enumerators who collected and assisted in the data entry and all farmers who participated in the survey. We are also grateful to the anonymous reviewers for their useful comments.

Competing interests

The authors declare that they have no competing interests.

Funding

This study had no external funding. It was fully funded by the authors.

References

1. Ghana Statistical Service. Ghana living standards survey round 6 (GLSS6): poverty profile in Ghana (2005–2013). Ghana: Accra; 2014.
2. Ghana Statistical Service. 2010 Population and housing census summary, report of final results. Ghana: Accra; 2012.
3. Asfaw S, Shiferaw B, Simtowe F, Lipper L. Impact of modern agricultural technologies on smallholder welfare: evidence from Tanzania and Ethiopia. Food Policy. 2012;37(3):283–95.
4. Ehiakpor SD, Danso-Abbeam G, Zutah J, Hamdiyah A. Adoption of cocoa farm management practices in Prestea Huni-Valley District, Ghana. Russ J Agric Soc Sci. 2016;5(53):117–24.
5. Wiredu AN, Zeller M, Diagne A. What determines the adoption of fertilizers among rice-producing households in Northern Ghana? Q J Int Agric. 2015;54(3):263–83.
6. Danso-Abbeam G, Setsoafia ED, Ansah IGK. Modelling farmers investment in agrochemicals: the experience of smallholder cocoa farmers in Ghana. Res Appl Econ. 2014;6(4):12–27.
7. Bruce KKA, Donkoh SA, Ayamga M. Improved rice variety and its effect on farmers' output in Ghana. J Dev Agric Econ. 2014;6(6):242–8.
8. Aneani F, Anchirinah VM, Owusu-Ansah F, Asamoah M. Adoption of some cocoa production technologies by cocoa farmers in Ghana. Sust Agric Res. 2012;1(1):103–17.
9. Gebrehiwot KG. The impact of agricultural extension on households' welfare in Ethiopia. Int J Soc Econ. 2015;42(8):733–48.
10. Christoplos I, Kidd A. Guide for monitoring, evaluation and joint analyses of pluralistic extension support. Lindau: Neuchâtel Group; 2000.
11. Food and Agriculture Organization (FAO) of the United Nations. Ethiopia Country Brief; 2010. Retrieved from www.fao.org/countries/55528/en/eth/.
12. Bonye SZ, Alfred KB, Jasaw GS. Promoting community-based extension agents as an alternative approach to formal agricultural extension service delivery in Northern Ghana. Asian J Agric Rural Dev. 2012;2(1):76–95.
13. Alemu AE, Maetens M, Deckers J, Bauer H, Mathijs E. Impact of supply chain coordination on honey farmers' income in Tigray, Northern Ethiopia. Agric Food Econ. 2016;4:9.
14. Swanson BE. Global review of good agricultural extension and advisory service practices. Rome: Food and Agriculture Organization of the United Nations; 2008.
15. Ghana Statistical Service. 2010 Population and housing census, districts report. Ghana: Accra; 2014.
16. Greene WH. Econometric analysis. 5th ed. New Jersey: Prentice-Hall; 2003.
17. Imbens G. Non-parametric estimation of average treatment effects under exogeneity: a review. Rev Econ Stat. 2004;96(1):4–29.
18. Maddalla GS. Limited dependent and qualitative variables in econometrics. Cambridge: Cambridge University Press; 1983.
19. Asres E, Makoto N, Kumi Y, Akira I. Effect of agricultural extension program on smallholders' farm productivity: evidence from three peasant associations in the highlands of Ethiopia. J Agric Sci. 2013;5(8):163–81.
20. Smith J, Todd P. Does matching overcome Lalonde's critique of non-experimental estimators? J Econ. 2005;125:305–53.
21. Heckman J. Sample selection bias as specification error. Econometrica. 1979;47:153–61.

22. Bushway S, Johnson BD, Slocum LA. Is the magic still there? The use of the Heckman two-step correction for selection bias in criminology. J Quant Criminol. 2007. https://doi.org/10.1007/s10940-007-9024-4.

23. Genius MG, Pantzios CJ, Tzouvelekas V. Information acquisition and adoption of organic farming practices. J Agric Res Econ. 2006;31(1):93–113.

24. Tiwari KR, Sitaula BK, Nyborg LP, Paudel GS. Determinants of Farmers' adoption of improved soil conservation technology in a middle mountain watershed of Central Nepal. Environ Manag. 2008;42(2):210–22.

25. Mendola M. Agricultural technology adoption and poverty reduction: a propensity score matching analysis for rural Bangladesh. Food Policy. 2007;32(33):372–93.

26. Gebregziabher G, Holden S. Does irrigation enhance and food deficits discourage fertilizer adoption in a risky environment: evidence from Tigray, Ethiopia. J Dev Agric Econ. 2011;30(10):514–28.

27. Feder G, Murgai R, Quizon JB. Sending farmers back to school: the impact of FFS in Indonesia. Rev Agric Econ. 2004;26(1):45–62.

28. Davis K, Nkonya E, Kato E, Mekonen DA, Odendo M, Miro R, Nkuba J. Impact of farmer field schools on agricultural productivity and poverty in East Africa. World Dev. 2012;40(2):402–13.

29. McGarry D. A synopsis of the measured outcomes of farmer field schools in integrated pest management for cabbage production, Timor Leste (2003–2005). FAO Brief, January 24, 2008.

30. Meti SK. Farmers field school strategies for effective diffusion of IPM technology for sustainable cotton yield: a critical analysis. Paper presented at the 23rd annual conference of the association for international agricultural and extension education, Polson, Montana, 20–24 May; 2007.

31. Geer T, Debipersaud R, Ramlall H, Settle W, Chakalall B, Joshi R. Introduction of aquaculture and other integrated pest management practices to rice farmers in Guyana and Suriname. FAN FAO Aquaculture Newsletter No. 35; 2006.

32. Kelemework, F. Impact evaluation of farmer field school: the case of integrated potato late blight management in the Central Highland of Ethiopia. University of Antwerp (Thesis); 2005.

33. Gujarati DN, Porter DC. Basic econometrics. 5th ed. Irwin: McGraw-Hill; 2009.

## 8

Evidence of rapid spread and establishment of *Tuta absoluta* (Meyrick) (Lepidoptera: Gelechiidae) in semi-arid Botswana

Honest Machekano, Reyard Mutamiswa and Casper Nyamukondiwa[*] [ID]

Abstract

Background: *Tuta absoluta* (Meyrick), a major invasive pest of Solanaceous plants, was recently detected in Botswana. Abiotic and biotic factors, together with a suite of population demographic traits are likely key for species propensity and invasion success. First, we determined the movement of *T. absoluta* from its core detection centre to new invasion areas using pheromone baiting and established likely biotic dispersal drivers. Second, we measured thermal tolerance vis critical thermal limits and lower and upper lethal limits to determine how these traits shape population establishment.

Results: We detected *T. absoluta* in all 67 pristine sites across nine districts of Botswana. Within-district trap catches varied between cultivated and wild hosts but were generally not statistically significant ($P > 0.001$). We report three major wild host plants for *T. absoluta* as biotic dispersal drivers: *Solanum coccineum* (Jacq.), *Solanum supinum* (Dunal) and *Solanum aculeatissimum* (Jacq.). *Solanum coccineum* and *S. supinum* were omnipresent, while *S. aculeatissimum* distribution was sporadic. Thermal tolerance assays showed larvae were more heat tolerant, with a higher critical thermal maxima (CT_{max}) than adults ($P < 0.001$), whereas the adults were more tolerant to cold with a significantly lower ($P < 0.001$) critical thermal minima (CT_{min}) compared to larvae. The upper lethal temperatures ranged from 37–43 °C, whereas the lower lethal temperatures ranged from − 1 to − 12 °C for 0–100% mortality, respectively. In the light of prevailing environmental (habitat) temperatures (T_{hab}), warming temperature (7.29 °C) and thermal safety margin (22.39 °C) were relatively high.

Conclusion: *Tuta absoluta* may not be under abiotic physiological or biotic constraint that could limit its geographical range extension within Botswana. The ubiquity of wild Solanaceous plants with the bridgehead of year-round intensive monocultures of Solanaceous crops within a favourable climatic framework may mean that environmental suitability aided the rapid spread of *T. absoluta*.

Keywords: Tomato leaf miner, Insect invasion, Thermal tolerance, Global change, Solanaceous plants

*Correspondence: nyamukondiwac@biust.ac.bw
Department of Biological Sciences and Biotechnology, Botswana
International University of Science and Technology (BIUST), Private Bag
16, Palapye, Botswana

Background

Invasive species are a major threat to agroecosystems and global change [1, 2] and increased global connectivity [3] has drastically increased the diversity and magnitude of such invasions especially in the hot-dry Afrotropical region. Tomato leaf miner, (*Tuta absoluta*) (Meyrick) (Lepidoptera: Gelechiidae), is one of the most destructive insect pests of tomatoes globally [2, 4]. It is of South American origin and was first detected in Spain in 2006 [5, 6] before rapidly spreading and establishing in novel environments in the Mediterranean Basin, Europe, Middle East, South Asia (India), north, east and west Africa [5–9] and recently Southern Africa [10–12]. Because of its high reproductive potential, multivoltinism and potential to acclimatize to different climatic conditions [1], *T. absoluta* is currently considered a key limiting phytosanitary factor affecting the global Solanaceous crops value chain [13].

The larvae of *T. absoluta* feeds on all aerial parts of the plants including the fruits, resulting in significant yield losses and cosmetic damages as well as secondary infection [14, 15]. Characteristic larval mesophyll mining also compromises photosynthetic capacity of crops significantly reducing yields [14]. In the absence of control, yield losses ranging 80–100% have been reported in open and protected tomato fields [5]. A cost–benefit analysis has shown a significant increase in cost of production through high use of insecticides [2, 16], increased tomato market prices as farmers try to recover the high production cost, spatial prohibition of tomato seedlings and fruits trade [17] culminating into overall increased food and nutrition insecurity [18].

In tropical sub-Saharan Africa, irrigated tomatoes are an essential component of horticulture, a major pillar of sustainable development, with a significant contribution to food and nutritional security as well as household source of income especially for resource-poor farmers [18]. However, a major constraint to growing field horticultural crops in Southern Africa is the reduction in yield and quality caused by insect pests [19]. The potential invasion of Southern Africa by *T. absoluta* has already been described [1, 3, 6] with models based on its invasion history and global warming [20]. However, there are no reports based on field data on its thermal fitness and how this correlates with availability and distribution biotic resources, e.g. wild host plant species. Although *T. absoluta* survival on wild Solanaceae, Amaranthaceae, Fabaceae, Chenopodiaceae and Asteraceae plant families was reported, no report has so far combined this knowledge with its on-going invasive movement in the light of prevailing climate data.

For an invasive species to be established, it first has to overcome several environmental barriers [21] including transport, introduction, population establishment and spread [22]. Upon introduction into a novel environment, high propagule pressure [23], species genetic and demographic characteristics [24] and physiological tolerance allow the establishment and habitat permeability [3]. Climate synchrony should exist between introduced areas' and species' environmental stress tolerance to allow successful spread during transience and niche occupation post-invasion [3, 25]. As such, physiologists often use species' thermal tolerance assays as proxy for determining potential for establishment of invasive species. Similarly, it has also been clear from modelling studies, that even when propagule material is high, environmental suitability remains an overriding factor for invasive species successful establishment [3, 23, 26]. Indeed, physiological assays have found use in niche modelling and invasive species risk assessments to determine critical risk invasion areas [27, 28]. *Tuta absoluta* is known to respond naturally to rapidly changing environments [29]. This is characteristic of successful invaders, which should inherently possess high basal and plastic physiological tolerance, including rapid genetic adaptive shifts [30]. Nevertheless [3], also show that native environmental heterogeneity may contribute to species invasive success. This means, species coming from a more heterogeneous environment may likely cope with a changing novel environment through phenotypic adjustment, compared to those coming from a more stable environment.

Temperature is the most important abiotic factor exerting direct and indirect effects on *T. absoluta* population dynamics [1] and consequently invasion success [31, 32]. Therefore, temperature forms a first abiotic 'ecological filter' [33] for successful invasion in a new environment [34], and failure to mount any compensatory mechanisms against it may result in the species failing to establish [35, 36]. The proximity of the environmental temperatures to species thermal physiological limits can therefore indicate species vulnerability and dispersal fitness [31]. Species introduced into habitats close to their thermal tolerance limits are more affected by environmental temperature [37] than those introduced into habitats far from their thermal tolerance limits.

Insects have been reported to experience multiple overlapping abiotic and biotic stressors such as temperature, starvation and desiccation in the wild [3, 38, 39]. Hence, an understanding of bioecology of invasive species is of paramount importance in enlightening mechanisms underlying the successful spread and establishment of invasive alien species [1, 40]. This will also involve determining how the invasive species may respond to native wild host plants. The availability and distribution of alternative wild (non-cultivated) host plants play a significant inoculum sink–source role across the novel landscapes

[6]. Since its detection in Zambia [41], South Africa [10, 42] and Botswana [12], no work has documented *T. absoluta* spread and establishment across the biotic and abiotic frontiers. Here, we ought to establish whether *T. absoluta* was indeed spreading and elucidate the major environmental drivers to successful establishment in Botswana. We measured its thermal tolerance vis limits to activity (critical thermal minima [CT_{min}] and critical thermal maxima [CT_{max}]) and lethal limits (lower and upper lethal limits [LLT] and [ULT], respectively) and compared this with prevailing ambient climatic environment. Second, we investigated wild Solanaceous host diversity and linked this to *T. absoluta* invasion. To date, data on *T. absoluta* invasion potential in tropical climates have only been derived from modelling [1, 2]. No studies have looked at *T. absoluta* physiological thermal tolerance limits with field climate data to test the possible role of climate on its range expansion and spread. Similarly, no study has coupled physiological tolerance and its interaction with host availability on *T. absoluta* invasion pathway. The objective of this study was therefore to investigate whether *T absoluta* has spread and established from its core detection site across other pristine districts of Botswana, since its first detection [12]. Such information is important for pest risk assessments, niche modelling and may aid in developing phytosanitary regulations for effective invasive pest management.

Methods

Insect trapping and sites

Following the detection of *T. absoluta* at Genesis farm (S21.14776; E27.64744), Matshelagabedi village in the North East District of Botswana December 2016 [12], a follow-up surveillance trapping was conducted across 9 of the 10 districts of Botswana (Fig. 1). Traps were not set in Kgalagadi as it is largely part of the Kalahari Desert with very minimal vegetation, human settlement and agricultural activity. A total of 201 (67 sites with 3 traps per site) yellow delta traps (Chempac-Progressive-Agricare®) (Suider Paarl, South Africa) equipped with sticky pads were placed ~ 1 m above ground in tomato fields (cultivated host) and open forests (wild hosts) in each of the study districts during the hot-rainy summer season when the wild hosts were flourishing. High temperature, high relative humidity [1] and presence of the host [2] were reported to possibly enhance its propensity to spread. The major male-attracting synthetic sex pheromone (3E,8Z, 11Z)-3,8,11-tetradecatrienyl acetate (TDTA) loaded on grey rubber dispensers at a dosage of 110 µg per lure, (*Tuta absoluta*-optima PH-937-OPTI Russel IPM, Flintshire, UK) was used. Trap

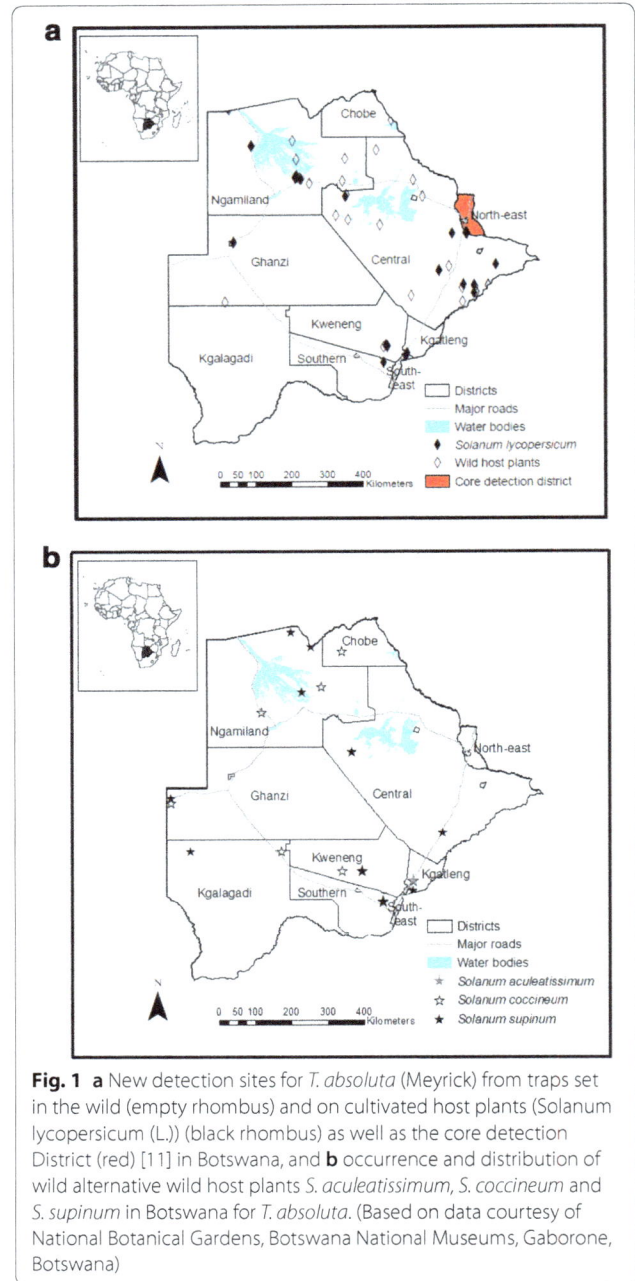

Fig. 1 a New detection sites for *T. absoluta* (Meyrick) from traps set in the wild (empty rhombus) and on cultivated host plants (Solanum lycopersicum (L.)) (black rhombus) as well as the core detection District (red) [11] in Botswana, and **b** occurrence and distribution of wild alternative wild host plants *S. aculeatissimum*, *S. coccineum* and *S. supinum* in Botswana for *T. absoluta*. (Based on data courtesy of National Botanical Gardens, Botswana National Museums, Gaborone, Botswana)

catch data were collected after ~ 30 days, and trapped moths were counted using dyed pointers (chopsticks dipped in insect dye) and mechanical (tally) counters following gross morphological identification [43]. Global Positioning System (GPS) points were recorded for each trapping site using a Garmin® (GPSMAP 62 model, Olathe, USA). Climate data for the sampling areas were obtained from the Meteorological Department, Ministry of Environment, Wildlife and Tourism (MEWT), Republic of Botswana.

Basal thermal tolerance experiments

Insect culture

Larvae were collected on damaged tomato fruits into insect cages (BugDorm®, MegaView Science Co., Ltd. Taiwan) from Noka farm (North East District) (S21.12860; E27.48830), with a general temperature range of 3.4–35.5 °C; mean, mean minimum and monthly temperature range of 20.5–22.6 °C, 11.9–13.3 and 29.1–30.4 °C, respectively [44]. These were allowed to pupate in the laboratory in climate chambers Memmert® climate chambers (HPP 260, Memmert GmbH – Co.KG, Germany) set at 25 ± 1 °C, $65 \pm 5\%$ relative humidity (RH) and 12L–12D photoperiod. This laboratory rearing temperature closely approximated mean annual temperature from the environment from which the specimens were collected. Eclosed T. absoluta adults were placed in 25-cm³ clean cages, where they fed on 10% sucrose solution ad libitum using the cotton dental wick source method (a feeding apparatus for liquid-feeding insects; insects suck the liquid from a wet cotton wick that draws the solution through capillarity) and provided with organically produced tomato-fruiting plants to lay eggs. Experiments were conducted using fourth instar F_1 generation larvae and freshly emerged unsexed adults (± 2 days old). Sex was not considered a factor in our experiments since it has been reported not to affect thermal tolerance traits in some related species (e.g. [45–47]).

Lethal temperature assays

Lethal temperatures were determined using established methods as outlined in [48]. Upper and lower lethal temperatures (ULTs and LLTs) were determined using direct plunge protocol at 2-h duration at temperatures that elicited 0–100% mortality. Ten insects were placed in 60-ml polypropylene vials with gauzed lids and placed in a 33 × 22 cm ziplock bag, replicated three times. This was then plunged into a Merck® water bath (Modderfontein, South Africa) filled with 99.9% circulating ethanol. For ULT, tiny wet filter paper was suspended in each vial to maintain benign humidity and prevent desiccation-related mortality. Following treatment (ULT and LLT), test insects were placed at 25 ± 1 °C and $65 \pm 5\%$ RH in Memmert® climate chambers for 24 h before scoring survival. All insects had access to food and water ad libitum during the 24-h recovery period. Survival was defined as the ability to coordinate muscle response to stimuli such as gentle prodding, or normal behaviours such as feeding, flying or mating [48, 49].

Critical thermal limits (CTLs)

CTLs were assayed using a programmable waterbath (LAUDA Ecogold® RE 2025, Lauda-Königshofen, Germany) connected to a transparent double-jacketed chamber as outlined by [45]. A thermocouple (type K 36SWG) connected to a digital thermometer (Fluke 54 series IIB) was inserted into the central organ pipe (control chamber) to record chamber temperature. A total of ten test insects replicated three times to yield 30 replications per treatment were used in these experiments. Test insects were individually placed into the organ pipes of the double-jacketed chamber connected to a programmable water bath filled with 1:1 water: propylene glycol to allow for subzero temperatures [50]. Both CT_{max} and (CT_{min} experiments started from an ambient set point temperature of 25 °C from which temperature was ramped up (CT_{max}) or down (CT_{min}) at 0.25 °C/min until CTLs were recorded. Although it is likely faster than natural diurnal heating or cooling rates in the wild [45], this ramping rate was chosen as a compromise between ecological relevance and maximum throughput (see also discussions in [45, 51]). In this study, we defined CTLs as the temperature at which each individual insect lost coordinated muscle function and the ability to respond to mild stimuli (e.g. prodding with a thermally inert object).

Data analyses

New detection sites and the distribution of wild Solanaceous host plants were presented on maps (ArcGIS, ArcMap 10.2.2). Trap catch and thermal tolerance data analyses were carried out in STATISTICA, version 13.2 (Statsoft Inc., Tulsa, Oklahoma) and R version 3.3.0 [52]. CTLs met the linear model assumptions of constant variance and normal errors; therefore, they were analysed using one-way ANOVA in STATISTICA. LLT and ULT assays results did not meet the assumptions of ANOVA, and thus, they were analysed using generalized linear models (GLM) assuming a binomial distribution and a logit link function in R. Tukey–Kramer's post hoc tests were used to separate statistically heterogeneous means.

Warming tolerance (WT) and the thermal safety margin (TSM) of T. absoluta under Botswana conditions were calculated as outlined by [53]:

$$WT = CT_{max} - T_{hab} \quad [53]$$

$$\text{and,} \quad TSM = T_{opt} - T_{hab} \quad [53]$$

where CT_{max} = critical thermal maximum for T. absoluta adult (the migratory stage), T_{hab} = habitat temperature—Botswana mean annual temperature for 2015/16. T_{opt} = optimum temperature for T. absoluta.

Results

The spread of T. absoluta in Botswana

Apart from North East District, the area of T. absoluta first detection [12], the species was recorded in eight other districts (Fig. 1a). Moths were detected both in the wild (forests, grazing lands and national parks distant from agroecosystems) and on cultivated solanaceous crops; mainly tomato *Solanum lycopersicum* (L.). We detected *T. absoluta* in areas such as Moremi Island (Okavango Delta) more than 200 km from the nearest human settlements and agricultural activities and bordered by Moremi and Chobe Game Reserves) (Fig. 1a). Surveillance results support our hypotheses that *T. absoluta* spread and successfully established across Botswana (Fig. 1a).

Tuta absoluta wild host plants belonging to the Solanaceae family showed a cosmopolitan distribution (Fig. 1b). Wild host species diversity showed three dominant species; *Solanum aculeatissimum* (Jacq.), *Solanum coccineum* (Jacq.) and *Solanum supinum* (Dunal). *Solanum supinum* was the most widely distributed, occurring in all districts except only in Chobe, North-East and South-East and was found on the Moremi Island of the Okavango Delta (Fig. 1b) giving credence to the occurrence of *T. absoluta* in such a remote area. *Solanum coccineum* had more sporadic distribution, occurring in Chobe, Ngamiland, Ghanzi, Kgalagadi, Kweneng districts and the surrounding areas of the Okavango Delta. However, *S. aculeatissimum* was only found in Kgatleng district (Fig. 1b).

Moths abundance in wild and cultivated hosts

Large numbers of *T. absoluta* moths were captured in all districts, in both cultivated and wild hosts. The cultivated host, *S. lycopersicum* hosted significantly higher numbers ($P < 0.001$) (Table 1) than the wild host plants within districts, especially in Kweneng and Central districts (Fig. 2). Inter-district populations were also generally not significantly different within the same host type (Fig. 2). Overall, in the wild host plants, we recorded a grand mean of 411.1 ± 13.38 moths/trap/month from the cultivated *S. lycopersicum* which was significantly higher ($P < 0.001$) (Table 1) than 187.4 ± 12.21 moths/

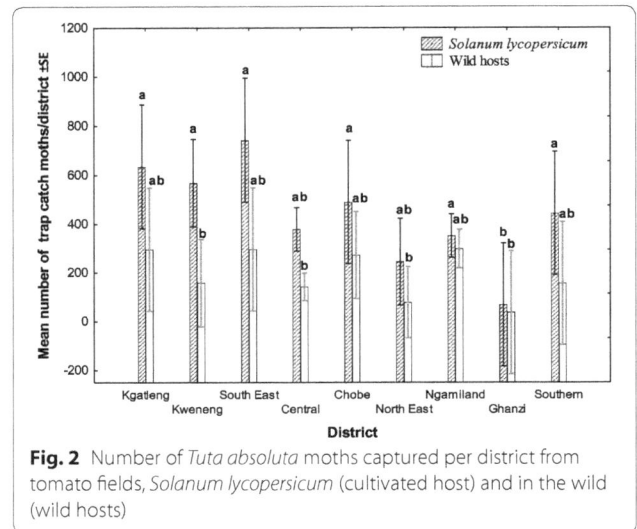

Fig. 2 Number of *Tuta absoluta* moths captured per district from tomato fields, *Solanum lycopersicum* (cultivated host) and in the wild (wild hosts)

trap/month recorded from the wild hosts. High numbers were recorded on *S. lycopersicum* in Central, South-East, Chobe, Kgatleng and Southern districts and in tunnels compared to open fields in other districts. There were no significant interaction effects between the host plant and the district ($P > 0.05$) (Table 1) signifying that in each district host type was not a significant factor affecting abundance (trap catches).

Basal thermal tolerance

Both life stages of *T. absoluta* showed relatively high temperature tolerance although the larvae had significantly higher (47.9 ± 1.25 °C) CT_{max} than the adult (44.1 ± 0.43 °C) (Fig. 3A). The highest temperature where *T. absoluta* could not survive (ULT_0) was 43.0 °C, while the highest temperature for 100% survival (ULT_{100}) was 37 °C (for a 2-h stressful high-temperature exposure). There were significant differences ($\chi^2 = 107.29$, df $= 4$, $P < 0.001$) in survival between test temperatures, again signifying the role of temperature severity and duration in its survival (Fig. 3B). However, on low temperature tolerance, the adult had a significantly lower CT_{min} (-5.2 ± 0.23 °C) than the larvae (3.5 ± 0.07 °C) (Fig. 3C), and the LLTs ranged from -12.0 to -1.0 °C for LLT_0 and LLT_{100}, respectively, based on a 2-h duration at stressful low temperature (Fig. 3D). There were significant differences ($\chi^2 = 163.73$, df $= 6$, $P < 0.001$) in survival between the low test temperatures, implying that survival was determined by both temperature severity and duration of exposure.

Climate data and basal thermal tolerance

Field temperature data from eight districts of Botswana in 2015/2016 seasons (period post-first detection prior

Table 1 Differences in mean moth trap catches between cultivated *S. lycopersicum* and wild hosts and different districts in Botswana

Effect	SS	DF	MS	F	P value
District	443,092	8	55,387	3.5151	< 0.001
Host type	406,792	1	406,792	25.8168	< 0.0001
District * host	206,503	8	25,813	1.6382	> 0.05
Error	756,328	48	15,757		

Tests of significance were done using factorial ANOVA and Tukey's HSD test was used to separate statistically significant means at 95% CI

DF Degrees of freedom, *SS* sum of squares, *MS* mean sum of squares, *F* Fisher–Snedecor test statistic

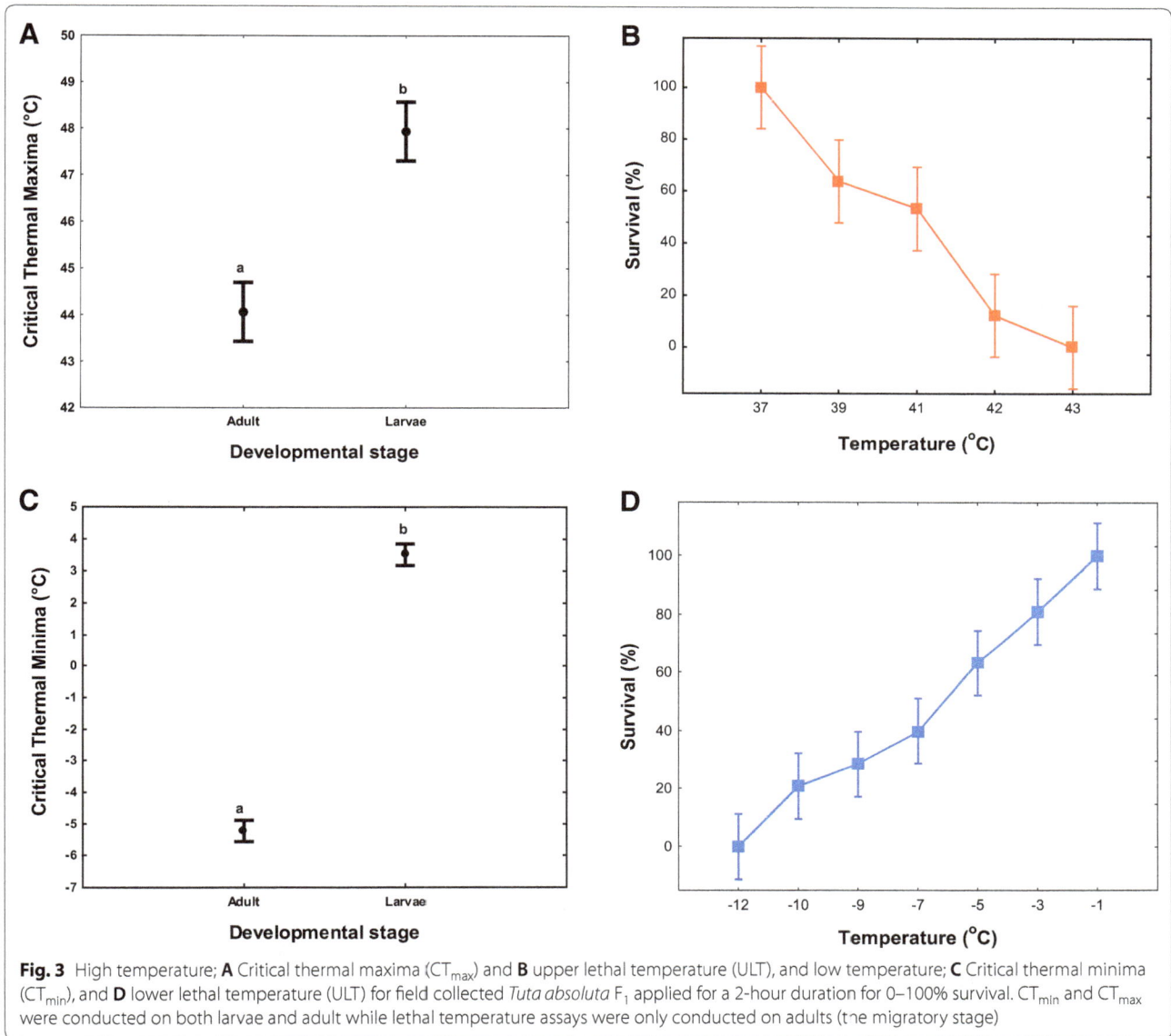

Fig. 3 High temperature; **A** Critical thermal maxima (CT_{max}) and **B** upper lethal temperature (ULT), and low temperature; **C** Critical thermal minima (CT_{min}), and **D** lower lethal temperature (ULT) for field collected *Tuta absoluta* F_1 applied for a 2-hour duration for 0–100% survival. CT_{min} and CT_{max} were conducted on both larvae and adult while lethal temperature assays were only conducted on adults (the migratory stage)

to and during establishment and spread of *T. absoluta*) are shown in Fig. 4a and b. The mean monthly maximum temperatures ranged from a low of 22.3 °C (Kweneng district) to a high of 37.4 °C. (South-East district) (Fig. 4a). Highest maximum field temperatures were below *T. absoluta* CT_{max} by about 6 °C (adults) and above 10 °C (larvae). Relating ULTs to the field maximum temperature data showed that the *T. absoluta* ULT_0 of 43 °C (Fig. 4a) was well above the highest maximum temperatures recorded in nature (37.4 °C; Fig. 4a), implying that *T. absoluta* was not under high-temperature-related physiological stress that could limit its spread and establishment.

The mean monthly minimum temperatures ranged from a low of 1.1 °C (Kweneng district) to a high of 21.3 °C in December 2015 (Ngamiland district) (Fig. 4b). The lowest minimum field temperatures were above *T. absoluta* adult CT_{min} by about 6 °C (adult) and below that of the larvae by about 2.4 °C (see Fig. 4b). This implied that the minimum field temperatures were not physiologically constraining survival of *T. absoluta* adult but the larvae. Adult *T. absoluta* LLT_0 was − 12.0 °C, while LLT_{100} was − 1.0 °C for a 2-h stressful low-temperature exposure (see Fig. 3D). Both temperatures fell well below the most extreme low temperatures recorded in the environment (see Fig. 4b), implying that adult *T. absoluta*

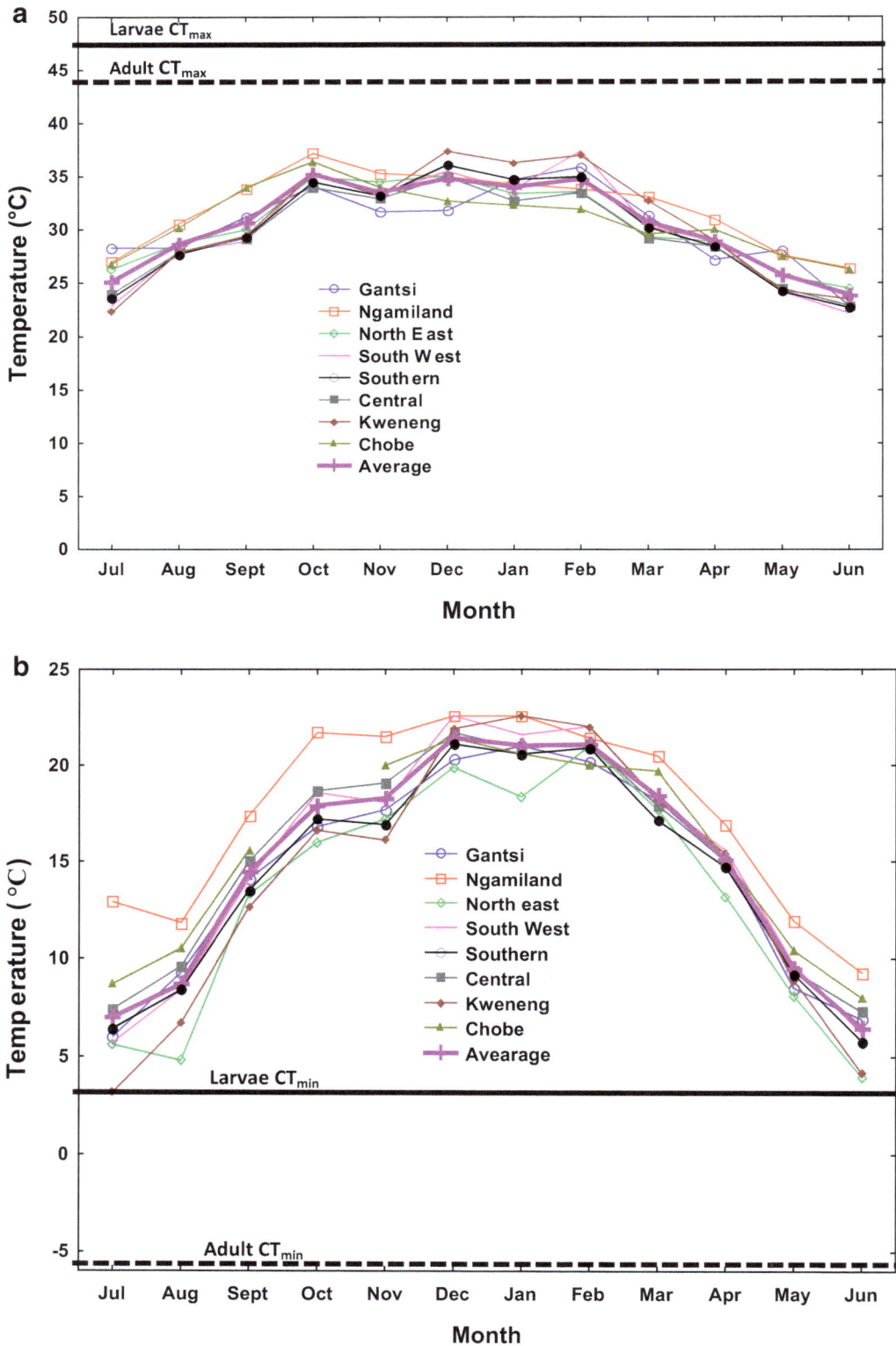

Fig. 4 Field temperatures, **a** mean monthly maximum temperatures and **b** mean monthly minimum temperatures for eight districts of Botswana in 2015/16 season related to *T. absoluta* CT_{max} and CT_{min}, respectively. Horizontal lines denote CT_{min} and CT_{max} for larvae (continuous) and adult (dotted)

may not be under diurnal low-temperature physiological stress.

Warming tolerance (WT) and thermal safety margin (TSM)

The annual mean temperature for Botswana in 2015/2016 was 22.71 °C considered in this study as the habitat temperature (T_{hab}), and the optimum temperature for *T. absoluta* performance and population growth is 30 °C [54] considered as T_{opt}. The adult CT_{max} was recorded as 44.1 °C (see "Basal thermal tolerance" section). Based on these data, the warming tolerance (WT) and the thermal safety margin (TSM) of *T. absoluta* under Botswana conditions were calculated according to [53]:

$$WT = 44.1 °C - 22.7 °C$$
$$WT = 22.39 °C$$

and similarly,

$$TSM = 30 °C - 22.71 °C$$
$$TSM = 7.29 °C$$

Discussion

Following its first invasion in Botswana in December 2016 [12], our results confirm that *T. absoluta* has spread and successfully established in almost all districts of Botswana, thus potentially eliciting widespread economic damage to Solanaceous crops. Although *T. absoluta* was first recorded in North east district of Botswana, evidence from this work suggest its rapid and wide extension of its distribution horizons with new records reported in various districts in the country within an 8-month period (January to August 2017) (Fig. 1). Indeed, this trend is not unusual for the species [2]. The species has been reported to spread at a rate of ~800 km/year aided through wind currents and plant material belonging to families Amaranthaceae, Convolvuceae, Fabaceae, Malvaceae and Solanaceae identified through volatile cues by female moths for egg laying [2]. Therefore, the observed rapid spread and successful niche establishment may be directly linked to the reported availability of host plants in the wild [3], climate suitability and physiological thermal tolerances [26]. These characteristics are consistent with other globally invasive economic insect pest species, e.g. *Chilo partellus* (Swinhoe) [55], *Bactrocera dorsalis* (Hendel) [56], *Ceratitis capitata* (Wiedemann) [57] and *Drosophila suzuki* (Matsumura) [58]. Our results associated *T. absoluta* with a wide range of cultivated and wild host plants [as in example 2, 59], consistent with polyphagic characteristic of many invasive species [22]. *Tuta absoluta* thermal activity physiological thresholds examined here also suggest that there is a conducive climate niche across the country and that species activity and

hence invasion may not be constrained by temperature. Our survey showed Botswana hosts wild solanaceous plants: *S. aculeatissimum, S. coccineum* and *S. supinum*, which are all suitable hosts to *T. absoluta* [see 2, 15]. Amongst these wild host plants, *S. supinum* was the most widely distributed, occurring in all districts of the country, while *S. coccineum* and *S. aculeatissimum* distribution was sporadic. Therefore, it is highly likely that these wild host plants provide biotic resources, (food and shelter) supporting the invasion pathway of *T. absoluta* in Botswana. Although tomato is the preferred host for *T. absoluta,* the species can switch hosts from cultivated to wild as a survival strategy, a notion supported by [5] and [60]. Such availability of biotic resources and suitable environmental conditions are also known to impair diapausing in *T. absoluta* larvae [1] resulting in increased breeding propagule pressure even under less favourable climate conditions, with implications on niche invasion success.

Short-distance dispersal (adjacent field to field or field to tunnels) of *T. absoluta* is known to be facilitated by wind especially soon after introduction [13] with moths capable of active flights of up to 100 km [59], a characteristic that may aid the species' dispersal [2, 32, 61]. Pressure distribution from the Indian Ocean was reported to traditionally create strong east-westerly air masses in the Southern African region [62]. This supports the possible movement of *T. absoluta* through wind currents from the north-east district (core detection district) to the central, southern and western parts of the country. On the other hand, long-distance dispersal may occur through open tomato trade, markets and other related activities [32]. These attributes together may to a larger extent have promoted the spread and establishment of *T. absoluta* propagule moths which could easily locate either cultivated or wild hosts during dispersal. However, the detection of *T. absoluta* in Moremi Island (Fig. 1a); (~200 km from human settlements and agroecosystems) suggests that wind and wild host plants might have played a more significant role in its invasion success.

High populations of *T. absoluta* were recorded on *S. lycopersicum* in Central, South-East, Chobe, Kgatleng and Southern districts (Fig. 2) where production of tomatoes is done in tunnels. The reason may be that the moths were contained within tunnels and hence highly concentrated resulting in the observed high trap catches. Amongst these districts, South-East recorded the highest moth catches. The district is a horticultural hotspot with high concentration and prolonged availability of the cultivated host plants (tomato, green and red pepper) which hosts *T. absoluta*. Since production of tomatoes is carried out throughout the year in this district, the tunnels also act as inoculum reservoirs that form bridgeheads for further introduction and reinfestation of outdoor culti-vated and wild host plants [1–3]. Similarly, relatively high *T. absoluta* moths were recorded in the wild (Fig. 2) sig-nifying its ability to survive outside the cultivated host plant ranges (agroecosystems). This therefore nullifies the possibility of controlled production of Solanaceous crops as a management measure against *T. absoluta*, as has been the case, e.g. *Pectinophora gossypiella* (Saunders) in cotton [63].

Native environmental heterogeneity may also con-tribute to invasion success [3]. Propagules from a more heterogeneously stressful environment are more adapt-able to multiple stressful conditions [64] and, together with other factors, may work synergistically towards the succession of ecological filters [reviewed in 3]. Climate matching between the native and novel environment is known to aid invasion success of invasive alien species [30, 65, 66]. Interestingly, African biotic and climatic con-ditions are closely related to *T. absoluta's* native region [1]. Insect species have specific optimum temperatures at which they optimally perform and develop [31, 37]. In addition, lower and upper developmental thresholds mark the temperatures beyond which they cannot per-form and develop [31, 37, 67]. As such, basal environ-mental stress tolerance, phenotypic plasticity and rapid genetic adaptive shifts are key to invasive species estab-lishment [30]. *Tuta absoluta* tolerance to temperature and relative humidity versus typical Botswana climate [68] may form the primary characteristics defining its range expansion [1, 37, 68]. Prior predictions using cli-matic suitability indices defined the eco-climatic index (EI) of Botswana to fall within 20–50, classified as high risk of establishment for *T. absoluta* [1]. Our results are thus in agreement with this prediction. Climatic condi-tions (chiefly temperature) are known to significantly influence generation numbers of multivoltine insects, with higher temperatures facilitating faster degree day accumulation and shorter generation times [69]. At an optimum temperature of 30 °C, the life cycle of *T. abso-luta* ranges approximately 26 days [53] accounting to ~ 12 generations per year [2]. Global warming comes with increased mean temperatures and variability thereof and is reported to increase insect metabolism [70]. With Afri-can temperatures projected to increase, future popula-tions of this pest may likely increase in tropical relative to temperate regions [1]. Botswana is arid to semi-arid with mean monthly maximum temperatures recorded in 2015/2016 season ranging 23.6 to 35.1 °C (Fig. 4a). Given that the optimum temperature for *T. absoluta* is 30 °C [54], a TSM of 7.29 °C was relatively high [37, 53]. This signifies that *T. absoluta* can tolerate an increase in atmospheric temperature of 7.29 °C from current Bot-swana ambient environmental temperatures of 22.71 °C

(T_{hab}) before its population growth and general perfor-mance can drop to critical levels. This is a considerably high TSM compared to most tropical species whose TSM is ~ 0 °C [53]. This, coupled with a wider WT (22.39 °C), further supports that Botswana environmental tempera-tures were conducive for the performance, rapid spread and establishment of *T. absoluta*. The lower and upper developmental threshold for *T. absoluta* is ~ 14 and 34.6 °C, respectively [54], translating to a wide thermal window (~ 20.6 °C) which is known to optimize key insect activity and life-sustaining behaviours such as develop-ment, mating and dispersal [3, 55], and may potentially facilitate the invasion pathway of *T. absoluta*. Thus, con-ducive climatic conditions might have chiefly facilitated the rapid accumulation of degree days hence culminat-ing into shorter generation time. This high reproductive capacity may also have contributed to its increased inva-sion success in novel environments in the country [as in 1].

Improved environmental tolerance and thermal plas-ticity are the key contributing factors towards invasion success of invasive alien species into a novel environ-ment [30, 39, 71]. Lower and upper lethal temperatures (LLT and ULTs) for *T. absoluta* adults ranged from − 1 to − 12 °C and 37 to 43 °C respectively for 2 h treat-ments. In addition, the CT_{max} for larvae and adults were 47.9 ± 1.25 and 44.1 ± 0.43 and CT_{min} were 3.5 ± 0.07 and $− 5.2 \pm 0.23$ respectively. Field temperature recorded dur-ing 2015/2016 season show that highest maximum tem-peratures were below both ULT and CT_{max} for both *T. absoluta* larvae and adults. In addition, LLTs and CT_{min} for adults were relatively lower than the lowest minimum field temperatures (Fig. 4b). This, added to the high TSM and WT, indicates that *T. absoluta* may not be at risk of cold and heat stress both of which has an implication on the invasion succession pathway. These results supports that *T. absoluta* is highly temperature tolerant at both extremes and may survive in arid/semi-arid sub-Saharan Africa whenever hosts plants are available. Its high basal thermal tolerance, coupled by favourable climates in Bot-swana (Fig. 4a and b), may mean that *T. absoluta* survives all-year-round temperature conditions, in the absence of diapause, a characteristic likely aiding successful estab-lishment. Furthermore [38], showed rapid cold harden-ing may also aid invasion success in insects, and indeed *T. absoluta* has been shown to rapidly cold-harden [see 2], a phenomenon likely aiding the invasion pathway. The absence of native coevolved natural enemies has also been reported to promote invasion success in novel environments [5, 64]. It is highly likely that the rapid spread and establishment of *T. absoluta* in Botswana may have been facilitated by the absence of biological con-trol agents. We thus recommend that native fortuitous

Evidence of rapid spread and establishment of Tuta absoluta (Meyrick) (Lepidoptera: Gelechiidae)...

107

natural enemies need be identified and promoted coupled with a campaign against the instinctive overuse of pesticides by small scale farmers [19] to preserve potential native natural enemies, reduce cost of production and protect public health. Further work needs to determine *T. absoluta* insecticide resistance to establish a controlled effective spraying program, coupled with the identified effective natural enemies to establish an efficacious tailor-made integrated pest management (IPM) program. Overall, an area-wide approach to *T. absoluta* management is recommended, and one that involves a coordinated Southern African region, to prevent further spread and establishment of the species [2].

Conclusion

Current results support the rapid spread and establishment of *T. absoluta* in Botswana following its first detection. This continued invasion by *T. absoluta* in tropical climates is a real concern for the horticultural industry, as well as African food and nutrition security. Host plant availability, climate suitability and high thermal tolerance may to a larger extent have contributed to the successful invasion, rapid spread and establishment of *T. absoluta* in the semi-arid tropical Botswana. In addition, intensive monocultures, continuous irrigation and unrestricted trade of Solanaceous crops coupled with strong winds and a lack of natural enemies may also be contributory factors. Furthermore, absence of efficient and coordinated area-wide management practices may have exacerbated the successful rapid invasion. A significant long-term management strategy would be necessary to optimize surveillance and monitoring of *T. absoluta* in the region for developing sustainable management options. Similarly, introduction of egg-targeting parasitoids (*Trichogramma spp.*) and predators as well as larval parasitoids (mostly belonging to Braconidae families) and predators (Miridae) [2, 14] could improve management of African suppression programmes, more especially in non-agroecosystem and natural environments.

Authors' contributions
HM and CN contributed to conceptualization and methodology; CN contributed to funding acquisition, project administration, resources and supervision; HM and RM contributed to investigation and writing of the original draft; and HM, RM and CN contributed to data curation, validation, formal analysis, writing, review and editing. All authors read and approved the final manuscript.

Acknowledgements
We acknowledge Botswana International University of Science and Technology for funding and Russell IPM for *T. absoluta* pheromone lures. We acknowledge assistance from the Department of Crop Protection (Ministry of Agriculture), on site selection, the Botswana National Botanical Gardens for data on distribution of wild Solanaceous plants and Department of Meteorological Services for climate data. We are also grateful to Dr. Tharina L. Bird for map drawing and assistance with trapping in the Okavango Delta and Mmabaledi Buxton and Mphoeng Ofitlhile for assistance with trap monitoring in some districts.

Competing interests
The authors declare that they have no competing interests.

Funding
The project was funded through Botswana International University of Science and Technology (BIUST) Research Office grant.

References
1. Tonnang HEZ, Mohamed SA, Khamis F, Ekesi S. Correction: identification and risk assessment for worldwide invasion and spread of *Tuta absoluta* with a focus on Sub-Saharan Africa: implications for phytosanitary measures and management. PLoS ONE. 2015;10:e0138319. https://doi.org/10.1371/journal.pone.0138319.
2. Biondi A, Guedes RNC, Wan FH, Desneux N. Ecology, worldwide spread, and management of the invasive south American tomato pinworm, *Tuta absoluta*: past, present and future. Ann Review Entomol. 2018;63:239–58.
3. Renault D, Laparie M, McCauley SJ, Bonte D. Environmental adaptations, ecological filtering, and dispersal central to insect invasions. Ann Rev Entomol. 2018;63:345–68.
4. Germain JF, Lacordaire AI, Cocquempot C, Ramel JM, Oudard E. Un nouveau ravageur de la tomate en France: *Tuta absoluta*. PHM-Revue Horticole. 2009;512:37–41.
5. Desneux N, Wajnberg E, Wyckhuys KAG, Burgio G, Arpaia S, Narváez-Vasquez CA, González-Cabrera J, Catalán Ruescas D, Tabone E, Frandon J, Pizzol J, Poncet C, Cabello T, Urbaneja A. Biological invasion of European tomato crops by *Tuta absoluta*: ecology, geographic expansion and prospects for biological control. J Pest Sci. 2010;83:197–215.
6. Brévault T, Sylla S, Diatte M, Bernadas G, Diarra K. *Tuta absoluta* Meyrick (Lepidoptera: Gelechiidae): A new threat to tomato production in sub-Saharan Africa. African Entomol. 2014;22:441–4.
7. Pfeiffer D, Muniappan R, Sall D, Diatta P, Diongue A, Dieng EO. First record of *Tuta absoluta* (Lepidoptera Gelechiidae) in Senegal. Fla Entomol. 2013;96:661–2.
8. Chidege M, Al-zaidi S, Hassan N, Julie A, Kaaya E, Mrogoro S. First record of tomato leaf miner *Tuta absoluta* (Meyrick) (Lepidoptera: Gelechiidae) in Tanzania. Agric. Food Sec. 2016;5:17. https://doi.org/10.1186/s4006 6-016-0066-4.
9. Tumuhaise V, Khamis FM, Agona A, Sseruwu G, Mohamed SA. First record of *Tuta absoluta* (Lepidoptera: Gelechiidae) in Uganda. Int J Trop Insect Sci. 2016;36:135–9. https://doi.org/10.1017/S1742758416000035.
10. Visser D, Uys VM, Nieuwenhuis RJ, Pieterse W. First records of the tomato leaf miner Tuta absoluta (Meyrick. (Lepidoptera: Gelechiidae) in South Africa. Biol Invas ons. 1917;2017:6.
11. Chidege M, Abel J, Afonso Z, Tonini M, Fernandez B. Tomato Leaf Miner, *Tuta absoluta* (Meyrick) (Lepidoptera: Gelechiidae) Detected in Namibe Province Angola. J. Appl. Life Sci. Int. 2017;12:1–5.
12. Mutamiswa R, Machekano H, Nyamukondiwa C. First Report of Tomato Leaf miner, *Tuta absoluta* (Meyrick) (Lepidoptera: Gelechiidae) in Botswana. Agric. Food Sec. 2017;6:49. https://doi.org/10.1186/s4006 6-017-0128-2.
13. Desneux N, Luna MG, Guillemaud T, Urbaneja A. The invasive South American tomato pinworm. *Tuta absoluta* continues to spread in Afro-Eurasia and beyond: the new threat to tomato world production. J Pest Sci. 2011;84:403–8.
14. Urbaneja A, Gonzalez-Cabrera J, Arno J, Gabarra R. Prospects for the biological control of *Tuta absoluta* in tomatoes of the Mediterranean Basin. Pest Manag Sci. 2012;68:1215–22.
15. Bawin T, Dujeu D, De Backer L, Francis F, Verheggen FJ. Ability of *Tuta absoluta* (Lepidoptera: Gelechiidae) to develop on alternative host plant species. Can Entomol. 2016;148:434–42.
16. Toševski I, Jović J, Mitrović M, Cvrković T, Krstić O, Krnjajić S. *Tuta absoluta* (Meyrick, 1917) (Lepidoptera, Gelechiidae): a new pest of tomato in Serbia. Pestic. Phytomed. 2011;26(3):197–204.
17. Abbes K, Harbi A, Elimem M, Hafsi A, Chermiti B. Bioassay of three solanaceous weeds as alternative hosts for the invasive tomato leafminer *Tuta absoluta* (Lepidoptera: Gelechiidae) and insights on their carryover potential. African Entomol. 2016;24(2):334–42.

18. FAO. The State of Food Insecurity in the World: Economic growth is necessary but not sufficient to accelerate reduction of hunger and malnutrition. Rome, FAO. 2012.

19. Machekano H, Mvumi BM, Nyamukondiwa C. Diamondback Moth, *Plutella xylostella* (L.) in Southern Africa: research trends, challenges and insights on sustainable management options. Sustainability. 2017;9:91. https://doi.org/10.3390/su9020091.

20. IPCC. Climate change 2014: synthesis report. In: Core Writing Team, Pachauri RK, Meyer LA, editors. Contribution of working groups I, II and III to the fifth assessment report of the intergovernmental panel on climate change. Geneva: IPCC; 2014. p. 151.

21. Richardson DM, Pysek P. Plant invasions-merging the concepts of species invasiveness and community invisibility. Prog Phys Geogr. 2006;30:409–31.

22. Blackburn TM, Pysek P, Bacher S, Carlton JT, Duncan RP, Jarosic V, Wilson JR, Richardson DM. A proposed unified framework for biological invasions. Trends Ecol Evolut. 2014;26:333–9.

23. Duncan RP, Blackburn TM, Rossinelli S, Bacher S. Quantifying invasion risk: the relationship between establishment probability and founding population size. Methods Ecol Evol. 2014;5:1255–63.

24. Szucs M, Melbourne BA, Tuff T, Hufbauer RA. The roles of demography and genetics in the early stages of colonization. Proc R Soc B. 2014;2014(28):20141073.

25. Dixon AF, Honek A, Kell P, Kotela MAA, Sizzling AL, Jarosik V. Relationship between the minimum and maximum temperature thresholds for development of insects. Funct Ecol. 2009;23:257–64.

26. Kelley AL. The role thermal physiology plays in species invasion. Conserv Physiol. 2014. https://doi.org/10.1093/conphys/cou04527.

27. Kumschick S, Gaertner M, Vila M, Essl F, Jeschke JM, Pysek P, Ricciardi A, Bacher S, Blackburn TM, Dick JT, Evans T. Ecological impacts of alien species: quantification, scope, caveats, and recommendations. Bioscience. 2015;65:55–63.

28. Nentwig W, Bacher S, Pysek P, Vila M, Kumschick S. The generic impact scoring system (GISS): a standardised tool to quantify the impact of alien species. Environ Monit Assess. 2016;188:1–13.

29. Biber-Freudenberger L, Ziemacki J, Tonnang HEZ, Borgemeister C. Future risks of pest species under changing climatic conditions. PLoS ONE. 2016;11(4):e0153237. https://doi.org/10.1371/journal.

30. Perkins LB, Leger EA, Nowak RS. Invasion triangle: an organizational framework for species invasion. Ecol. Evol. 2011;1:610–25.

31. Hoffmann AA. Physiological climatic limits in Drosophila: patterns and implications. J Exp Biol. 2009;213:870–80.

32. Karadjova O, Ilieva Z, Krumov V, Petrova E, Ventsislavov V. *Tuta absoluta* (Meyrick) (Lepidoptera: Gelechiidae): Potential for entry, establishment and spread in Bulgaria. Bulg J Agric Sci. 2013;19:563–71.

33. Crowl TA, Crist TO, Parmenter RR, Belovsky G, Lugo AE. The spread of invasive species and infectious disease as drivers of ecosystem change. Front Ecol Environ. 2008;6:238–46.

34. Olyarnik SV, Bracken ME, Byrnes JE, Hughes AR, Hultgren KM, Stachowicz JJ. Ecological factors affecting community invasibility. In: Rilov G, Crooks J, editors. Biological invasions in marine ecosystems. Berlin: Springer; 2009. p. 215–38.

35. Chown SL, Terblanche JS. Physiological Diversity in Insects: ecological and Evolutionary Contexts. Adv. Insect Physiol. 2007;33:50–152.

36. Gerhardt F, Collinge SK. Abiotic constraints eclipse biotic resistance in determining invasibility along experimental vernal pool gradients. Ecol Appl. 2007;17:922–33.

37. Andrew NR, Hill SJ. Effect of climate change on insect pest management. In: Coll M, Wajnberg E, editors. Environmental pest management: challenges for agronomists, ecologists, economists and policymakers. Oxford: Wiley; 2017.

38. Nyamukondiwa C, Kleynhans E, Terblanche JS. Phenotypic plasticity of thermal tolerance contributes to the invasion potential of Mediterranean fruit flies (*Ceratitis capitata*). Ecol Entomol. 2010;35:565–75.

39. Gotcha N, Terblanche JS, Nyamukondiwa C. Plasticity and cross-tolerance to heterogeneous environments: divergent stress responses co-evolved in an African fruit fly. J Evol Biol. 2017;31:98–110.

40. Kelley AL. The role thermal physiology plays in species invasion. Conserv Physiol. 2014. https://doi.org/10.1093/conphys/cou045.

41. IPPC. Reporting pest presence: Preliminary surveillance reports on *Tuta absoluta* in Zambia. International Plant Protection Convention (IPPC). 2016a https://www.ippc.int/en/countries/zambia/pestreports/2016/09/reporting-pest-presence-preliminary-surveillance-reports-on-tuta-absoluta-in-zambia/. Accessed 14 Sep 2016.

42. IPPC. First detection of *Tuta absoluta* in South Africa. International Plant Protection Convention (IPPC). 2016b. https://www.ippc.int/en/countries/south-africa/pestreports/2016/09/first-detection-of-tuta-absoluta-in-south-africa. Accessed 1 Sept 2016.

43. Hayden JE, Lee S, Passoa SC, Young J, Landry JF, Nazari V, Mally R, Somma LA, Ahlmark KM. Digital Identification of Microlepidoptera on Solanaceae. USDA-APHIS-PPQ.2013.

44. Machekano H, Mvumi BM, Nyamukondiwa C. Loss of coevolved basal and plastic responses to temperature may underlie trophic level host-parasitoid interactions under global change. Biol Control. 2017. https://doi.org/10.1016/j.biocontrol.2017.12.005.

45. Nyamukondiwa C, Terblanche JS. Thermal tolerance in adult Mediterranean and Natal fruit flies (*Ceratitis capitata* and *Ceratitis rosa*): effects of age, gender and feeding status. J Therm Biol. 2009;34:406–14.

46. Bertoli CI, Scannapieco AC, Sambucetti P, Norry FM. Direct and correlated responses to chill-comma recovery selection in *Drosophila buzzatii*. Entomol Exp Appl. 2010;134:154–9.

47. Chang XQ, Ma CS, Zhang S, Lu L. Thermal tolerance of Diamondback moth, *Plutella xylostella*. J Appl Ecol. 2012;23:772–8.

48. Chidawanyika F, Terblanche JS. Rapid thermal responses and thermal tolerance in adult codling moth *Cydia pomonella* (Lepidoptera:Totricidae). J Insect Physiol. 2011;57:108–17.

49. Nyamukondiwa C, Weldon CW, Chown SL, le Roux PC, Terblanche JS. Thermal biology, population fluctuations and implications of temperature extremes for the management of two globally significant insect pests. J Insect Physiol. 2013;59:1199–211.

50. Chown SL, Nicolson SW. Insect physiological ecology: mechanisms and patterns. Oxford: Oxford University Press; 2004.

51. Terblanche JS, Deere JA, Clussella-Trullas S, Janion C, Chown SL. Critical thermal limits depend on methodological context. Proc R Soc B. 2007;274:2935–42.

52. R Development Core Team. R: A language and environment for statistical computing. Vienna. Austria R Foundation for Statistical Computing. 2016.

53. Deutsch CA, Tewksbury JJ, Huey RB, Sheldon KS, Ghalambor CK, Haak DC, Martin PR. Impacts of climate warming on terrestrial ectotherms across latitude. Proc Natl Acad Sci USA. 2008;108:6668–72.

54. Martins JC, Picanço MC, Bacci L, Guedes RNC, Santana PA Jr, Ferreira DO, Chediak M. Life table determination of thermal requirements of the tomato borer *Tuta absoluta*. J Pest Sci. 2016;89:897–908.

55. Mutamiswa R, Chidawanyika F, Nyamukondiwa C. Dominance of spotted stemborer *Chilo partellus* Swinhoe (Lepidoptera: Crambidae) over indigenous stemborer species in Africa's changing climates: ecological and thermal biology perspectives. Agric For Entomol. 2017. https://doi.org/10.1111/afe.12217.

56. Lux SA, Copeland RS, White IM, Manrakhan A, Billah MK. A new invasive fruit fly species from the *Bactrocera dorsalis* (Hendel) group detected in East Africa. Insect Sci Appl. 2003;23:355–61.

57. Carey JR. Biodemography of the mediterranean fruit fly: aging, longevity and adaptation in the wild. Exp Gerontol. 2011;46:404–11.

58. dos Santos LA, Mendes MF, Krüger AP, Blauth ML, Gottschalk MS, Garcia FRM. Global potential distribution of *Drosophila suzukii* (Diptera, Drosophilidae). PLoS ONE. 2017. https://doi.org/10.1371/journal.pone.0174318.

59. Ferracini C, Ingegno BL, Navone P, Ferrari E, Mosti M, Tavella L, Alma A. Adaptation of indigenous larval parasitoids to *Tuta absoluta* (Lepidoptera: Gelechiidae) in Italy. J Econ Entomol. 2012;105:1311–9.

60. Cocco A, Deliperi S, Lentini A, Mannu R, Delrio G. Seasonal phenology of *Tuta absoluta* (Lepidoptera: Gelechiidae) in protected and open-field crops under Mediterranean climatic conditions. Phytoparasitica. 2015;43:713–24.

61. CFIA. *Tuta absoluta* (Tomato Leafminer)—Fact Sheet. Ontario. Government of Canada Publications. 2016.

62. Kruger AC, Goliger AM, Retief JV. Strong wind climatic zones in South Africa. Wind Struct. 2010;13:1.

63. Henneberry TJ. Integrated systems for control of the pink bollworm *Pectinophora gossypiella* in cotton 567–579. In: Vreysen MJB, Robinson AS, Hendrichs J, editors. Area-wide control of insect pests. Arizona: Western Cotton Research Laboratory USDA/ARS; 2007.

64. Manenti T, Sørensen JG, Loeschcke V. Environmental heterogeneity does not affect levels of phenotypic plasticity in natural populations of three Drosophila species. Ecol Evol. 2017;7:2716–24.

65. Farji-Brener AG, Corley JC. Successful invasions of hymenopteran insects into NW Patagonia. Ecol Austral. 1998;8:237–49.

66. Jarošík V, Kenis M, Honěk A, Skuhrovec J, Pyšek P. Invasive insects differ from non-invasive in their thermal requirements. PLoS ONE. 2015;10:e0131072. https://doi.org/10.1371/journal.pone.0131072.

67. Khadioli N, Tonnang ZEH, Ongamo G, Achia T, Kipchirchir I, Kroschel J, Le Ru B. Effect of temperature on the life history parameters of noctuid lepidopteran stemborers, *Busseola fusca* and *Sesamia calamistis*. Ann Appl Biol. 2014;165:373–86. https://doi.org/10.1111/aab.12157.

68. Son D, Bonzi S, Somda I, Bawin T, Boukraa S, Verheggen F, Francis F, Legreve A, Schiffers B. First record of Tuta absoluta (Meyrick, 1917) (Lepidoptera: Gelechiidae) in Burkina Faso. African Entomol. 2017;25:259–63.

69. Bale JS. Implications of cold-tolerance for pest management. In: Denlinger DL, Lee RE, editors. Low temperature biology of insects. Cambridge: Cambridge University Press; 2010. p. 342–72.

70. Dillon ME, Wang G, Huey RB. Global metabolic impacts of recent climate warming. Nature. 2010;467:704–6.

71. Chown SL, Slabber S, McGeoch MA, Janion C, Leinaas HP. Phenotypic plasticity mediates climate change responses among invasive and indigenous arthropods. Proc R Soc Lond B Biol Sci. 2007;274:2661–7.

## 9

Agricultural history nexus food security and policy framework in Tanzania

Msafiri Yusuph Mkonda[1]* and Xinhua He[2,3]

Abstract

Background: Understanding the production trend of the major food crops is an important step for any nation that evaluates her agricultural progress. This evaluation should mostly focus on the yields per unit area. So far, it can also earmark the expansion of farms to determine the general yields trend. The main objective of this paper is to assess the production trend of the major food crops and their efficacy to food security in Tanzania. This is particular important because for the past three decades, the country has failed to control food security (especially food availability and accessibility).

Results: Here, crop data from 1980 to 2015 were gathered from the Ministry of Agriculture, Livestock and Fishery (MALF), and in the respective regions. In some incidences, the regional data were averaged to elicit their preciseness. To determine the objectivity of this study, agricultural policy, programs, and plans from MALF were reviewed for similar purpose. Mostly, the Mann-Kendal Test and Microsoft Excel were used for data analyses. The results show that the production of the total yields had a positive trend (*i.e., growing at $R^2 = 0.4$ and 0.8*), while that of the yields (ton/ha) had a negative trend (*i.e., declining at $R^2 = 0.02$ and 0.3*). It was further realized that the total yields mostly boomed due to farm expansion.

Conclusions: Despite the efforts from various agricultural stakeholders, the country has not yet achieved a sustainable crop yield and food security. Explicitly, this situation has been affecting peoples' livelihoods, and other sectors either directly or indirectly. Therefore, there is a need to improve the production strategies and approaches (i.e., more especially technology and marketing) to limit this problem.

Keywords: Agricultural production trend, Crop yields, Food security, Policy, Programs, Tanzania

Background

Since the inception of Tanzania in 1964, the country's population has increased to about five times, from approximately 11.7 million in 1965 to more than 53.4 million in 2015 [1–3]. The demand placed on national agricultural production arising out of population and economic growth has enormously increased [4–7]. Despite the increase in both land expansion and agricultural development, the problem of food security has remained unresolved [8, 9]. Reports from both domestic and international agencies such as FAO have reinstated

the same. In Tanzania, the agricultural sector has a substantially outstanding contribution to socioeconomic development as it supports the GDP (Gross Domestic Product) with 28%, provides 95% of the food and employs over 75% of the national labor forces [3].

In that respect, agricultural growth has a direct contribution to food security and economic growth of the country and livelihoods of the people. However, some regions have favorable biophysical characteristics (i.e., soils, water sources and climate just to mention a few) than others. Potentially, this differential biophysical endowment has significant implications to crop yields. Ultimately, this situation leads to differential agricultural outputs among the regions within the country [9, 10].

Generally, the country has a wide range of agricultural potentials such as 44 million hectares of arable land, numerous rivers for irrigation, labor force, policies and

*Correspondence: msamkonda81@yahoo.co.uk
[1] Department of Geography and Environmental Studies, Solomon Mahlangu College of Sciences and Education, Sokoine University of Agriculture, Morogoro 3038, Tanzania
Full list of author information is available at the end of the article

programs just to mention a few. Nevertheless, less than 24% and 4% of the arable land and irrigation potentials has been harnessed, respectively. This inefficient utilization of resources has been attributed by low investment in the sector in terms of finance and technology and ultimately resulted to low productivity. Preferably, maize is a staple food to about 70% of the Tanzanians, while sorghum, millet, paddy, banana, cassava, wheat and diverse varieties of potato just to mention a few are dominant food crops depending on the regional preferences.

Although the assessment of crop production and associated challenges is progressing rapidly, a variety of knowledge gaps still exist. Despite the implementation of programs and initiatives, yet there is high inconsistence of crop yields that always lead to enormous food shortage and insecurity in Tanzania [2, 3, 10, 11]. This study aims to assess the efficacy of the major food crop production and its implications to food security, economic development and policy framework. It intends to depict the balance between the efforts placed in the production process and its return. By determining such a balance, this study will be capable to propose strategies at local and policy level that would optimize yields to sustain the livelihoods of over 80% smallholder farmers who entirely depend on rain-fed agriculture.

To meet the study objective, this paper attempts to answer the following questions: (1) How did the production trends of the major food crops behave in the past three decades? (2) Were the obtained yields enough for food requirement? (3) What were the main factors that influenced the production trend? (4) How policy framework and other associated programs influenced the efforts for yield optimizations? (5) What should be done to intensify crop production with higher yields that can ensure food security and economic development?

Methods
Profile of the study site
Tanzania is located on the eastern coast of Africa, south of the Equator between latitudes 1°00′S and 11°48′S and longitudes 29°30′E and 39°45′. Eight countries: Kenya and Uganda in the north, Rwanda, Burundi, Democratic Republic of Congo and Zambia in the west, Malawi and the Republic of Mozambique to the south share boundaries with Tanzania. The eastern side of Tanzania is a coastline of about 800 km long marking the western side of the Indian Ocean. Tanzania has a total of 945 087 km², and out of this area, water bodies cover 61, 495 km² which is equivalent to 6.52% of the total area.

The mean annual rainfall varies considerably from place to place ranging from less than 400 mm to over 2500 mm per annum. Rainfall in about 75% of the country is erratic, and only 21% of the country can expect an annual rainfall of more than 750 mm with a 90% probability. Intergovernmental Panel on Climate Change (IPCC) [12] informed that Tanzania and other sub-Saharan African countries will continue to be vulnerable to the impacts of climate change (i.e., excessive droughts). This is caused by their weak adaptive capacities. The current report by IPCC [13] slotted Tanzania as among the thirteen countries that area worst affected by the impacts of climate change and vulnerability and has weak adaptive capacities to cope or recover from the stress. The report further described that if proper adaptation measures are not virtually taken, more significant impacts will stress the country.

Since soil is the major determinant of agricultural production as it acts as a mother factor for the whole process, it was liable to understand the dominant types of the soil in the country. According to World Reference Base of Soil Resource (WRB), Tanzania has 19 dominant soil types and they are grouped into two groups, namely organic soil and mineral soils. The structure, concepts and definitions of the WRB are strongly influenced by (the philosophy behind and experience gained with) the FAO-UNESCO Soil Classification System [14, 15]. Literally, Tanzania has different types of soils such as clay, loam and sand as identified by the normal farmers. These soil types have different potentials in terms of fertility and moisture conservation for crop production.

Ecologically, the *Rufiji, Ruvu, Wami, Ruaha, Kilombero, Malagarasi and Pangani* basins form numerous hydro-ecological zones in Tanzania that provide fruitful potentials for crop production. Similarly, the existence of diverse agro-ecological zones in the country has significant potentials to sustainable crop production. These biophysical potentials including the 44 million hectares of arable land are convenient for the progression of agricultural industry in the country [3].

Data collection and analyses
This study was designed to typically entail the long-term (1980–2015) crop data from authentic sources. The yield data for major food crops were collected from March to December 2016 at the Ministry of Agriculture, Livestock and Fishery (MALF), and at regional and district levels. The data were particularly gathered from documentaries (statistic unit) of the ministry. Interviews with some agriculture officials at the ministry were employed to collect qualitative data and cross-check some quantitative data.

In some instances, it happened that some years had no data. Other data were collected from ten (10) regions earmarking the regions which are best and least producers of food crops. Among these regions are *Iringa, Mbeya, Kigoma, Tabora, Kilimanjaro, Kagera, Lindi, Ruvuma, Singida* and *Dodoma*. Then, the data from these areas were cross-checked with those documented at MALF. Then, the mean was calculated to harmonize some raised discrepancies and missing data.

To correct this, we searched the missing data at regional and district level to obtain the same. Where necessary, we estimated the data below 5% to avoid erroneous. National Agricultural Policy of 2013 and its allied programs from the Ministry were equally consulted during data collection. Since most of data were quantitative in nature, we employed the Mann–Kendall Test and Microsoft Excel to plot the yield trends. Qualitative data especially those from interviews were analyzed through content method, and the results were inserted in the main text during discussion.

Results

Trends of crop production

This empirical study has succeeded to perform the temporal analyses for yields of the major crops. The results show that the yields have had high inconsistence throughout the time frame (1980–2015). This is evident in Figs. 1a, b, 2a–c, 3a–c and 4a–c where the yields of maize, paddy, millet, sorghum, cassava, banana, potatoes and wheat have been plotted against time and thus far, have shown unreliability. This plotting entailed two major dimensions that would determine the quantity of yields. Firstly, it involved the total yields in tons and secondly was about the yields per hectare in tons. On the basis of total yields, the production trend for maize, banana, beans and sorghum exhibited a positive slope (significant growth at $R^2 = 0.8$), as well, that of paddy, cassava, millet, wheat and potatoes showed a positive slope (between $R^2 = 0.4$ and 0.7), as seen in Fig. 1a, b. Besides, the yields per hectare in tons exhibited a negative slope (between $R^2 = 0.02$ and 0.3) as seen in Figs. 2a–c, 3a–c and 4a–c. These results depict the widest context of crop yields at national level.

Further, the present study has revealed that there has been a temporal expansion of crop land as presented in Table 1. Explicitly, this has been a major cause for the total yield to boom. For example, within a period of two decades the crop land under maize increased for about three times, while that of sorghum, millet and cassava have more than doubled. While the land under crop production has experienced significant increase from 1996 to 2015, the population has almost doubled within such a time frame and thus, increasing more food demand.

Implications to food security

The results from analyses have a message to convey to agricultural and social livelihood analysts. This is about the implications of what has been gotten from farms (yields) and the particular food requirements at household and national level. According to FAO, food security encompasses three major aspects, namely availability, accessibility and use. In the Tanzanian context, all the three aspects are dominant. Along these aspects, poverty (entitlement failure) has also exacerbated the magnitude of food insecurity. Since more than 90% of the food requirement is produced in the country; thus, whatsoever come from the farm determines the food security status. Overall, the production turbulence as seen in Figs. 1, 2, 3 and 4 indicates the unsustainability of food security in the country.

There has been a correlation between the amount of obtained yields and the level of food security; as regions with low self-sufficient ratio have been experiencing intensive food shortage. For example, most regions found in arid and semiarid zones have been suffering from all three aspects of food security. The food produced has always been outsmarted by the requirements; thus, food availability is not ensured. On the other hand, people (i.e., especially the destitute) have weak purchasing capacity to access food in market and ultimately, bringing about food shortage and insecurity. These results are in agreement with the government report as recently in 2013/2014, there was a food deficit of 43,452 tons [11]. That deficit happened because the total food requirement was 7,656,673 tons, but the production was placed at 7,613,221 tons.

Hindrances for crop production

There have been numerous challenges contributing to agricultural unsustainability in the country. These challenges range from natural to human-induced problems. Reports from IPCC, FAO and other international organizations have revealed that climate change has posed severe impacts to agricultural sector in most developing countries [6, 12, 13]. In this era of global climate change, the poor nations like Tanzania are most vulnerable and are weak to cope or recover from such dreadful condition. Besides, the study by URT (United Republic of Tanzania) [2, 3, 10], Kangalawe [16], Rowhani et al. [17], Neufeldt et al. [18] and Paavola [19] further asserted that climate change has already affected and will continue affecting poor households in most developing countries and thus, increasing the vulnerability of the dependable livelihoods.

In some ways, the government has contributed to such unsustainability due to low investment in the sector. It is understandable that for the sector to progress, at least 10% of the developmental budget should be placed in it.

However, good step has left unimplemented despite establishing some initiatives to overcome the crisis. At farmer's level, Tanzanian agriculture is predominated by small-scale holders for over 70%. This dominance is

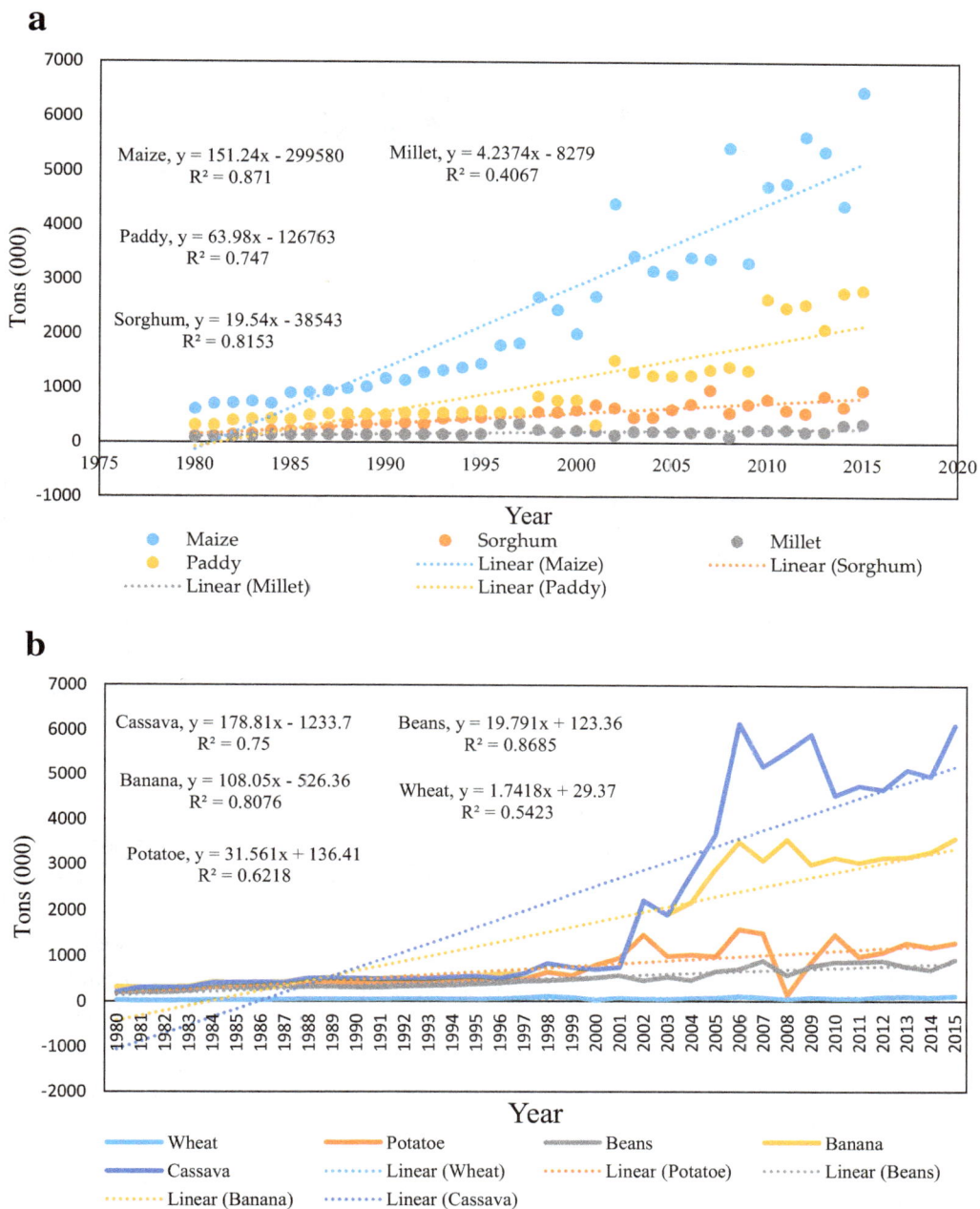

Fig. 1 Yields trend of the major food crops in Tanzania from 1980 to 2015. **a** Yields trend for maize, sorghum, millet and paddy, **b** yields trend for wheat, potatoes, beans, banana and cassava *Source*: Analyses from the data obtained from MALF

even higher in rural areas where it exceeds 80%. Unfortunately, this farming scale is meant for subsistence and not for commercial purpose. However, despite targeting for subsistence, these farmers have in most cases failed to meet their minimum requirement due to meager yields obtained from their farms. Correspondingly, it is this vulnerability that has contributed to decline in

yields per hectare in tons as seen in Figs. 2a–c, 3a–c and 4a–c). Besides, the market constraints and increased poverty among the rural households are some of the salient factors for agricultural dwindling [3]. Overall, the human-induced factors have been increasing over time because the rate of solving them is surpassed by that of its creation.

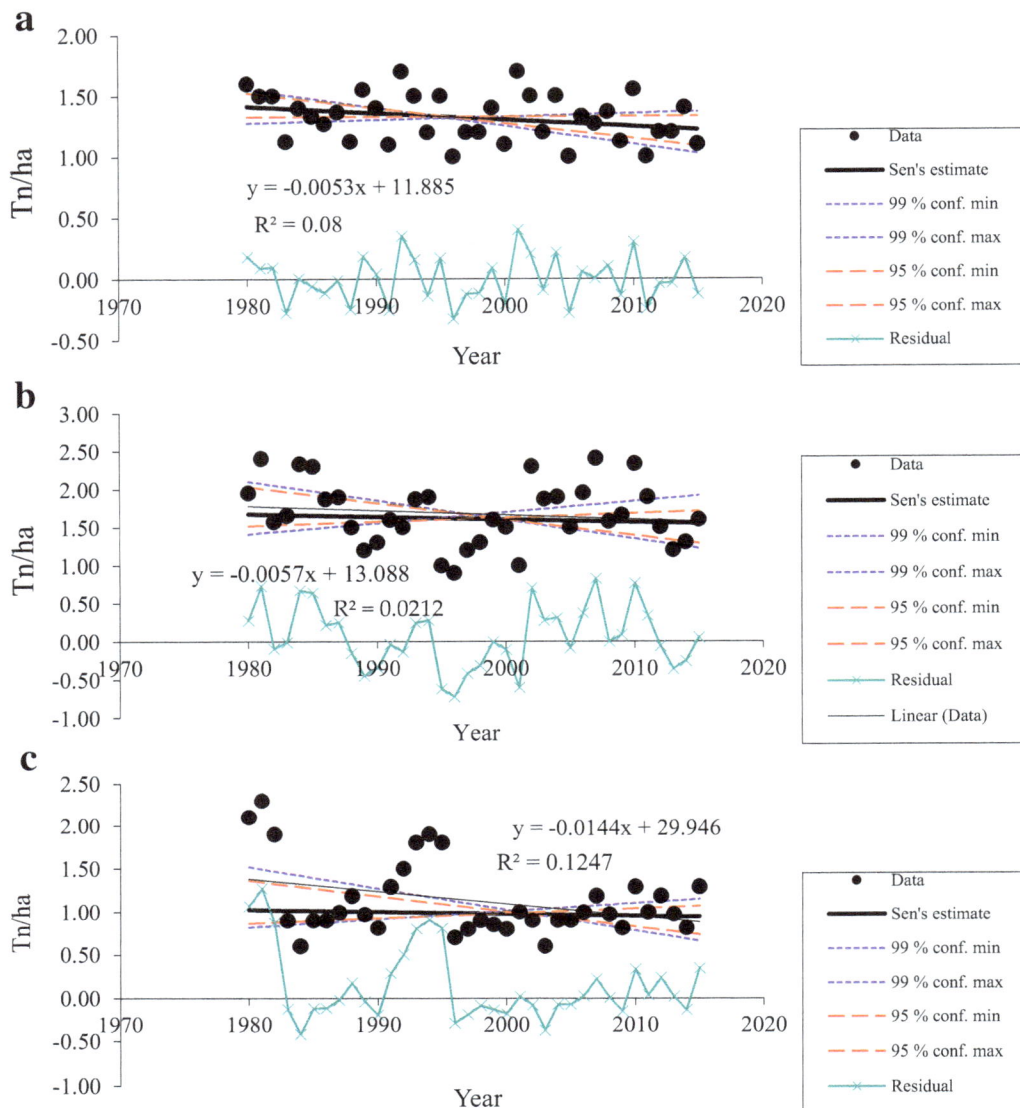

Fig. 2 Yields of the major food crops per hectare in tons in Tanzania from 1980 to 2015. **a** Maize yields per hectare in tons, **b** paddy yields per hectare in tons, **c** sorghum yields per hectare in tons *Source*: Analyses from the data obtained from MALF

Policy framework entailed in crop optimizations

National Agricultural Policy (NAP) [3] is the main document responsible for agricultural sector in Tanzania. The main aim of this policy is to optimize crop production for food security and economic development. Its implementation is done in a series of programs, initiatives, strategies, plans and projects. Among these, the Agricultural Sector Development Program (ASDP) was launched in 2005 to implement the Agricultural Sector Development Strategy (ASDS) of 20013.

In addition, PADEP—Participatory Agricultural Development and Empowerment Project (2003),

TAFSIP—Tanzania Agriculture and Food Security Investment Plan (2011-2021) and Agricultural First (locally, *Kilimo Kwanza*) Initiative of 2009 were equally introduced to spearhead agricultural production. Externally, the Comprehensive Africa Agricultural Development Program (CAADP) is an initiative to improve food security in most African countries in which Tanzania is inclusive. Altogether, these programs and initiatives meant to raise agricultural produce, especially for rural households in order to optimize crop production. However, despite these programs and initiatives, little has been achieved.

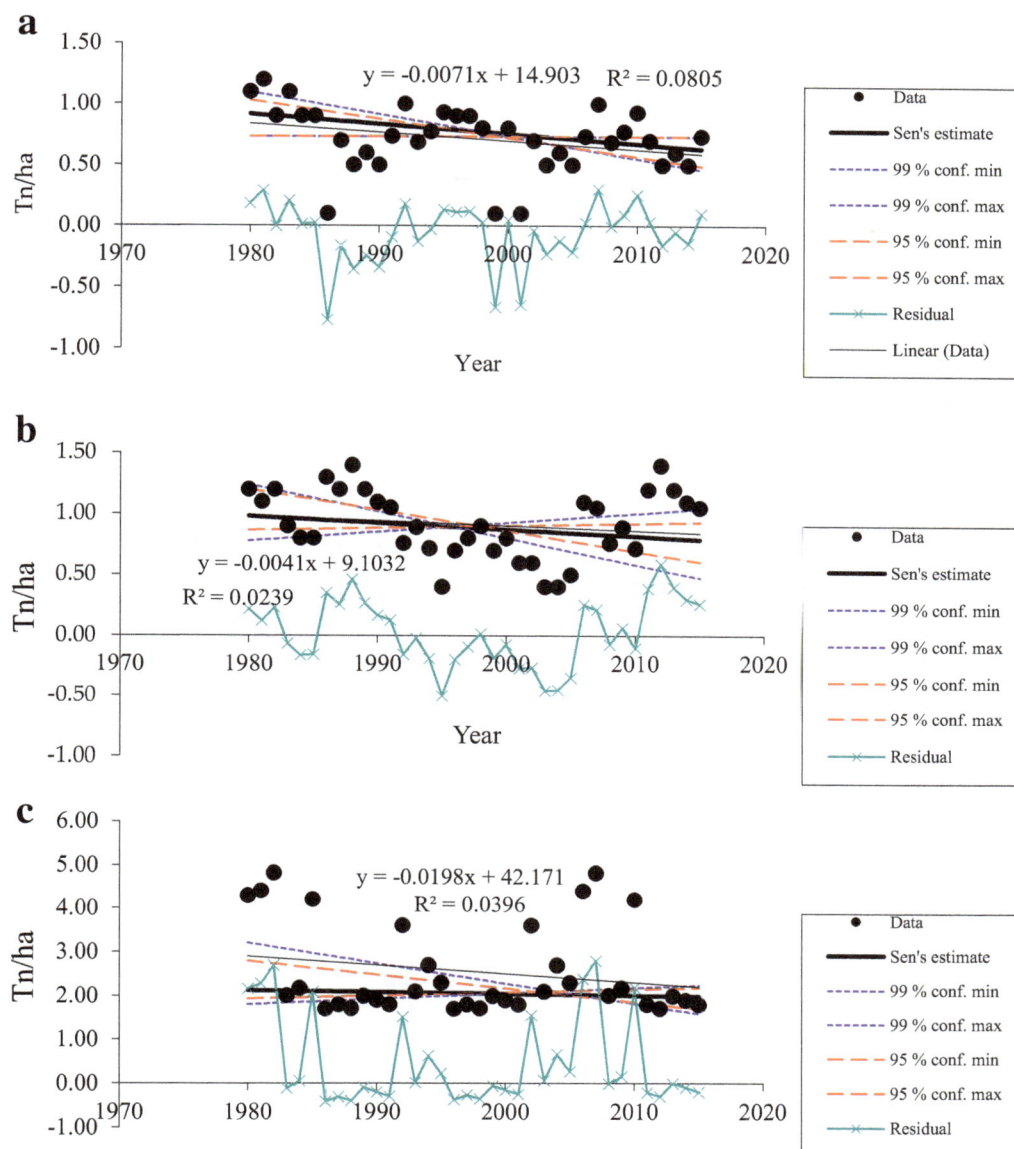

Fig. 3 Yields of the major food crops per hectare in tons in Tanzania from 1980 to 2015. **a** Millet yields per hectare in tons, **b** beans yields per hectare in tons, **c** potatoes yields per hectare in tons *Source*: Analyses from the data obtained from MALF

Discussion

The results of this empirical study can be among the best platforms to discuss where and why we do not do much better as a nation. This is because for the past three decades (study time coverage), lots have been reported as food shortage. More unfortunately, this crisis has been almost randomly scattered throughout the country although the semiarid areas have been more susceptible. As confirmed in Figs. 1a, b, 2a–c, 3a–c and 4a–c, and Table 1, Tanzanian agriculture is inconsistent, and thus, as a nation, has still a long way heading to productive

and sustainable industry. While the production trend of total yields for maize, sorghum, banana and beans had a positive slope (at $R^2 = 0.8$), the remaining crop yields had a positive slope (between $R^2 = 0.4$ and 0.7) as seen in Fig. 1a, b.

However, this growth was highly influenced by expansion of crop land (Table 1). For example, the areas under maize, sorghum, paddy, millet and cassava cultivation have increased from 1,343,300, 566,700, 434,500, 243,200 and 265,500 in 1996 to 3,854,600, 1,389,600, 1,364,300, 467,900 and 1,254,300 in 2015, respectively (Table 1).

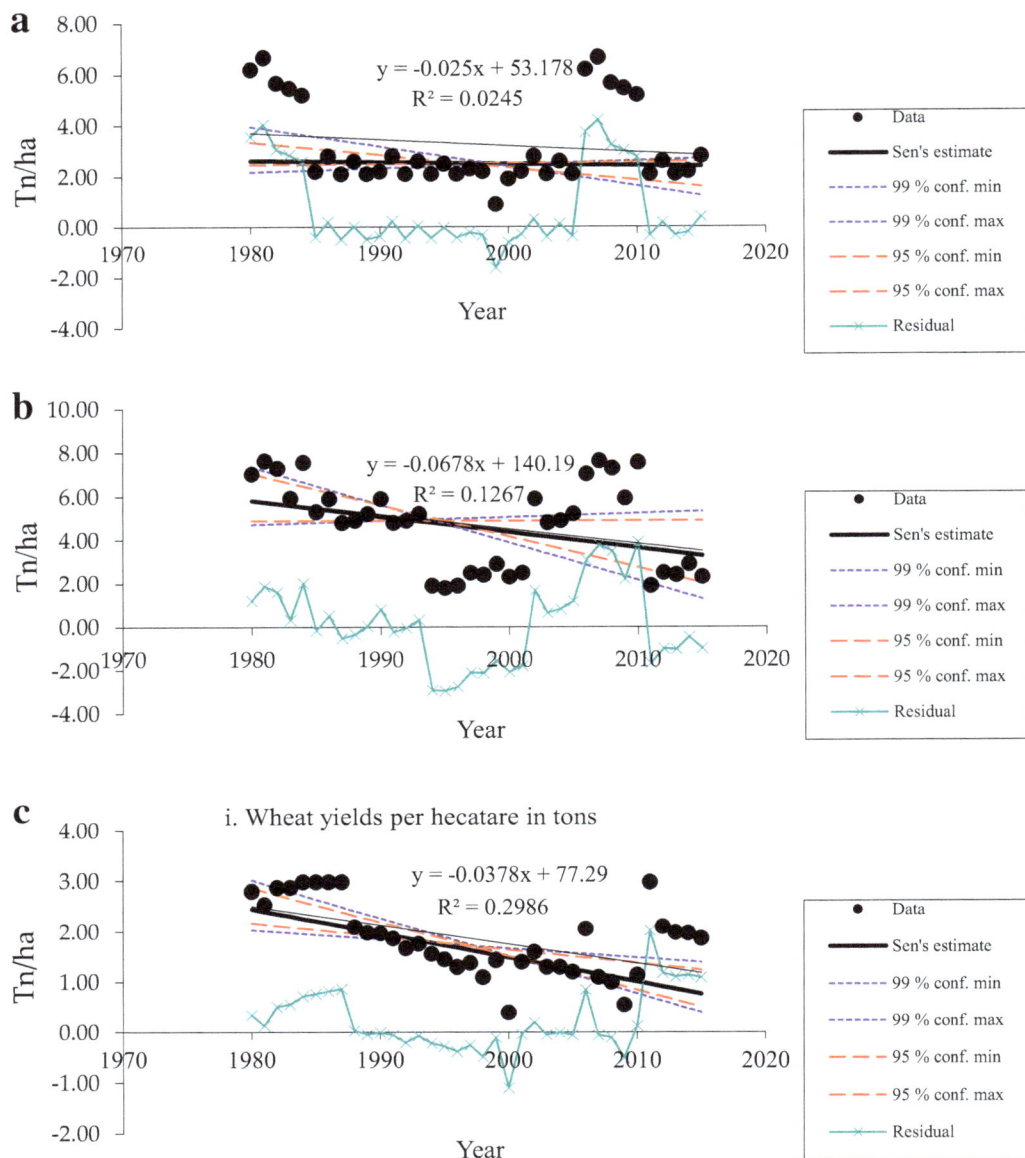

Fig. 4 Yields of the major food crops per hectare in tons in Tanzania from 1980 to 2015. **a** Cassava yields per hectare in tons, **b** banana yields per hectare in tons, **c** wheat yields per hectare in tons *Source*: Analyses from the data obtained from MALF

Despite this, the country has not yet fully harnessed the potential arable land as less than 24% of the 44 million hectares of parable land has been under effective utilization 3. This, therefore, indicates that we have still lots to do if need to attain sustainable crop yields and food security at both local and national level.

In due fact, this farm expansion has on one side been accrued by population increase. In other words, there has been insignificant increase in the yields even in regions called the food-producing zones or food basket regions

[2, 3]. Frankly speaking, different regions have been experiencing different crop yields. This spatial differentiation is mainly influenced by diverse biophysical characteristics of the locality, i.e., climate, soil and agro-ecological zones just to mention a few. However, the gotten yields (i.e., increased) have not curbed the food demand, especially at national level [20], and more particularly on the food availability and accessibility. For that case, the food demands have continued to widen over time. Thus, in most cases the growth of the total yields has

Table 1 Area under crop production in '000' hectares by region *Source*: **Extracted from MALF, 2016**

Year	Maize	Sorghum	Millet	Paddy	Cassava
1996	1343.3	566.7	243.2	434.5	565.5
1997	1564.1	874.2	253.6	439.3	559.4
1998	2088.3	618.4	268.1	654.5	599.4
1999	1764.4	685.5	295.8	473.9	779.3
2000	1870.5	817.9	251.8	517	1824.5
2001	1572.2	566.7	201.1	323.7	905.5
2002	2956.7	874.2	227.3	642.7	752.7
2003	2852.3	618.4	225.2	688.5	1191.9
2004	2854.2	715.8	225.8	689.6	1313.1
2005	2854.5	817.9	227.9	691.2	1345.4
2006	2570.9	715.8	338.6	633.7	993.2
2007	2600.3	817.9	346.8	557.9	779.1
2008	3980.9	1239.3	355.7	837.7	876.9
2009	2961.3	1323.4	396.5	805.6	1081.4
2010	3050.7	1239.3	452.7	1136.3	872.9
2011	3056.7	1323.4	456.5	1143.3	890.5
2012	3058.8	1148.3	457.6	1174.5	954.4
2013	3730.5	1170.8	421.2	1276.4	1034.5
2014	3729.5	1285.7	453.2	1254.7	1154.8
2015	3854.6	1389.6	467.9	1364.3	1254.3

not necessarily implied agricultural development. This is because there are other aspects that determine such development apart from the total yields.

The exploration of crop yields per hectare in tons is far most the best way of determining the actual agricultural development. The analyses in this aspect had a real basis of determining agricultural progress. This is because it can supply more food to sustain the rapidly growing population. However, the results from analyses of yields per hectare in tons indicate that almost each crop experienced negative growth as seen in Figs. 2a–c, 3a–c and 4a–c. From 1980 to 2015, the yields for these crops declined between $R^2 = 0.02$ to 0.3 (Figs. 2a–c, 3a–c and 4a–c). This means, the production trend per hectare has been deteriorating over time due to multiple reasons such as soil degradation, increased droughts, poor farming methods, eruption of diseases and less investment in agricultural sector, especially research. Apparently, the increased incidences of climate change impacts have contributed to worsen the situation [13, 17].

Apart from establishing the correlation between the components under study, the present study elicits or/and annotates specific aspects that categorizes the country as food insecure [21–23]. The production systems seem to be less monitored as anyone can join or quit agricultural production. This reminds us that there is a possibility of the majority abandoning the sector or opting

transformative adaptation due to climate change. Thus, all the suddenly, the situation can be at its worst and little can be done by the government and/or donors to rescue the victims in such massive mess.

In addition to that, despite employing over 80% of the labor forces, especially in rural areas, the sector has failed to attract more young generation because it is perceived as the least paying. This has been evidenced in various occasions where the farmer has remained with very little control on marketing his/her yields. For example, the government may come abruptly to solve immediate problem at the expenses of the farmer who had anticipated to sell their yields in the market of their favor.

For example, exporting maize or selling in its raw form to get more income. The discussion with undisclosed agricultural officers had some blames being directed to some politicians who always don't adhere to the technical aspects. Most of this people from political cadre do this reckless tendency deliberately to win political credit.

Given the production turbulent which the country experiences, it is evident that the efforts that have been placed in the sector have not born enough fruits. The episode of making agriculture as a backbone of the national economy that started (back in 1970s) over four decades ago using different approaches has never been fruitful. For example, in 1980s the government involved directly in agriculture by making herself a funder, facilitator and supplier of agricultural inputs that would boost its development; however, this approach was fruitless. One of the major reasons for this failure was weak supervision of the government firms, infrastructures and assets. Ultimately, government companies and factories ran into bankrupt and the approach ended unsuccessfully.

Recently (2000s), numerous participatory approaches have been adopted in the planning and implementation of various agricultural practices as a substitute of the previous approach. The PADEP, ASPS, ASDP and *Kilimo Kwanza* are among these programs and initiatives that were geared to boost agriculture industry in the country 11, 29. The NAP 2013 serves to give guidance on how agriculture sector should go about. Mainly it implies to optimize yields from agriculture for the betterment of people's social welfare. Originally, this policy was enacted in 1997, whereas amendment and new draft was in 2013. However, despite giving explicit concept on the available resources for sustainable agricultural development, there are some agricultural aspects that are not well addressed. Most of these are the new challenges that emanates/comes as response from human-induced factors.

As a response to agricultural uncertainty, the NAP 2013 [3] has been emphasizing on the adaptation measures to climate change impacts, intensification of agriculture and the use of improved crop seeds among the farmers.

However, despite those responses the state of food security has not been stabilized as in the series of years from 2000 to 2014; there have been frequent food shortages in various area of the country (either due to food unavailability or inaccessibility the latter being caused by financial shenanigans). Thus, there is a need to adopt sustainable and exhaustive approaches that could adequately serve on diverse agro-ecological zones of the country.

On the other hand, there are numerous factors that impede agricultural sector in the country. Adjustment to climate change impacts has been a major challenge among the smallholder farmers 10. This is because they have weak adaptive capacity either to cope or recover from such a dreadful condition. Since more than 70% of the agricultural industry is dominated by the smallholder farmers, it is evident that the impact of climate change to them is worst compared to medium and large scale. This has ultimately elevated poverty levels in rural areas [2, 11]. This observation is in agreement with the reports by UNDP [24, 25] which informs that most developing countries have elevated poverty levels mainly due to poor yields.

Market constraint is another challenge that has been affecting agriculture. This refers to uncontrolled market systems that involve goods and inputs related to agriculture. In terms of inputs, the flow of fertilizers and seeds to farmers has been inconvenient [26]. In most cases, these inputs have been reaching people out of their farming calendars. In so doing, the provided inputs become useless/fruitless. In this aspect of marketing, the distribution of food within the country has been less convenient. Among the causes for this inconvenience is poor infrastructure such as roads and storage facilities.

For example, in 2014 there were optimal yields gotten in the main producing regions (Iringa, Mbeya, Morogoro, Rukwa and Kigoma); however, shortage of storage facilities was the main cause for yield destruction. While the yields were destroyed in those regions, other parts of the country had food shortage.

As a survival strategy, some farmers intended to sell their produce to other countries, but were forbidden by the government as it could arise food shortage within the country. Ultimately, this situation brought about enormous loss to some farmers, who some of them decided not to further engage in agriculture in the next season. Thus, there is a need to streamline all market components so that it benefits the farmers and related stakeholders.

Inadequate technology is among the major obstacle for agricultural development in the country as it limits agricultural intensification. This encompasses low use of improved inputs such as seed, fertilizers and other farm implements. Harnessing irrigation potentials is significantly impacted by poor technology [2, 27, 29]. It is only less than 4% of the potential irrigable land that has been harnessed. This exploitation is quite different from that in the developed countries where over 50% of such potential is exploited. Sufficient budget could be of help in exploiting the irrigation potentials, especially by hiring innovative machines and considerable technology.

Sparingly, the government through TAFSIP aimed to comprehensively transform the sector to achieve food and nutrition security, and reduce poverty by allocating 10% of its budget in agriculture sector [28, 29]. However, this financial allocation has not been made. It is this financial discrepancy that impedes the implementations of various agricultural programs and projects. If this budget were allocated, among other things it could facilitate the provision of agricultural subsidies to farmers. So, as long as the allocation was not made, the farmers have continued to be economically powerless due to such insignificant services.

The solicitation of agricultural loans among the farmers has also been a challenge. The easiest way of acquiring such a loan is through banks. However, most of the farmers have no entitlement to loans access due to lack of relevant collaterals required by most financial institutions. Even for those who access, they are not sure if they could manage to pay back given the market constraints of their prospective yields.

Agricultural turbulence has also affected the national GDP throughout the time frame. For example, from the year 2000 to 2009, the agricultural contribution to GDP fell from 29 to 25%, respectively. While that happened, the nutrition level fell from 25 to 23.5% within the same time frame. Specifically, the malnutrition is more pronounced to children under 5 years [30–33].

To reverse this trend, there is a need to employ the newest techniques in agricultural industry that can transform agriculture from subsistence to commercial [30–33]. This is because numerous programs, initiatives and project have been established and implemented; however, increased incidences of food shortage and poverty have been reported. Since 1980 to date, a series of initiatives have been in place with either less fruits or short-terms impacts.

In this aspect, the over dependency on foreign support appeared to increase the unsustainability of the sector. In most cases, when the project phases out, the operation of all agricultural activities which were formerly done by the project is suspended. The surest solution to curb this uncertainty is to increase the generation of national income and allocate more funds in agricultural sector.

Alternatively, the government through the Ministry of Agriculture should adhere to the demands of authentic and potential challenges through appropriate review of its policy to make it more vigorous and friendly to small holder farmers. As usual, flawless consideration of different agro-ecological zones in the reviews should be adhered. This will have long-term significant contributions to socioeconomic development of the people. Likewise, the provision of raw materials to domestic industries will be ensured from the local production.

Conclusions

This study has laid down some empirical conclusions that when adopted can be a good basis for agricultural intensification in the country. Among others, it revealed that despite the temporal increase in the total yields of the major food crops, the food demand has been widening due to sudden population increase and other allied factors. The observed increase in total yields has happened due to expansion of crop land. Besides, the study noticed that the yields per hectare in tons have been fluctuating at the decreasing trend. The enormous and ever-expanding food demand has necessitated usual policy responses to overcome the associated challenges. This has involved the implementation of various programs and projects that could upsurge crop yields. On the same basis, the NAP 2013 has been identifying some arising issues that impede agricultural production. Among others, the impacts of climate change and soil degradations have been adequately identified.

Given to this vulnerability, the impetus, tireless and improved instruments (including more robust policies) should be employed in agriculture sectors. Determination should be well set to overcome the huddles. To curb these challenges, this study proposes the adoption of drought-resistant crops and increased fertilizations have been proposed. However, the technology and knowledge of curbing the same have not well tricked down at local scale. The assessment on the utilization of various research findings can be the further research priorities.

Abbreviations

FAO: Food and Agriculture Organization; GDP: Gross Domestic Product; IPCC: Intergovernmental Panel on Climate Change; MALF: Ministry of Agriculture, Livestock and Fishery; NAP: National Agricultural Policy; URT: United Republic of Tanzania; WRB: World Reference Base of Soil Resource.

Authors' contributions

Both authors (MYM and XH) designed the study, while MYM carried out the majority data collection portion. XH cooperated in the research, providing input throughout and reviewing details. Both authors wrote, reviewed and commented on the manuscript. Both authors read and approved the final manuscript.

Author details

[1] Department of Geography and Environmental Studies, Solomon Mahlangu College of Sciences and Education, Sokoine University of Agriculture, Morogoro 3038, Tanzania. [2] Centre of Excellence for Soil Biology, College of Resources and Environment, Southwest University, Chongqing 400715, China. [3] School of Plant Biology, University of Western Australia, Crawley 6009, Australia.

Acknowledgements

We give thanks to the College of Resources and Environment of Southwest University for supporting this study. We also wish to thank research assistants who involved in the data collection process.

Competing interests

The authors declare that they have no competing interests.

Funding

This study was funded by the College of Resources and Environment of Southwest University of China.

References

1. URT. National population policy. Dar es Salaam: Government Publishing Press; 2006.
2. URT. Poverty and human development report "research and analysis working group. Dar es Salaam: Government Publishing Press; 2009.
3. URT. National agriculture policy. Dar es Salaam: Government Publishing Press; 2013.
4. Food and Agriculture Organization of the United Nations. Soil fertility management in support of food security in Sub-Saharan Africa. Rome: FAO; 2001.
5. FAO. FAOSTAT, Rome. Food and Agriculture Organization, Rome. 2012. http://faostat.fao.org/. Accessed 10 Jan 2017.
6. FAO: Declaration of the world summit on food security. World Summit on Food Security. Neufeldt H, Jahn M, Campbell BM. Beyond climate smart agriculture: toward safe operating spaces for global food systems. Agric Food Secur. 2013; 2:12.
7. Reddy AA. Food security indicators in India compared to similar countries. Curr Sci. 2016;111(4):632–40.
8. Lobell DB, Burke MB. On the use of statistical models to predict crop yield responses to climate change. Agric For Meteorol. 2010;150(11):1443–52.
9. Mkonda MY. Rainfall variability and its association to the trends of crop production in Mvomero District, Tanzania. Eur Sci J. 2014;10(20):263–73.
10. URT. United Republic of Tanzania, national adaptation programme of action (NAPA). Dar es Salaam: Division of Environment, Vice President's Office; 2007.
11. URT. Review of food and agricultural policies in the United Republic of Tanzania. MAFAP country report series. Rome: FAO; 2014.
12. Intergovernmental Panel on Climate Change. Climate change 2013: the physical science basis. In: Stocker TF, Qin D, Plattner GK, Tignor M, Allen SK, Boschung J, Nauels A, Xia Y, Bex V, Midgley PM, editors. Contribution of working group I to the fifth assessment report of the intergovernmental panel on climate change. Cambridge: Cambridge University Press; 2013.
13. Intergovernmental Panel on Climate Change. In: Field CB, Barros VR, Estrada YO, Genova RC, Girma B, Kissel ES, Levy AN, MacCracken S, Mastrandrea PR, White LL, editors. Climate change 2014 impacts, adaptation, and vulnerability. Part A: global and sectoral aspects. Contribution of working group II to the fifth assessment report of the intergovernmental panel on climate change. Cambridge: Cambridge University Press; 2014.
14. Food and Agriculture Organization. FAO/Unesco soil map of the World, revised legend, with corrections and updates. (World Soil Resources Report 60) FAO, Rome (reprinted with updates as Technical Paper 20, ISRIC, Wageningen, 1988.
15. Food and Agriculture Organization: FAOSTAT Agriculture Data. FAO, Rome, Italy. 2003. http://apps.fao/cgi-binnph-db.pl?suset=agriculture.
16. Kangalawe RYM. Climate change impacts on water resource management and community livelihoods in the southern highlands of Tanzania. Climate Dev. 2016. https://doi.org/10.1080/17565529.2016.1139487.
17. Rowhani P, Lobell DB, Linderman M, Ramankutty N. Climate variability and crop production in Tanzania. Agric For Meteorol. 2011;15:449–60.
18. Neufeldt H, Jahn M, Campbell B. Beyond climate smart agriculture: toward safe operating spaces for global food systems. Agric Food Secur. 2013;2(12):16–8.
19. Paavola J. Livelihoods, vulnerability and adaptation to climate change in Morogoro, Tanzania. Environ Sci Policy. 2008;11:642–54.
20. Wik M, Pingali P, Broca S. Global agricultural performance: past trends and future prospects. Washington DC: World Bank; 2008.
21. Sen AK. Development as freedom. In: Knopf A, editor. Agricultural Development. Oxford: Oxford University; 1999.

22. Harvey CA, Chacón M, Donatti CI. Climate-smart landscapes: opportunities and challenges for integrating adaptation and mitigation in tropical agriculture. Conserv Lett. 2014;7:77–90.

23. Conceição P, Levine S, Lipton M, Warren-Rodríguez A. Toward a food secure future: ensuring food security for sustainable human development in Sub-Saharan Africa. Food Policy. 2016;60:1–9.

24. UNDP. World population prospects: the 2000 revision. New York: Department of Economic and Social Affairs; 2000.

25. UNDP. Africa human development report 2012: towards a food secure future. New York: United Nations Development Programme (UNDP); 2012.

26. Cornia G, Deotti L, Sassi M. Sources of food price volatility and child malnutrition in Niger and Malawi. Food Policy. 2016;60:20–30.

27. ECA (Economic Commission for Africa). Harnessing technologies for sustainable development. Addis Ababa: ECA; 2002.

28. URT. Tanzania agriculture and food security investment plan (TAFSIP). Dar es Salaam: Government Publishing Press; 2011.

29. URT. Participatory agricultural development and empowerment project (PADEP). Dar es Salaam: Government Publishing Press; 2003.

30. Mkonda MY, He XH. Climate variability, crop yields and ecosystems synergies in Tanzania's semi-arid agro-ecological zone. Ecosyst Health Sustain. 2018;1:1. https://doi.org/10.1080/20964129.2018.1459868.

31. Mkonda MY, He XH, Festin ES. Comparing smallholder farmers' perception of climate change with meteorological data: experiences from seven agro-ecological zones of Tanzania. Weather Clim Soc. 2018. https://doi.org/10.1175/wcas-d-17-0036.1.

32. Mkonda MY, He XH. Yields of the major food crops: implications to food security and policy in Tanzania's semi-arid agro-ecological zone. Sustainability. 2017;9(8):1490. https://doi.org/10.3390/su9081490.

33. Mkonda MY, He XH. Are rainfall and temperature really changing? Farmer's perceptions, meteorological data, and policy implications in the Tanzanian semi-arid zone. Sustainability. 2017;1:1. https://doi.org/10.3390/su9081412.

## 10

Correlations of cap diameter (pileus width), stipe length and biological efficiency of *Pleurotus ostreatus* (Ex.Fr.) Kummer cultivated on gamma-irradiated and steam-sterilized composted sawdust as an index of quality for pricing

Nii Korley Kortei[1*], George Tawia Odamtten[2], Mary Obodai[3], Michael Wiafe-Kwagyan[2] and Deborah Louisa Narh Mensah[3]

Abstract

Background: Consumption patterns of mushrooms have increased in Ghana recently owing to its acknowledgement as a functional food. Different mushroom cultivation methods and substrate types have been linked to the quality of mushrooms produced, thereby affecting its pricing.

Methods: A comparative regression analysis was carried out to assess the correlation of stipe lengths, cap diameters and biological efficiencies of mushroom fruit bodies of *Pleurotus ostreatus* cultivated on steam-sterilized and gamma-irradiated sawdust after exposure to ionizing radiations of doses 0, 5, 10, 15, 20, 24 and 32 kGy from a ^{60}CO source (SL 515, Hungary) at a dose rate of 1.7 kGy/h. Steam sterilization of composted substrates was also done at a temperature of 100–105 °C for 2 h.

Results: Cap diameters of the mushrooms ranged 41–71.5 and 0–73 mm for gamma-irradiated samples depending on dose and steam-sterilized composted sawdust, respectively. Stipe lengths ranged between 4.4–61 and 0–58.1 for gamma-irradiated samples depending on dose and steam-treated substrates, respectively. Total yields of *P. ostreatus* grown on the gamma irradiation-treated composted sawdust ranged between 8.8 and 1517 g/kg, while mushrooms from steam sterilized recorded 0–1642 g/kg. Biological efficiencies of mushrooms grown on irradiated sawdust ranged 3–93.3%, while steamed sawdust ranged 0–97%. Good linear correlations were established between the cap diameter and biological efficiency ($r^2 = 0.70$), stipe length and biological efficiency ($r^2 = 0.91$) for mushrooms cultivated on gamma-irradiated sawdust. Similarly, good correlations were established between cap diameter and biological efficiency ($r^2 = 0.89$) stipe length and biological efficiency ($r^2 = 0.95$) for mushrooms cultivated on steam-sterilized sawdust.

Conclusion: These correlations provide the possibility to use only the cap diameter and stipe lengths to predict their biological efficiency and also use this parameter for grading and pricing of mushrooms earmarked for the consumer market.

Keywords: Correlation, Biological efficiency, Cap diameter, Stipe length, *P. ostreatus*, Steam, Gamma-irradiated compost

*Correspondence: nkkortei@uhas.edu.gh
[1] Department of Nutrition and Dietetics, School of Allied Health Sciences, University of Health and Allied Sciences, PMB 31, Ho, Ghana
Full list of author information is available at the end of the article

Background

Oyster mushrooms (*Pleurotus* sp.) have been widely cultivated in many different parts of the world. These mushrooms have the ability to grow at a wide range of temperatures utilizing various lignocelluloses [1] due to its powerful degrading cellulolytic and pectinolytic enzymes produced in vivo. Enzymatic efficiency of mushrooms makes their nutrition one of the most proficient biotechnological processes for lignocellulosic and organic waste recycling [2] by converting plant waste residues from agricultural activities as well as forestry debris. The technology used for its cultivation is eco-friendly since it exploits the natural ability of the fungus to degrade these complex polysaccharides to generate much simple compounds useful for human nutrition.

The current state-of-the-art research shows the usage of fungal biotechnology in fields as restoration of damaged environments (mycorestoration) via mycofiltration (i.e. use of mycelia to filter water), mycoforestry (i.e. use of mycelia to restore forests), mycoremediation (using mycelia to ameliorate heavily polluted soils), myconuclear bioremediation (the use of mycelia to sequester soils of radioactive materials), mycopesticide (use as biopesticide to control pests), and also spent composts could be used as biofertilizers to enhance the fertility of the soil [3, 4]. These methods represent fungal ability to restore the ecosystem where there are no adverse effects after fungal application.

The global economic value of fungi is now well known, the reason for the rise in consumption of mushrooms is a combination of their value as food, and their medicinal and nutraceutical properties [5] hence are gaining so much popularity in Ghana.

Several studies [2, 4, 6, 7] suggest that the methodology described for substrate preparation consists of composting agricultural residues followed by pasteurization and this could be achieved in diverse ways [8, 9]. The steam method is the most popular used for its artificial cultivation [10]. Nonetheless, the conventional method of steam sterilization of substrates has some disadvantages. It is cumbersome since factors such as precise sterilization time and temperature also depend on the residual micro- and mycoflora in a given substrate material [11]. There are also the associated discomfort and health hazards of standing by the heat for long hours not excepting the use of fuel wood for heating when natural cooking gas is not readily available. Fire wood burning depletes the forests of environmentally useful plant species.

Compost substrate sterilization by autoclaving at 121 °C or hot water dipping (pasteurization) in steel drum at 60 °C for 2.3 h has been reported by [12]. Chemical pretreatment of substrates has been reported by [13]. Oyster mushrooms have also been traditionally produced using the outdoor log technique, thereby excluding substrate sterilization [14].

In recent times, gamma irradiation has been used successfully to sterilize different substrates including mushroom compost to cultivate oyster mushrooms in Ghana [4, 15–17].

This study was to show correlations of growth parameters (cap diameter and stipe length) and biological efficiency in relation to the use of gamma irradiation for sterilization for substrates for mushroom cultivation.

Materials and methods

Preparation of pure culture

Pure culture was prepared using the modified method of [18]. Malt extract agar (Oxoid, Basingstoke, Hampstead, England) was prepared according to manufacturer's instructions. The media were sterilized in an autoclave for 15 min at 121 °C with 1.5 kg/cm^2 pressure. The sterilized media in the test tubes were kept in slanting/sloping positions. The stipe of the mushroom was surface-sterilized with 0.1% sodium hypochlorite. A scalpel was then dipped in alcohol and flamed until it was red hot. It was then cooled for 10 s. The stipe was cut lengthwise from the cap to the tip of the stipe downwards. Small piece of the internal tissue of the broken mushroom was cut and removed with a sterile scalpel. Using an inoculation needle, the cut tissue was then immediately inserted into test tube slant and the tissue laid on the agar surface. The mouth of the test tube was flamed before the needle was inserted. The mouth of the test tube was plugged with cotton wool and was incubated at 28±2 °C for mycelia growth. After 7 days, the tissue was covered with a white mycelium that was spread on the agar surface.

Preparation of spawn

The stock culture substrate was prepared by using good quality sorghum grains and CaCO$_3$ packed tightly in 25 × 18 cm polypropylene bag. These packets were sterilized in an autoclave for 1 h at 121 °C and 1.5 kg/cm^2 atmospheric pressure and were kept 24 h for cooling before inoculation. After 7–9 days, there was a complete coverage of running mycelium. The spawn then was ready for inoculation of substrate bags.

Substrate preparation

The substrate consisted of 'wawa' (*Triplochiton. scleroxylon*) sawdust 80–90%, 1–2% of CaCO$_3$ and 5–10% wheat bran. Moisture content was adjusted to 65–70% [19]. The sawdust was mixed thoroughly, heaped to a height of about 1.5 and 1.5 m base and covered with polythene and made to undergo natural fermentation for 28 days. Turning was made every 4 days to ensure homogeneity.

Bagging

Composted sawdust was compressed into 0.18 m × 0.32 m heat-resistant polyethylene bags. Each bag contained approximately 1000 g (1 kg) of substrate and replicated six times.

Sterilization/pasteurization

Bagged composted sawdust substrates were sterilized with either moist heat at a temperature of 100–105 °C for 2.5 h or treated with gamma irradiation.

Gamma irradiation

Bagged composted sawdust substrates were subjected to radiation doses of 0, 5, 10, 15, 20, 24 and 32 kGy at a dose rate of 1.7 kGy per hour in air from a cobalt-60 source (SLL 515, Hungary) batch irradiator. Doses absorbed were confirmed using the conventional ethanol-chlorobenzene (ECB) dosimetry system. Radiation was carried out at the Radiation Technology Centre of the Ghana Atomic Energy Commission, Accra, Ghana.

Each treatment dose was replicated six times.

Innoculation and incubation

Each bag was closed with a plastic neck and plugged in with cotton and inoculated with 5 g sorghum spawn. The bags were then incubated at 26–28 °C and 60–65% relative humidity for 20–34 days in a well-ventilated, semi-dark mushroom growth room.

Calculations of growth and yield parameters of mushrooms

Growth and yield parameters were estimated according to methods prescribed by [20–22] as follows [6]:

Stipe length = length of cap base to end of stalk

$$\text{Average cap diameter} = \frac{\text{longest} + \text{shortest cap diameters}}{2}$$

Biological efficiency (B.E%)

$$= \frac{\text{Weight of fresh mushrooms harvested (g)}}{\text{Dry substrate weight (g)}}$$

$$\times 100$$

Yield (g/kg) = Weight of fresh mushrooms

Statistical and regression analysis

Regression analysis of correlation was employed using Microsoft Excel (Windows 10 version). Means of yield and growth parameters were subjected to analyses of variance (one-way ANOVA).

Differences between the means were determined using the Least Significant Difference test. All findings were considered statistically significant at $P < 0.05$.

Results

The various substrate treatments resulted in different degrees of growth and yield: stipe length (mm), cap diameter (mm), yields (g/kg) and biological efficiencies (%) (Figs. 1, 2, 3, 4, 5, 6, 7, 8). Cap diameters of *P. ostreatus* from gamma-irradiated composted sawdust ranged between 41 and 71.5 mm, while steam sterilized ranged between 0 and 73 mm (Fig. 1). Non-sterilized substrates produced poor growth of mushrooms and were significantly ($P < 0.05$) low. Cap diameters of mushrooms obtained on steam sterilized were similar with those obtained on 15, 20 and 32 kGy doses. They, however, differed ($P > 0.05$) statistically.

Stipe lengths of *P. ostreatus* also ranged between 44 and 61 mm for gamma-irradiated composted sawdust, while that of steam sterilized ranged between 0 and 58.1 mm. Generally, there was no significant ($P > 0.05$) difference in stipe length growth for both 5 and 10 kGy likewise 20 and 24 kGy doses. Interestingly, steam sterilized and 32 kGy dose produced similar growth and were not significantly ($P > 0.05$) different.

Biological efficiencies of *P. ostreatus* from the gamma-irradiated composted sawdust ranged between 3 and 93.3%. Gamma radiation doses of 5, 10, 24 and 32 kGy produced comparable ($P > 0.05$) biological efficiencies. Similarly, doses of 15 and 20 kGy treatments were also comparable. Steam-sterilized mushrooms recorded a range of 0–97%.

Generally, the average yields of mushrooms recorded were of range 8.8–1517 g/kg for mushrooms depending on dose applied to the substrate, and higher doses tended to yield more mushrooms. A similar trend was observed for biological efficiency and was recorded for the yield. Steam-treated substrates produced a range of 0–1642 g/kg. Although the overall greatest average total yield of mushrooms (1642 g/kg) was obtained from the 5 kGy, the 10 kGy-treated composted sawdust produced comparable results. Yield of steam-sterilized mushrooms recorded was in a similar range as 15 and 20 kGy and showed no significant difference ($P > 0.05$).

The correlations between biological efficiencies (%) and stipe lengths (mm) as well as cap diameters (mm) of *P. ostreatus* produced on both gamma irradiation and steam sterilization substrates are presented in Figs. 5, 6, 7 and 8.

There were good correlation coefficients of determinations of $r^2 = 0.699$ (cap diameter) and $r^2 = 0.914$ (stipe length) and biological efficiencies for *P. ostreatus* from gamma-irradiated sawdust compost (Figs. 5, 6).

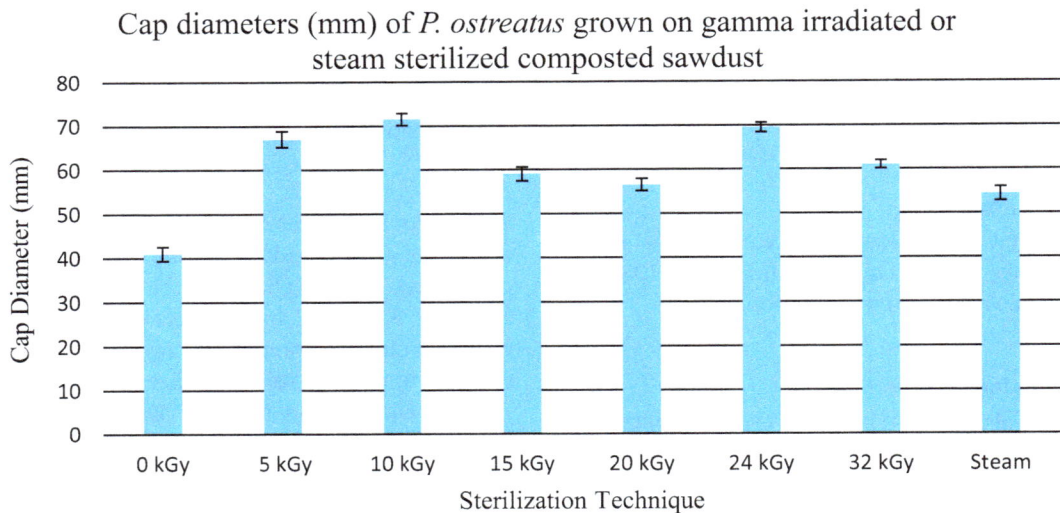

Fig. 1 Cap diameters of *P. ostreatus* grown on gamma-irradiated and steam-sterilized composted sawdust

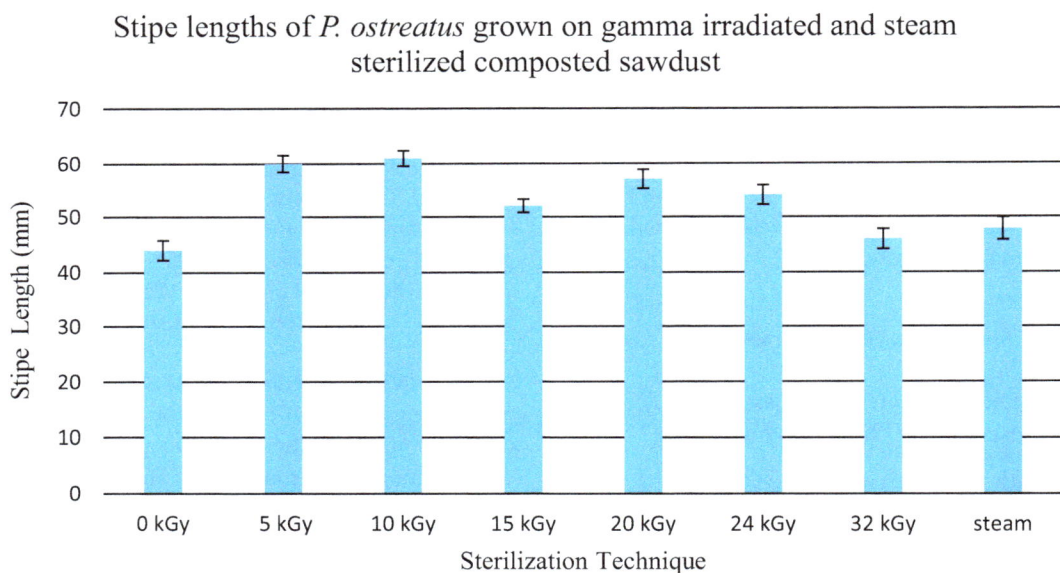

Fig. 2 Stipe lengths of *P. ostreatus* grown on gamma-irradiated and steam-sterilized composted sawdust

Furthermore, the highest scattered points were obtained from correlation of stipe length and biological efficiency (Figs. 5, 6). Very good correlation coefficients were also obtained for cap diameter ($r^2 = 0.886$) and stipe length ($r^2 = 0.951$) on steam-sterilized substrates with biological efficiency (Figs. 7, 8).

Generally, greater stipe lengths and cap diameters (pileus) were attended by high values of biological efficiencies (Figs. 5, 6, 7, 8). Figures 5, 6, 7 and 8 show how the regression line fits the data. The highest

coefficients of determination r^2 (0.95) were obtained from the regression line between biological efficiency and cap diameter (pileus width). The highest scattered points were obtained from the correlation of biological efficiency and stipe length.

Discussion

Cap diameter (pileus width) and stipe length

Growth parameters of cap diameter and stipe lengths of *P. ostreatus* obtained on the various treatments of gamma

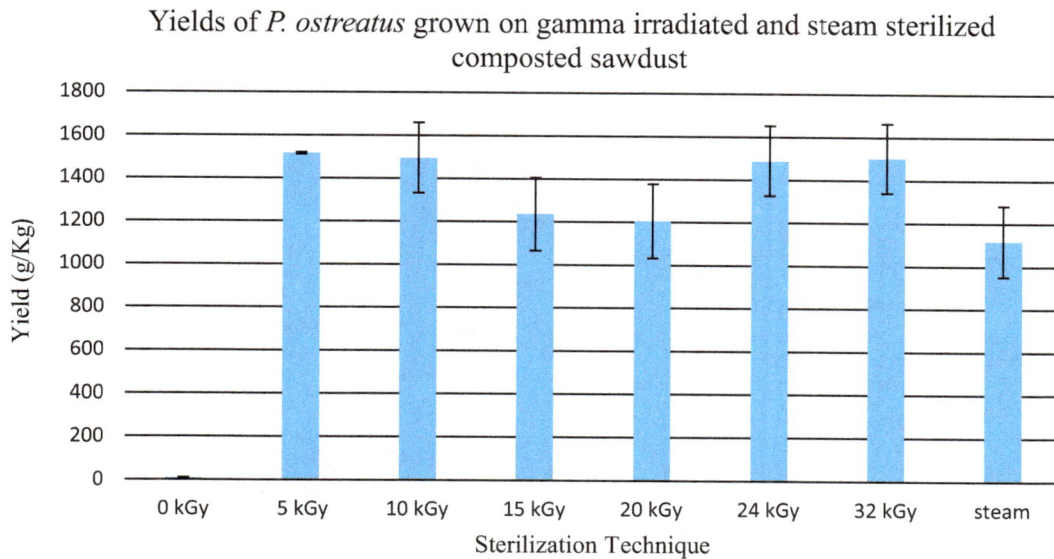

Fig. 3 Average yields of *P. ostreatus* grown on gamma-irradiated and steam-sterilized composted sawdust

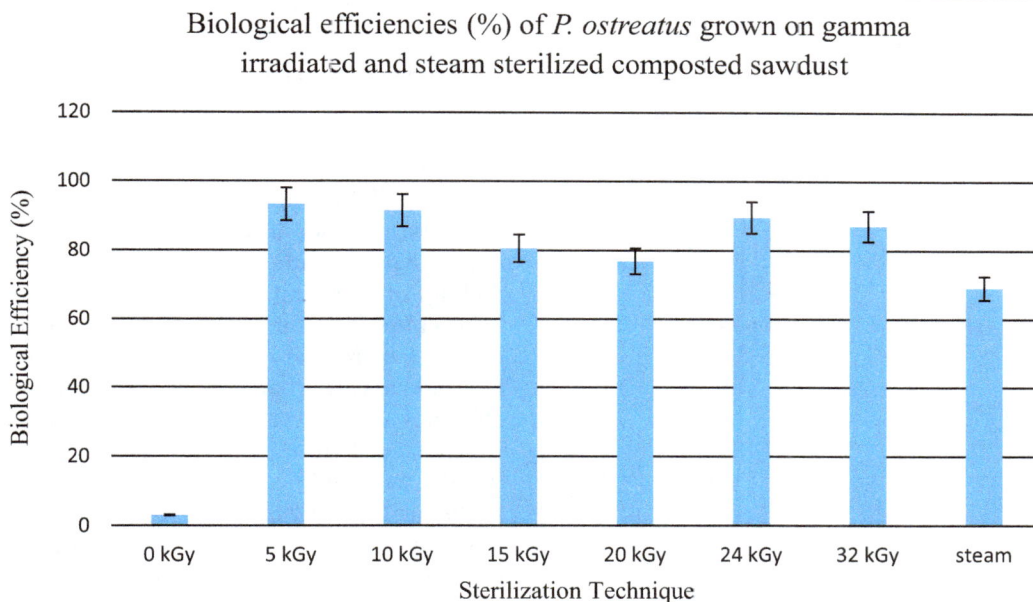

Fig. 4 Biological efficiencies (%) of *P. ostreatus* grown on gamma-irradiated and steam-sterilized composted sawdust

irradiation and steam sterilization varied presumably due to the extent release of nutrients from depolymerization of substrate and subsequent mobilization by hyphae in the substrate for growth.

The performance of oyster mushroom grown on composted lignocellulosic substrates with respect to stipe length and cap diameter (pileus width) also depended on the structure, compactness and physical properties of the substrate which in turn was influenced by the type

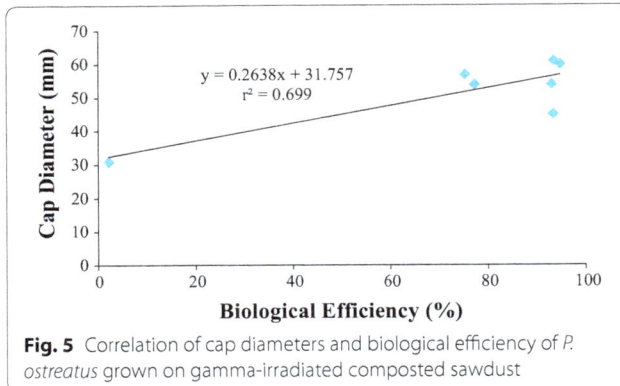

Fig. 5 Correlation of cap diameters and biological efficiency of *P. ostreatus* grown on gamma-irradiated composted sawdust

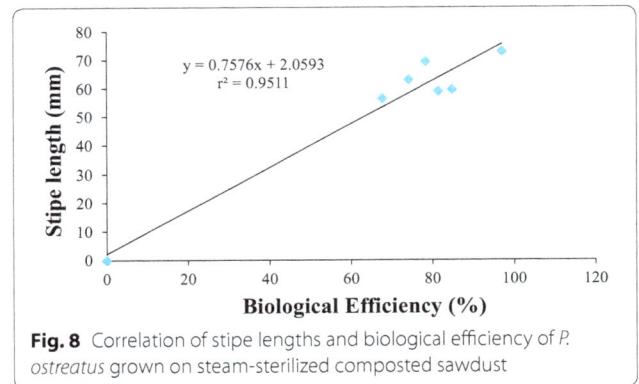

Fig. 6 Correlation of stipe lengths and biological efficiency of *P. ostreatus* grown on gamma-irradiated composted sawdust

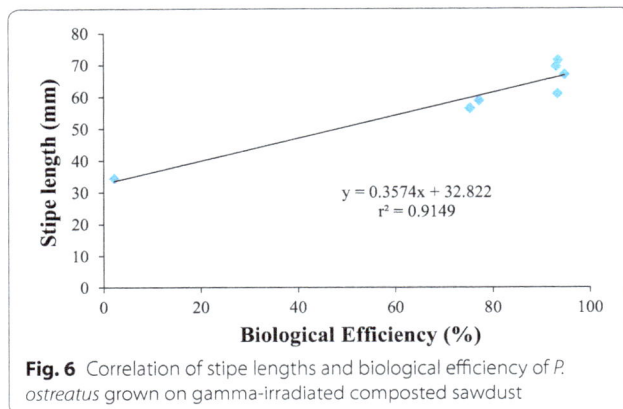

Fig. 7 Correlation of cap diameters and biological efficiency of *P. ostreatus* grown on steam-sterilized composted sawdust

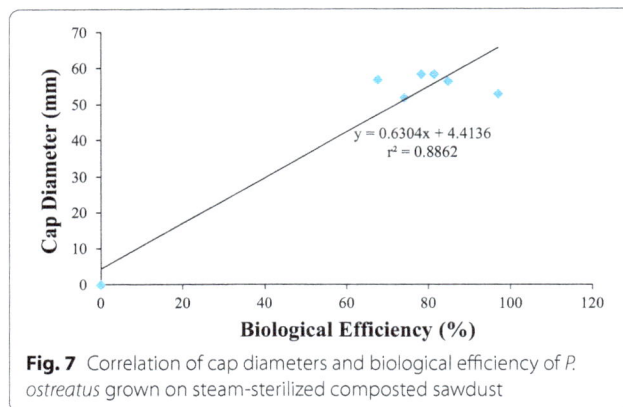

Fig. 8 Correlation of stipe lengths and biological efficiency of *P. ostreatus* grown on steam-sterilized composted sawdust

Results of stipe lengths and cap diameters obtained in this experiment agree with findings of some researchers [6, 23, 24]. Conversely, Kortei and Wiafe-Kwagyan [17] reported higher values of 95 and 80 mm for cap diameter and stipe lengths, respectively, for P-31 strains cultivated on gamma-irradiated corncobs when they investigated the growth of P-31 strain on eight different gamma-irradiated substrates. Kortei [24] and Sarker et al. [25] reported a lower range of 1–5 and 1.85–6.57 cm for stipe lengths and cap diameters, respectively.

Yield and biological efficiency
According to [26], biological efficiency (B.E) is an expression of the bioconversion of dry substrate to fresh fruiting bodies and indicates the fructification ability of the fungus exploiting the substrate. Biological efficiencies (65–98%) recorded in this present study were relatively higher than literature values, and our data agree with the findings of Garo and Girma [27] who reported range of 31.98–146% from the study of responses of oyster mushrooms (*P. ostreatus*) as influenced by different substrates in Ethiopia. Gitte et al. [28] also reported high biological efficiency of milky mushrooms on different substrates which ranged from 51.57 to 146.3%. On the contrary, low B.E (%) values were reported by [6] ranging from 61 to 0% for *P. ostreatus* on different lignocellulosics. Raymond et al. [24] reported a B.E range of 8.95–62.8% for cultivated oyster mushrooms (*Pleurotus* HK37) on sisal wastes fractions supplemented with cow dung manure in Tanzania.

Both sterilization methods were effective in decontamination and depolymerisation of substrates as evidenced in the yield as well as biological efficiency since the two are directly linked. Yields were moderately higher for mushrooms cultivated from gamma-irradiated substrates than for those from steam treatment. This observation could be attributed to the higher degree of splitting of complex polysaccharides of the substrates into smaller

of agricultural wastes and method used in preparing the substrates. Chukwurah et al. [23] reported that substrates with higher moisture retaining capacity grow better than those with lower moisture retaining capacity and that substrates which contained mixtures of different types of agricultural wastes performed better than those with single agricultural waste [23].

utilizable units by gamma rays than what obtained for steam. Secondly, gamma rays have a high lethal effect on the genome of the micro-organisms resident in the substrate either directly or indirectly during pasteurization. It can be deduced from our results that 5 kGy produced the overall best yield. This study has confirmed that the use of different doses (5, 10, 15, 20, 24 and 32 kGy) of gamma irradiation to sterilize sawdust substrate produced similar results for that of steam treatment. Although the overall maximum yield was marginally higher for steam-treated substrates, the cost of heating (either with wood, gas or electricity) with steam will be economically more expensive and capital intensive and laborious than the use of gamma irradiation.

Correlations of cap diameter, stipe length and biological efficiency

Varied degrees of growth of *P. ostreatus* fruit body parts have been observed on different substrates as well as different cultivation methods by some researchers [1, 12]. In this study, there were high coefficients of correlations of stipe lengths and biological efficiency (0.91–0.95) was better as compared to that of cap diameter and biological efficiency (0.69–0.88) for both methods. Ajonina and Tatah [29] reported that the size of the stalk and pileus is positively correlated with yield and with carbohydrate and protein, respectively. They noticed variations in stalk length (2.43–3.24 cm) for *P. ostreatus*. Ahmed et al. [30] suggested that in the case of yield, the larger the pileus size, the higher the yield. Fruiting body weight is to a large extent influenced by the thickness and diameter of the pileus. Large-sized fruit bodies are widely perceived to be of superior quality and hence highly ranked in mushroom cultivation. In grading oyster mushroom for pricing, the use of correlation of stipe length, cap diameter (pileus width) and weight will be good criteria for grading quality [31].

Shen and Royse [32] and Mondal et al. [18] reiterated that fruit bodies were susceptible to breakage during packing and transport and so reduced their storage quality. Harvested mushrooms packaged in a rigid appropriately determined packaging material with good aeration, humidity and compactness can have extended shelf life even after grading for the market.

Conclusion

Our results showed that there was a positive correlation between cap diameters, stipe lengths and biological efficiencies of *P. ostreatus* cultivated on both steam-sterilized and gamma-irradiated composted sawdust which gives a clear indication of the possibility to predict yield or biological efficiency with growth determinants. Yields of both methods were also comparable. Gamma radiation could be used as a decontaminating agent of substrates for mushroom cultivation in countries that have access to gamma irradiation facilities to augment the dreary processes associated with the conventional steam sterilization method. Pricing can be based on a quality parameter such as yield, stipe length and pileus width.

Authors' contributions
NKK, GTO and MO involved in the conception of the research idea, design of the experiments and data analysis and also drafted the paper. MW-K and DLNM involved in the design of the experiments and data collection. GTO and MO provided guidance, corrections and supervision to the entire research and critically reviewed the manuscript. NKK and GTO read, reviewed and amended the manuscript. All authors read and approved the final manuscript..

Author details
[1] Department of Nutrition and Dietetics, School of Allied Health Sciences, University of Health and Allied Sciences, PMB 31, Ho, Ghana. [2] Department of Plant and Environmental Biology, College of Basic and Applied Sciences, University of Ghana, P. O. Box LG 55, Legon, Ghana. [3] Food Microbiology Division, Council for Scientific and Industrial Research - Food Research Institute, P. O. Box M20, Accra, Ghana.

Acknowledgements
We are grateful to Messers Abaabase Azinkaba, Godson Agbley and Moses Mensah of Mycology Unit, CSIR-FRI, for under taking the steam sterilization and maintenance of the farm house. Messers S.N.Y. Annan, J.N.O. Armah, S.W.N.O. Mills and S.A. Acquah of the Radiation Technology Centre, Ghana Atomic Energy Commission (G.A.E.C), Kwabenya, Accra, carried out the irradiation process.

Competing interests
The authors declare that they have no competing interests.

Funding
Not applicable.

References
1. Sanchez C. Cultivation of *Pleurotus ostreatus* and other edible mushrooms. Appl Microbiol Biotechnol. 2010;85:1321–37.
2. Mandeel QA, Al-Laith AA, Mohamed SA. Cultivation of oyster mushroom (*Pleurotus* sp.) on various lignocellulosic wastes. World J Microbiol Biotechnol. 2005;21:601–7.
3. Stamets P. Mycelium running: how mushroom can help save the world. Berkeley: Ten Speed Press; 2005. p. 574.
4. Kortei NK. Comparative effect of steam and gamma irradiation sterilization of sawdust compost on the yield, nutrient and shelf life of *Pleurotus ostreatus* (Jacq.ex. Fr) Kummer stored in two different packaging materials. Ph.D. thesis, Graduate School of Nuclear and Allied Sciences, University of Ghana, Legon. 2015.
5. Kortei NK, Wiafe-Kwagyan M. Comparative appraisal of the total phenolic content, flavonoids, free radical scavenging activity and nutritional qualities of *Pleurotus ostreatus* (EM-1) and *Pleurotus eous* (P-31) cultivated on rice (*Oryzae sativa*) straw in Ghana. J Adv Biol Biotechnol. 2015;3(4):153–64.

6. Obodai M, Cleland-Okine J, Vowotor KA. Comparative study on the growth and yield of *Pleurotus ostreatus* mushroom on different lignocellulosic by-products. J Ind Microbiol Biotechnol. 2003;30:146–9.

7. Raymond P, Mshandete AM, Kivaisi AM. Cultivation of oyster mushroom (*Pleurotus* HK-37) on sisal waste fractions supplemented with cow dung manure. J Biol Life Sci. 2013;4(1):274–86.

8. Stamets P. Growing gourmet and medicinal mushrooms. Berkley: Ten Speed Press; 1993. p. 552.

9. Balasubramanya RH, Kathe AA. An inexpensive pre-treatment of cellulosic materials for growing edible oyster mushrooms. Biol Resour Technol. 1996;57:303–5.

10. Apetorgbor MM, Apetorgbor AK, Nutakor E. Utilization and cultivation of edible mushrooms for rural livelihood in southern Ghana. 17th Commonwealth Forestry Conference. Sri-Lanka. 2005.

11. Kwon H, Sik Kim B. mushroom growers' handbook: shiitake cultivation. Mushworld: Seoul; 2004. p. 260.

12. Oseni TO, Dlamini SO, Earnshaw DM, Masarirambi MT. Effect of substrate pre-treatment methods on oyster mushroom (*Pleurotus ostreatus*) production. Int J Agric Biol. 2012;14(2):251–5.

13. Contreras EP, Sokolov M, Mejia G, Sanchez J. Soaking of substrate in alkaline water as a pretreatment for the cultivation of *Pleurotus ostreatus*. J Hortic Sci Biotechnol. 2004;79(2):234–40.

14. Royse DJ. Cultivation of shiitake on natural and synthetic logs. 2009. p. 2. http://www.americanmushroom.org/wp-content/uploads/2014/05/Shiitake_how_to_grow_PSU.pdf. Accessed 22 Sept 2017.

15. Gbedemah C, Obodai M, Sawyer LC. Preliminary investigations into the bioconversion of gamma irradiated agricultural wastes by *Pleurotus* spp. Radiat Phys Chem. 1998;52(6):379–82.

16. Kortei NK, Odamtten GT, Obodai M, Appiah V, Annan SNY, Acquah SA, Armah JO. Comparative effect of gamma irradiated and steam sterilized composted 'wawa' (*Triplochiton scleroxylon*) sawdust on the growth and yield of *Pleurotus ostreatus* (Jacq.Ex.Fr.) Kummer. Innov Rom Food Biotechnol. 2014;14:69–78.

17. Kortei NK, Wiafe-Kwagyan M. Evaluating the effect of gamma irradiation on eight different agro-lignocellulose waste materials for the production of oyster mushrooms (*Pleurotus eous* (Berk.) Sacc. Strain P-31). Croat J Food Biotechnol Nutr. 2014;9(3-4):83–90.

18. Mondal SR, Rehana MJ, Noman SM, Adhikary SK. Comparative study on growth and yield performance evaluation of oyster mushroom (*Pleurotus florida*) on different substrates. J Bangladesh Agric Univ. 2010;8:213–20.

19. Buswell JA. Potentials of spent mushroom substrates for bioremediation purposes. Compost. 1984;2:31–5.

20. Amin RSM, Alam N, Sarker NC, Hossain K, Uddin MN. Influence of different amount of rice straw per packet and rate of inocula on the growth and yield of oyster mushroom (*Pleurotus ostreatus*). Bangladesh J Mushroom. 2008;2:15–20.

21. Tisdale TE, Susan C, Miyasaka-Hemmes DE. Cultivation of the oyster mushroom (*Pleurotus ostreatus*) on wood substrates in Hawaii. World J Microbiol Biotechnol. 2006;22:201–6.

22. Morais MH, Ramos AC, Matos N, Santos-Oliveira EJ. Production of shiitake mushroom (*Lentinus edodes*) on ligninocellulosic residues. Food Sci Technol Int. 2000;6:123–8.

23. Chukwurah NF, Eze SC, Chiejina NV, Onyeonagu CC, Okezie CEA, Ugwuoke KI, Ugwu FSO, Aruah CB, Akobueze EU, Nkwonta CG. Correlation of stipe length, pileus width and stipe girth of oyster mushroom (*Pleurotus ostreatus*) grown in different farm substrates. J Agric Biotechnol Sustain Dev. 2013;5(3):54–60.

24. Kortei NK. Determination of optimal growth and yield parameters of *Pleurotus ostreatus* grown on composted cassava peel based formulations. M.sc.Thesis, Kwame Nkrumah University of Science and Technology, Kumasi, Ghana. 2008.

25. Sarker NC, Hossain MM, Sultana N, Mian IH, Karim AJMS, Amin SMR. Performance of different substrates on the growth and yield of *Pleurotus ostreatus* (Jacquin ex Fr.) Kummer. Bangladesh J Mushroom. 2007;1:9–20.

26. Fan L, Pandey A, Mohan R, Soccol CR. Comparison of coffee industry residues for production of *Pleurotus ostreatus* in solid state fermentation. Acta Biotechnol. 2000;20(1):41–52.

27. Garo G, Girma G. Responses of oyster mushroom (*Pleurotus ostreatus*) as influenced by substrate difference in Gamo Gofa zone, Southern Ethiopia. IOSR J Agric Vet Sci. 2016;9(4):63–8.

28. Gitte VK, Priya J, Kotgire G. Selection of different substrate for the cultivation of milky mushroom (*Calocybe indica* P&C). Indian J Tradit Knowl. 2014;13(2):434–6.

29. Ajonina AS, Tatah LE. Growth performance and yield of oyster mushroom (*Pleurotus ostreatus*) on different substrates composition in Buea South West Cameroon. Sci J Biochem. 2012. https://doi.org/10.7237/sjbch/139.

30. Ahmed M, Abdullah N, Ahmed KU, Borhannuddin Bhuyan MHM. Yield and nutritional composition of oyster mushroom strains newly introduced in Bangladesh. Pesqui Agropecu Bras. 2013;48(2):197–202.

31. Onyango BO, Palapala VA, Arama PF, Wagai SO, Gichimu BM. Suitability of selected supplemented substrates for cultivation of Kenyan native wood ear mushrooms (*Auricularia auricula*). Am J Food Technol. 2011;6:395–403.

32. Shen Q, Royse D. Effect of nutrient supplement on biological efficiency, quality and crop cycle time on maitake (*Griofola frondosa*). Appl Microbiol Biotechnol. 2001;57:74–8.

## 11

Promoting sustainable agriculture in Africa through ecosystem-based farm management practices: evidence from Ghana

Caesar Agula[1]* [ID], Mamudu Abunga Akudugu[2], Saa Dittoh[3] and Franklin Nantui Mabe[1]

Abstract

Background: The type of farming practices employed within an agro-ecosystem have some effects on its health and sustainable agricultural production. Thus, it is important to encourage farmers to make use of ecosystem-friendly farming practices if agricultural production is to be sustainable and this requires the identification of the critical success factors. This paper therefore examined the factors to consider in promoting sustainable agriculture production in Africa through ecosystem-based farm management practices (EBFMPs) using Ghana as a case study. The study employed mixed methods—qualitative and quantitative techniques. Data were collected through key informant interviews, focus group discussions and a semi-structured questionnaire administered to 300 households. The Poisson and negative binomial models were employed to determine the factors that influence farmers' intensity of adoption of EBFMPs. Eight (8) EBFMPs were used in the paper as the dependent variable, which are organic manure application, conservation of vegetation, conservative tillage, mulching, crop rotation, intercropping with legumes, efficient drainage system and soil bunding.

Results: The paper found that the intensity of adoption of EBFMPs is significantly determined by the age of farmers, distance to farms, perception of soil fertility, knowledge of EBFMPs, number of extension visits and the type of irrigation scheme available to farmers.

Conclusions: To promote sustainable agricultural production in Ghana and elsewhere in Africa using EBFMBs, these factors must be considered.

Keywords: Africa, Agriculture, Ecosystems, Farm management practices, Ghana

Background

Agriculture production contributes to sustaining the livelihoods of many households, particularly in Africa. Despite the important roles it plays, some of the modern farming practices adopted by most farmers pose a threat to the environment [1, 2]; sustainable agricultural production [3] and the health and functional capacity of the agro-ecosystems [4]. In other words, unfriendly ecosystem farming practices create a condition that makes agricultural production costly and this traps future generations in the vicious poverty cycle [5] and the rural poor are the most disadvantaged. It is for this reason that sustaining the fertility of farmlands and maintaining ecosystems resilience has been of interest to many programmes and policies including the Comprehensive Africa Agriculture Development Programme (CAADP) and ECOWAS Agricultural Policy (ECOWAP) [6, 7]. Unfortunately, the response to these policies and programmes, especially in Sub-Saharan Africa has been low [8].

Most interventions on crop production in Ghana and elsewhere in Africa place greater emphasis on high yields with little concern on how to sustain farmlands for future benefits. For example, the focus of the Ministry of Food and Agriculture (MoFA) in Ghana has been on improving yields through dissemination of yield enhancing

*Correspondence: caesaragula@yahoo.com
[1] Department of Agricultural and Resource Economics, University for Development Studies, Tamale, Ghana
Full list of author information is available at the end of the article

technologies [8]. These yield enhancing technologies seek to improve food availability [9], which often derail the biological functioning of the agro-ecosystems [10]. Again, a lot of studies (e.g. [11, 12]) that have been carried out in Ghana on the adoption of sustainable farm practices have paid little attention to farmers' knowledge of indigenous sustainable farm practices and how this might affect farmers' intensity of adoption.

Meanwhile, sustainable farm practices termed as eco-system-based farm management practices (EBFMPs) can help maintain the fertility of agricultural lands and balance nutrients requirement of crops [3]. Ecosystem-based management farm practices (EBMFPs) within the context of this paper is the traditional farm-based practices (such as mulching, compost application, crop rotation, efficient drainage systems, and vegetation conservation among others) that aim at balancing agricultural output and maintaining agro-ecosystems resilience. According to [4], EBMFPs averagely conserve and boost the functional capacity of the ecosystems services through natural and biological means as well as intensive, high inputs systems.

Considering the varied functions that EBFMPs play, this paper sought to examine the factors that must be considered in promoting the adoption of EBFMPs by farmers. The paper thus provides evidence to policy formulators and implementers on some of the factors that enhance or inhibit the adoption of EBMFPs by farmers for sustainable agricultural production.

Methods
The study setting and sampling process
The study was conducted in the Upper East Region of Ghana. Ghana has varied agro-ecological zones which include the Rain Forest Zone, Coastal Savannah Zone, Semi-deciduous Forest Zone, Transitional Zone, Guinea-Savannah Zone and Sudan-Savannah Zone. The varied nature of the agro-ecological zones in Ghana makes her fairly representative of Africa, which has similar agro-ecological zones. The selection of Upper East Region was due to the fragile nature of its ecosystem that makes the need for EBFMPs to ensure sustainability in agricultural production imperative. Specifically, the study was conducted in two districts (Kassena-Nankana West District and Kassena-Nankana East District). The study districts (Fig. 1) fall within the Sudan-Savannah Vegetation Zone and has a total population of about 181,000 with about 61% from the Kassena-Nankana East District and 39% from the Kassena-Nankana West District [13]. The predominant economic activity in the area is farming with about 69% of the total population in agriculture [14].

A three-stage sampling technique was used to select study communities and households. In the first stage of the sampling, because of the critical role of irrigation in ensuring sustainable agricultural production, communities in the districts were divided into strata of community-managed and government-managed irrigation schemes of which three (3) communities each were randomly selected (Fig. 2).

In the second stage, a simple random sampling technique was used to select the required number of irrigated households from each community. According to [16], for any meaningful and more precise comparisons to be made, then a constant sample from each group, in this case community is critical. From a sample frame of 1813 households, 300 households (about 17% of the sample frame) were randomly selected for the study with each community having fifty (50) households as shown in Fig. 2. The 50 households from each community was more than 20% of the total number of households from each community, and thus representative of the communities from the view point of [16].

Theoretical and empirical review of models on sustainable farm practices
In social sciences, most studies usually deal with outcomes that are measured in counts such as number of soil conservative management practices, number of Integrated Pest Management (IPM) practices adopted, number of children as an indicator of fertility, and number of doctor visits as an indicator of health care demand among others [17]. Such studies are traditionally analysed with econometric models such as the binomial Probit or Logit models, which usually divide the dependent variable into two categories (1 = full adoption, 0 = no adoption at all) [18]. However, this might not be the true picture in most cases since technologies have different components, which could either be fully or partially adopted and binary choice models (e.g. Probit or Logit) cannot properly capture such situations. Thus, the Poisson regression or negative binomial regression models have been developed to handle such situations [18]. The count models (Poisson and negative binomial models) have the capacity to estimate the effect of a policy intervention either on the average rate or the probability of no event, a single event, or multiple events [17].

The Poisson model assumes that the response variable Y has a Poisson distribution and the logarithm of its expected value can be modeled by a linear combination of unknown parameters [19]. From [20], the model looks at the probability that the dependent variable Y (in this case the number of EBFMPs used) will be equal to a certain number y, and is represented mathematically as follows:

$$\text{Prob}(Y = y) = \frac{e^{\lambda} \lambda^{y}}{y!}, \quad y = 0, 1, 2, 3 \ldots n \qquad (1)$$

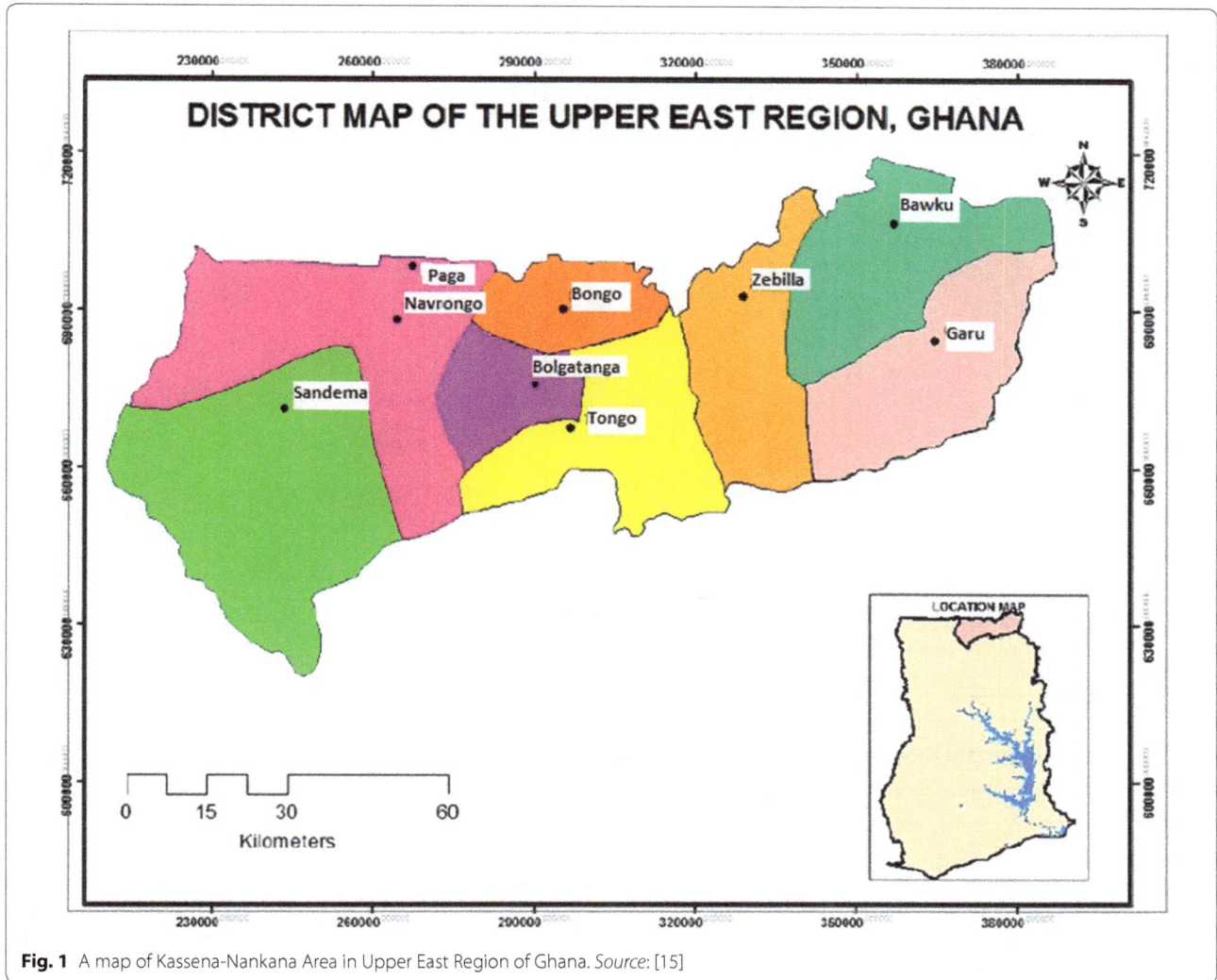

Fig. 1 A map of Kassena-Nankana Area in Upper East Region of Ghana. *Source*: [15]

where $\lambda =$ is the intensity or rate parameter, $\lambda = \exp\left(X_i'\beta\right)$, $\beta =$ a vector of unknown parameters to be estimated.

The intensity parameter (λ) is assumed to be log-linearly related to the explanatory variables [8]. This is because the parameter (λ) is expressed as an exponential function of the explanatory variables. From the Poisson distribution assumption, the intensity of y is determined by the mean. This suggests that the intensity of adoption of EBFMPs is determined by the mean.

The log-likelihood function is given by the equation:

$$\ln L = \sum_{i=1,2,...n} \left[-\lambda + y_i\beta' - \ln y_i!\right] \quad (2)$$

The interpretation of the coefficient is that, one unit increase in X_i will increase or decrease the average number of Y_i by the coefficient expressed as a percentage [20].

The marginal effect of a variable on the average number of events is:

$$\partial E(y_i/x_i)/\partial x_j = \beta_j \exp\left(X_i'\beta\right) \quad (3)$$

The interpretation of marginal effect is that one unit increase in X_i will increase/decrease the average number of the dependent variable by the marginal effect [20].

The key assumption is that the Poisson model has equi-dispersion property of the Poisson distribution. That is the equality of the mean and the variance specified as:

$$E\left(y/x\right) = Var(y/x) = \lambda \quad (4)$$

This property is much restrictive and often fails to hold in practice if there is 'over dispersion' in the data. This is common in developing countries like Ghana where farmers tend to recall agricultural information with a lot of discrepancies. According to [21], the Poisson model

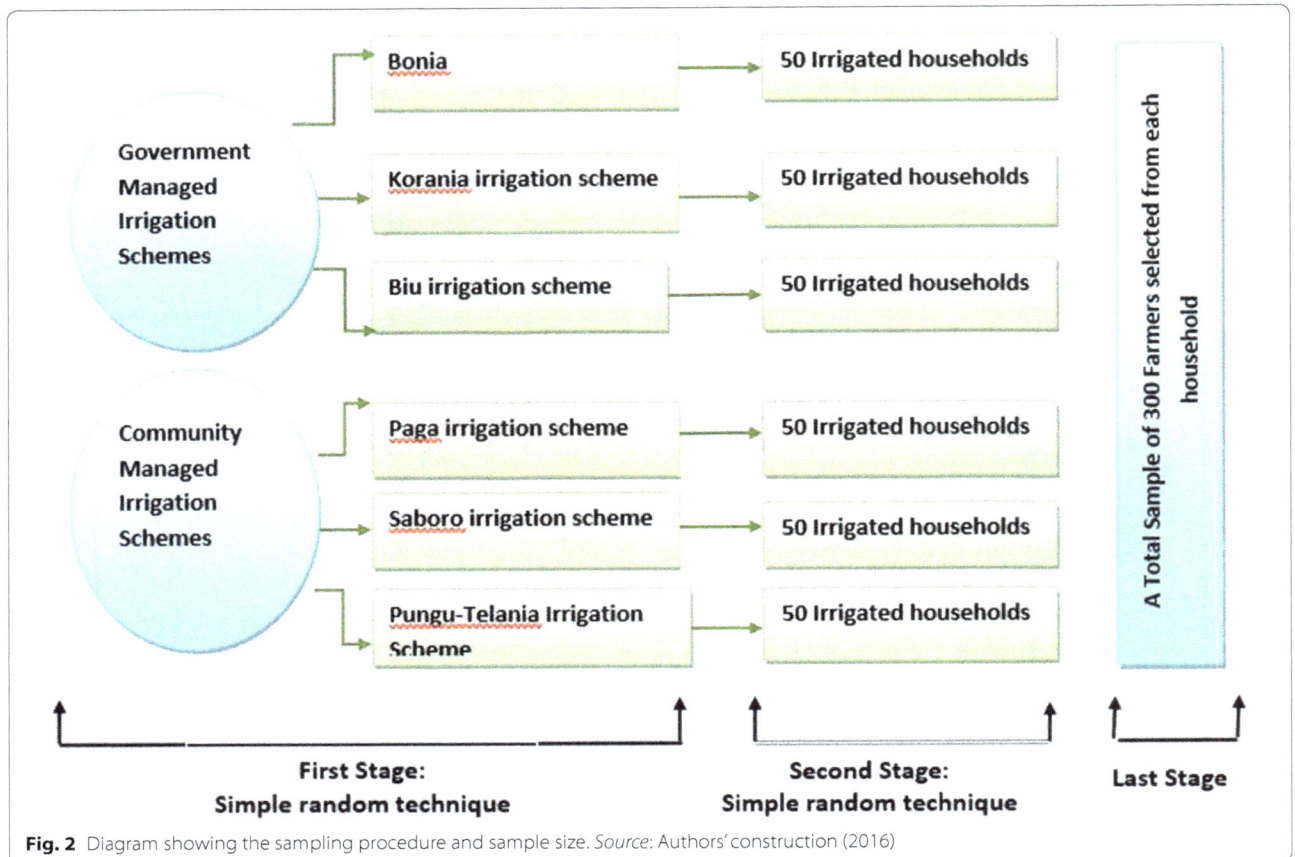

Fig. 2 Diagram showing the sampling procedure and sample size. *Source*: Authors' construction (2016)

relies heavily on an assumption that the conditional mean of outcome is equal to the conditional variance. But in practice, the conditional variance often exceeds the conditional mean. The negative binomial regression model however, deals with this problem by allowing the variance to exceed the mean [21]. Unlike the Poisson model, the negative binomial model (NBM) has a less restrictive property that the variance is not equal to the mean (μ) [22]. This is represented mathematically as follows:

$$Var(y/x) = \lambda + \alpha\lambda^2 \qquad (5)$$

The negative binomial model also estimates the over-dispersion parameter α. Therefore, there is the need to test for over-dispersion. To test for the over-dispersion, the negative binomial model (NBM) which includes the over-dispersion parameter α is estimated and tested to see if α is significantly different from zero [21]. When $\alpha = 0$, it comes back to the Poisson model estimates. When $\alpha > 0$; there is over-dispersion (which frequently holds with real data). When $\alpha < 0$; there is under-dispersion (which is not very common).

These two models (Poisson and negative binomial regression models) have shown to be very simple for analysing

count data and straightforward in interpretation. As a result, they are gaining greater usage by many researchers on current studies involving count data [17]. Thus, there are a number of current studies (e.g. [8, 12, 23]) on the adoption of sustainable practices that used count models.

The study by [18] was one of the first to explore the use of Poisson count regression models to analyse technology adoption. It was used to evaluate three technology transfer projects in Central America: Integrated Pest Management in Costa Rica, Agro-forestry systems in Panama, and Soil Conservation in El Salvador. However, the study by [18] has direct connection with this paper, which examined the factors to consider in promoting sustainable agriculture. Another study that employed one of the count regression models is [23]. Following [23], the adoption behaviour of farm households on farm management practices in three agro-biodiversity hotspots in India were investigated using the negative binomial count data regression model. The regression outcome revealed that farmers who received agricultural extension are more likely to use improved farm management practices. It also showed a negative relationship between cultivation of local varieties and adoption of farm management practices.

Again, in the work of [24], the Poison regression model was used to analyse the impact of farmers' experiences and perceptions of health risks of pesticides on the adoption of Integrated Pest Management (IPM) and pesticide use among small scale vegetable farmers in Nicaragua. Using the Poisson model, the authors were able to consider two levels of adoption process in that study (1) the count of IPM practices tested and (2) the count of practices actually used. The results revealed that previous experience with pesticide poisoning incidents has significant positive effect on the number of IPM practices tested by a farmer, but not on the adoption. Other factors, which showed significance, include school education, characteristics of cropping system, whether or not farmers had attended training in IPM and farmers who pay wage premiums to workers for application of pesticides.

In Ghana, the use of the Poisson and negative binomial regression models is equally gaining prominence. Classical examples include that of [8] and [12]. Nkegbe and Shankar [12] employed the Poisson model in the study to analyse the intensity of adoption of the sustainable soil and water conservation practices-composting, cover crops, agro-forestry, grass strip, soil bund and stone bund. The Gamma count was also used to further correct for over-dispersion in the data. From the empirical results of that study, access to information, social capital, per capita landholding and wealth play a crucial role in determining farmers' decision to intensively adopt sustainable soil and water conservation practices.

Again, [8] also closely tied with that of [12] except that the former had a broader scope, as it went beyond the factors that determine the adoption of the sustainable farming practices to consider the factor productivity. The study equally employed the Poisson model coupled with the stochastic frontier. From the study, credit, farm size, group membership and proximity to input sale points positively influence the adoption of conservation techniques. The covariates included gender, age, age square, education, farm size, household size, group membership, number of extension visits, credit obtained by the farmer and distance to input stores. The limitation of this study is on its inability to test for over-dispersion for the necessary corrections.

From the foregoing, the Poisson and the negative binomial regressions models are considered appropriate for this paper. It can also be deduced that all the above-mentioned studies have failed to consider farmers' knowledge of the ecosystem services as one of the factors that can influence their adoption of sustainable farm practices or EBFMPs. This paper therefore contributes to adoption studies literature on agro-ecosystems with a blend of indigenous farming practices (ecosystem-based farm management practices) knowledge and how it affects farmers' intensity of using the practices.

Empirical model specifications

To determine the factors that influence the adoption of EBFMPs, data were collected on the farm practices employed by each farmer in irrigation and rain-fed farming. These practices were then grouped into EBFMPs and non-EBFMPs. The total number of EBFMPs adopted by farmers in irrigation was then used as the dependent variable. The Poisson and negative binomial models were employed to examine the factors that influence the number of EBFMPs adopted by farmers. Below are the empirical models for the study and descriptions of the variables in Table 1:

Empirical model for community managed irrigation schemes (CIS)

$$Logy_{ci} = \beta_{c0} + \beta_{c1}Age_{ci} + \beta_{c2}sex_{ci}$$
$$+ \beta_{c3}Educ\_d._{ci} + \beta_{c4}Ext.serv._{ci}$$
$$+ \beta_{c5}Fm.distance_{ci} + \beta_{c6}Soil.perceptn_{ci} \quad (6)$$
$$+ \beta_{c7}Fsize_{ci} + \beta_{c8}Knw.EBFMP_{ci} + \varepsilon_{ci}$$

Empirical model for government managed irrigation scheme (GIS)

Table 1 Definition of variables and *apriori* expectations for adoption models

Variable	Variable definition	Units of measurement	Expected sign
Y	EBFMPs	Number of EBFMPs used	
Age	Age	Years	±
Sex	Sex	Dummy (1 = female, 0 = male)	±
Educ_d	Education	Dummy (1 = had formal education (at least JSS/JHS education), 0 = below JSS/JHS)	+
Ext.serv_d	Extension services	Dummy (1 = received at least 2 extension services last season, 0 = received 1 or no extension service)	+
Fm.distance.irr	Distance to irrigated farm	Kilometers	–
Soil.perceptn	Perception of soil fertility	Dummy (1 = fertile, 0 = not fertile)	–
Fsize.irr	Irrigable farm size	Acres	–
Knw.EBFMP	Perceived Knowledge of EBFMPs	Indexed on each EBFMP importance stated	+
Irrig_type	Category of irrigation	Dummy (1 = CIS, 0 = GIS)	+

Source: Authors' construction, 2016

$$Logy_{gi} = \beta_{g0} + \beta_{g1}Age_{gi} + \beta_{g2}sex_{gi}$$
$$+ \beta_{g3}Educ\_d._{gi} + \beta_{g4}Ext.serv._{gi}$$
$$+ \beta_{g5}Fm.distance_{gi} + \beta_{g6}Soil.perceptn_{gi} \quad (7)$$
$$+ \beta_{g7}Fsize_{gi} + \beta_{g8}Knw.EBFMP_{gi} + \varepsilon_{gi}$$

Empirical model for both government and community managed irrigation schemes

$$Logy_{cgi} = \beta_{cg0} + \beta_{cg1}Age_{cgi} + \beta_{cg2}sex_{cgi}$$
$$+ \beta_{cg3}Educ\_d._{cgi} + \beta_{cg4}Ext.visits._{cgi}$$
$$+ \beta_{cg5}Fm.distance.irr_{cgi} + \beta_{cg6}Soil.perceptn_{cgi}$$
$$+ \beta_{cg7}Fsize.irr_{cgi} + \beta_{cg8}Knw.EBFMP_{cgi}$$
$$+ \beta_{cg9}Irig_{cgi} + \varepsilon_{cgi} \quad (8)$$

Results and discussion

Socio-demographic characteristics of farmers

The survey found that farming is dominated by males in Ghana and likewise other parts of Africa in a broad scope (Table 2). This development emanates from the cultural and social setting of the people of Ghana and other countries in Africa, where resources (particularly productive agricultural lands) are controlled and owned by men. Until recently, farming was culturally seen as a male dominated economic activity in many parts of Africa while women were basically in-charge of sales of farm produce and other petty trading. It was revealed from the focus group discussions that agriculture is still labour intensive, which constraint women who are already preoccupied with domestic chores to engage themselves in it. Table 3 also shows that the average age of farmers is about 42 years with a standard deviation of 11 years. This suggests that averagely, the farmers in Ghana fall within the productive age cohort. Irrigation farming in several regions of Africa (e.g. Sub-Sahara Africa) has become an attractive force for most youth to engage in agriculture. The reason being that farm produce from irrigation (such as pepper, onions, tomatoes, rice among others) offer good prices relative to produce from rain-fed agriculture. Again, most of the agricultural lands owned by government-managed irrigation schemes (GIS) are operated as an open access system where the youth have an equal chance of securing lands for farming.

From Table 2, majority of the farmers had no formal education or had only basic education. The respondents' level of education shows that approximately 34% had at least Junior High School (JHS) education from the pooled data. This implies that only a few of the farmers might be able to read and understand new agricultural technologies and interventions. Like the agricultural sector in Ghana, agriculture is yet to acquire the needed level of investment in other parts of Africa, which can

Table 2 Summary statistics of categorical variables

Variables	Percentages		
	CIS	GIS	Pooled
Sex			
Females	42.00	16.00	29.00
Males	58.00	84.00	71.00
Marital status			
Married	58.00	72.67	65.33
Otherwise (single, separated and widowed)	42.00	27.33	34.67
Household head			
Yes	69.33	72.67	71.00
No	30.67	27.33	29.00
Perception of soil fertility			
Fertile	44.67	17.33	31.00
Otherwise	55.33	82.67	69.00
Education			
Had formal education (JHS education and above)	31.33	37.33	34.33
Below Junior High School (JHS) education	68.67	62.67	65.67
Extension services			
Received (at least two in the past season)	60.00	40.67	50.33
Otherwise	40.00	59.33	49.67
	N = 150	N = 150	N = 300

Source: Field survey, 2016

attract graduates from the tertiary level. As such, it is characterised by farmers with greater weakness in reading and understanding new agricultural interventions or programmes. This tend to affect farmers understanding of the nexus between new agricultural interventions and agro-ecosystems sustainability, hence they adopt practices that are not ecosystem-friendly. The survey also revealed that the mean household size of the respondents is about 6 with a standard deviation of 2 (Table 3). This means that averagely households have large potential labour force to help in farming activities. It can be observed in Table 2 that about 65% of the respondents are married while 35% otherwise (single, separated and widowed). Table 2 also shows that 71% of the respondents are household heads while 29% are not. Some household heads lost their spouses and some are staying with their children alone because of broken homes. Details of the statistics for the socio-demographic characteristics of farmers are shown in Tables 2 and 3.

Factors influencing ecosystem-based farm management practices adoption

The paper sought to determine the factors to be considered in promoting the use of ecosystem based farm

Table 3 Summary statistics of continuous variables

Variables	CIS		GIS		Pooled	
	Mean	SD	Mean	SD	Mean	SD
Age	45.19	11.10	38.07	10.01	41.63	11.14
Household size	6.17	2.70	5.39	1.93	5.78	2.37
Farm size for irrigation (acres)	0.61	0.39	1.64	1.09	1.12	0.97
Irrigated farm distance (km)	0.95	0.59	1.53	0.94	1.24	0.84
Knowledge of EBFMPs (indexed)	16.99	3.86	15.39	3.25	16.19	3.65

Source: Field survey, 2016

management practices (EBFMPs) for sustainable agriculture in Africa. The paper focused on the adoption of EBFMPs by farmers in Ghana who are into irrigation farming because it represents the hope for the future under the current trends in climate change and variability. Besides, irrigation farming was targeted because of the critical role it plays in ensuring sustainable production in agriculture and the effect of it on the various ecosystems within a landscape. Ghana was used as a case study for Africa because of the varied nature of the country's agro-ecological zones (six types of agro-ecological zones), making her fairly representative of the continent. The paper used eight (8) EBFMPs for its analysis, which are organic manure application, conservation of vegetation, conservative tillage, mulching, crop rotation, intercropping with legumes, efficient drainage system and soil bunding (Table 4). These practices formed the basis as the dependent variable for the analysis with the Poisson and negative binomial models.

The results (Table 5) indicate that there is no over-dispersion since the test for alpha is not statistically different from zero. As such, there is sufficient evidence that the conditional mean is equal to the conditional variance and hence, the negative binomial model reduces back to the Poisson model (check "Appendix" for Poisson estimates). Even though the negative binomial regression Pseudo R^2

Table 4 Distribution of EBFMPs adopted by farmers

EBFMPs adopted	Percentages		
	CIS	GIS	Pooled
Organic manure or compost application	72.00	46.67	59.33
Conservation of vegetation	76.67	52.67	64.67
Conservative tillage	81.33	36.00	58.67
Mulching	60.00	24.00	42.00
Crop rotation	28.67	38.00	33.33
Intercropping with legumes	46.00	28.67	37.33
Efficient drainage systems	47.33	22.67	35.00
Soil bunding	18.00	36.67	27.33
	$N = 150$	$N = 150$	$N = 300$

for the pooled data is low (about 10%), the overall significance of the model is high as indicated by the likelihood ratio Chi-square (significant at 1%). This implies that farmers' intensity of adoption of EBFMPs is determined by the set of covariates modelled in this paper. The regression results showed that farmers' age, distance to farm, perception of soil fertility, knowledge of EBFMPs, extension visits and the type of irrigation scheme the farmer cultivates significantly influence the adoption of EBFMPs.

The results from the pooled data (Table 5) indicate that age of a farmer influence the adoption of EBFMPs in farming. Specifically, the results show that as farmers' age increases by 1 year, the intensity of adopting EBFMPs on farms increases and this is statistically significant at 5%. Generally, the finding suggests that old people in farming within the Ghanaian society and extensively in some parts of the African continent adopt more sustainable practices (or EBFMPs) than younger ones. Most aged farmers are still traditional with regards to agriculture production and as such, used more of the EBFMPs because they are indigenous practices learnt from forefathers. Even though, most of them cannot explain the biological functioning of the indigenous practices (which are mostly EBFMPs), they acknowledged the importance of these practices in minimising cost of production and sustaining soil fertility. This finding is however contrary to the finding of [12] which reported that the age of farmers do not influence adoption of soil and water conservation practices in northern Ghana.

The services that farmers receive from extension officers specifically, smallholder farmers under community-managed irrigation schemes (CIS) have an influence on the level at which they adopt EBFMPs. From the marginal effect regression on farmers under CIS, it suggests that those who received extension education in the previous season have greater intensity of using EBMFPs than those who had no extension education (Table 5) and this is statistically significant at 10%. Agricultural extension agents in Ghana and other parts of Africa provide information and education on agricultural production, especially new interventions. The education enlightens farmers on the

Table 5 Coefficient estimates for factors that influence EBMFPs adoption

Variables	Estimates of negative binomial model (NBM)					
	CIS		GIS		Pooled	
	Coeff.	SE	Coeff.	SE	Coeff.	SE
Constant	1.016	0.296	0.044	0.318	0.498	0.213
Age	0.002	0.004	0.011	0.005**	0.006	0.003**
Sex	0.016	0.085	− 0.022	0.132	− 0.000	0.070
Educ_d	0.040	0.091	− 0.025	0.105	0.025	0.067
Ext_visits_d	0.160	0.094*	0.048	0.102	0.095	0.067
Fm_dist. (km)	− 0.234	0.083***	− 0.103	0.055*	− 0.146	0.046***
Fm_size (Acres)	0.040	0.106	0.005	0.046	0.005	0.042
Soil_perception	0.159	0.089*	0.239	0.123*	0.186	0.070***
Knw_EBFMPs	0.016	0.011	0.040	0.015***	0.026	0.009***
Irrig_type					0.161	0.081**
Number of observations	150		150		300	
LR chi^2(9)	42.230***		27.160***		106.860***	
Prob > chi^2	0.000		0.000		0.000	
Pseudo R^2	0.076		0.053		0.096	
Dispersion = mean						
Log likelihood	− 255.107		− 242.655		− 501.122	
Alpha	0.000		0.000		0.000	
Chibar^2(01)	0.000		0.000		0.000	
Prob ≥ chibar^2	1.000		1.000		1.000	

Source: Field survey, 2016

*, **, ***Represent 10, 5 and 1% levels of significance respectively

choice of activities at farm level and help them better understand the side effects of the practices they employ on farms. The significance and direction (positive) of the number of extension contacts are consistent with the finding of [12].

Farmers also consider the distance of their farms from places of abode in their adoption of sustainable farming practices or EBFMPs. Thus, distance to farms was found to have a negative influence on the number of EBFMPs that farmers adopt and this is statistically significant at 1, 10 and 1% for CIS, GIS and pooled models respectively. In other words, overall, when the distance to farms increases by 1 km, the intensity of adopting EBFMPs reduces in all the models. One of the major problems in terms of distance is that most farmers usually find it difficult to transport organic manure (one of the EBFMPs identified) from family compounds to farm sites. As such, only few farmers can apply organic manure on farms that are far from places of abode. It was also revealed from the focus group discussions that farms that are at the outskirts of communities or in the forests zones are usually very fertile and require little or no organic manure for greater yields. Such farms also have dense vegetation, which most farmers usually clear for farming activities.

Another factor that determines the intensity of adoption of EBFMPs is farmers' perception of soil fertility. From Table 5, perception of soil fertility is statistically significant at 10% in both CIS and GIS models. It is highly significant at 1% in the pooled model. In all the three models, it has positive effect on the intensity of adoption of EBFMPs. Farmers who perceived their farm plots to be fertile have a greater expected intensity of using EBFMPs than those who perceived their farm plots are infertile, all other things being equal. This finding is inconsistent with that of [9] who reported that farming on better soils decreases the adoption of soil improving practices. The reason given by farmers to support this finding is that those who perceived their soil fertility is low rather resort to the use of more inorganic measures to improve their soil fertility instead of the indigenous ecosystem friendly practices. Again, farmers who perceived that their soil fertility is high try to save cost by adopting organic practices to maintain the fertility of the soil. Another reason that accounts for this finding is that, farmers especially those under government-managed irrigation schemes think their soils are degraded to a non-responsive level for organic manure application. Thus, they rely on the usage of inorganic manure to improve their soils since it works faster than the organic manure.

Knowledge of farmers on the usefulness of EBFMPs affect their level of adoption of such EBFMPs (Table 5). Farmers who have more insights on the biological functions and benefits of ecosystem-based farm management practices tend to adopt more compared to those without adequate knowledge on the usefulness of EBFMPs. The result indicates that as farmers' knowledge on EBFMPs improves, the intensity of adopting EBFMPs increases and this was found to be statistically significant at 1% in both the GIS and the pooled models. Most farmers in Ghana and elsewhere in Africa, especially young and uneducated farmers focus more on yields at the expense of sustainability and this does not make them adopt EBFMPs. From the focus group discussions, most of the farmers attributed the current prevalence of strange pests and diseases in agriculture to the failure of this generation and the previous ones in maintaining some of the indigenous agricultural practices that could sustain the resilience of the agroecosystems (Table 6).

Lastly, the type of irrigation scheme or facility available to farmers influence the adoption of EBFMPs. The results in Table 5 indicate that farmers who cultivate under the community-managed irrigation schemes (CIS) have a greater intensity of adoption of EBFMPs than those under the government-managed irrigation schemes (GIS), *ceteris paribus* and the difference is statistically significant at 5%. Even though farmers in the CIS aim at maximizing yield as per their counterparts in the GIS, they are more conscious about the sustainability of their fields. This is probably because unlike the GIS where the land is publicly owned, farmers producing on CIS own the land upon which production takes place and hence have primary interest of maintaining the fertility of the farmlands even for future generations.

Conclusions and policy recommendations

The study sought to examine the factors that promote the adoption of ecosystem-based farm management practices (EBFMPs) in Africa, taking farmers in Ghana as a case study. Ghana became an ideal place because of its varied nature of agro-ecological zones. The agro-ecological zones in Ghana are into six (6) types and fairly representative of the agro-ecological zones in Africa. The Poisson and negative binomial models were employed for the analyses. The paper found that the intensity of EBFMPs adoption is significantly determined by age of farmers, distance to farms, perception of soil fertility, knowledge of EBFMPs, number of extension visits and the type of irrigation scheme available to farmers. Based on the results, it is concluded that to promote the use of EBFMPs in Ghana and other parts of Africa, it is important to focus on these factors. In other words, a focus on these factors is needed to bring about a shift from the current production system that relies heavily on intensive use of agrochemicals with negative consequences on ecosystem resilience and sustainability to a production system that is more ecosystem friendly using EBFMPs. It is therefore recommended that policy makers and implementers in Ghana and Africa generally come out with interventions that are generation specific (i.e. for the old and the young), distance neutral (i.e. not affected by distance to farmer residence), knowledge sensitive (i.e. literate and illiterate farmers) and production context specific (i.e. irrigation versus rain-fed; smallholder versus medium to large scale farmers). In all this, there is the need for policies that aim at building and sustaining robust agricultural extension systems that have at the centre ecosystem resilience and sustainability. Specifically, there is the need to review agricultural extension

Table 6 Marginal effects for factors that influence EBFMPs adoption

Variables	NBM's marginal effects					
	CIS		GIS		Pooled	
	dy/dx	SE	dy/dx	SE	dy/dx	SE
Age	0.012	0.016	0.032	0.013**	0.023	0.010**
Sex	0.070	0.357	− 0.061	0.362	− 0.001	0.238
Educ_d	0.169	0.386	− 0.069	0.290	0.088	0.232
Ext_visits_d	0.656	0.380*	0.135	0.285	0.326	0.228
Fm_dist. (km)	− 0.974	0.345***	− 0.286	0.152*	− 0.499	0.158***
Fm_size (Acres)	0.167	0.441	0.014	0.130	0.019	0.145
Soil_perception	0.669	0.378*	0.716	0.400*	0.658	0.258**
Knw_EBFMPs	0.070	0.049	0.112	0.042***	0.089	0.031***
Irrig_type					0.548	0.279**

Source: Field survey, 2016

*, **, ***Represent 10, 5 and 1% levels of significance respectively

policies to refocus them on how to expand agricultural production without compromising the biological functioning of the agro-ecosystems. Participatory approaches should be employed in the formulation and implementation of such policies to ensure community acceptability and ownership which will guarantee sustainability.

Abbreviations

CAADP: Comprehensive Africa Agriculture Development Programme; CGIAR: Consultative Group on International Agriculture Research; CIS: community managed irrigation schemes; EBFMPs: ecosystem-based farm management practices; ECOWAP: ECOWAS Agricultural Policy; GSS: Ghana Statistical Service; GIS: Government Managed Irrigation Schemes; MoFA: Ministry of Food and Agriculture; NBM: negative binomial model; IPM: Integrated Pest Management; IWMI: International Water Management Institute; WIAD: Women in Agriculture Development; WLE: Water, Land and Ecosystems; UDS: University for Development Studies.

Authors' contributions

CA designed data collection instruments, gathered data, analysed data and wrote the first draft of the manuscript. MAA, SD and FNM provided guide, corrections, inputs and supervision to the entire research study. All authors read and approved the final manuscript.

Author details

[1] Department of Agricultural and Resource Economics, University for Development Studies, Tamale, Ghana. [2] Institute for Interdisciplinary Research and Consultancy Services (IIRaCS), University for Development Studies, Tamale, Ghana. [3] Department of Climate Change and Food Security, University for Development Studies, Tamale, Ghana.

Acknowledgements

This paper is extracted from a larger study that was partly sponsored by a CGIAR Water, Land and Ecosystems (WLE) Project titled "Giving 'latecomers' a head start: Reorienting irrigation investment in the White Volta basin to improve ecosystem services and livelihoods of women and youth" led by Ghana Irrigation Development Authority (GIDA) with International Water Management Institute (IWMI), University for Development Studies (UDS) and Women in Agriculture Development (WIAD) as collaborators.

Competing interests

The authors declare that they have no competing interests.

Funding

This study was partly sponsored by a CGIAR Water, Land and Ecosystems (WLE) Project titled "Giving 'latecomers' a head start: Reorienting irrigation investment in the White Volta basin to improve ecosystem services and livelihoods of women and youth" led by Ghana Irrigation Development Authority (GIDA) with International Water Management Institute (IWMI), University for Development Studies (UDS) and Women in Agriculture Development (WIAD) as collaborators.

Appendix: Poisson estimates for factors that influence EBFMPs adoption by farmers

```
. poisson T_EBMFP_irr Age Sex Educ_d Ext_visits_d Fm_dist_km_irr Fm_size_irr Soil_percptn EBMFP_k
> nw Irrig_type

Iteration 0:   log likelihood = -501.12213
Iteration 1:   log likelihood = -501.12206
Iteration 2:   log likelihood = -501.12206


Poisson regression                              Number of obs   =        300
                                                LR chi2(9)      =     106.86
                                                Prob > chi2     =     0.0000
Log likelihood = -501.12206                     Pseudo R2       =     0.0963
```

T_EBMFP_irr	Coef.	Std. Err.	z	P>|z|	[95% Conf.	Interval]
Age	.0067793	.003103	2.18	0.029	.0006976	.0128611
Sex	−.0004684	.0700202	−0.01	0.995	−.1377055	.1367687
Educ_d	.0258911	.0676576	0.38	0.702	−.1067154	.1584977
Ext_visits_d	.0959868	.0670127	1.43	0.152	−.0353557	.2273294
Fm_dist_km_irr	−.14678	.0466758	−3.14	0.002	−.238263	−.055297
Fm_size_irr	.0055801	.0428365	0.13	0.896	−.078378	.0895381
Soil_percptn	.1864361	.0706725	2.64	0.008	.0479206	.3249516
EBMFP_knw	.0262394	.009294	2.82	0.005	.0080234	.0444554
Irrig_type	.1610305	.0818754	1.97	0.049	.0005578	.3215033
_cons	.4989228	.2132471	2.34	0.019	.0809661	.9168794

.

References

1. Dale VH, Polasky S. Measures of the effects of agricultural practices on ecosystem services. Ecol Econ. 2007;64(2):286–96.
2. Önder M, Ceyhan E, Kahraman A. Effects of agricultural practices on environment. Biol Environ Chem. 2011;24:28–32.
3. Rezvanfar A, Samiee A, Faham E. Analysis of factors affecting adoption of sustainable soil conservation practices among wheat growers. World Appl Sci J. 2009;6(5):644–51.
4. Thiaw I, Kumar P, Yashiro M, Molinero C. Food and ecological security: identifying synergy and trade-offs. UNEP Policy Series. 2011;1-12.
5. Millennium Ecosystem Assessment. A toolkit for understanding and action. Protecting nature's Services. Protecting ourselves. 2007. https://www.unpei.org/sites/default/files/PDF/ecosystems-economicanalysis/MEA-A-Toolkit.pdf. Accessed 30 Jan 2016.
6. ECOWAS Commission. Regional Partnership Compact for the implementation of ECOWAP/CAADP. 2009. http://www.oecd.org/swac/publications/44426979.pdf. Accessed 30 Jan 2016.
7. Zimmermann R, Bruntrüp M, Kolavalli S. Agricultural policies in Sub-Saharan Africa. Understanding CAADP and APRM policy processes. Bonn: Deutsches Institut für Entwicklungsforschung; 2009.
8. Abdul-Hanan A, Ayamga M, Donkoh SA. Smallholder adoption of soil and water conservation techniques in Ghana. Afr J Agric Res. 2014;9(5):539–46.
9. Nata JT, Mjelde JW, Boadu FO. Household adoption of soil-improving practices and food insecurity in Ghana. Agricul Food Secur. 2014;3(1):17.
10. Sterve H. Factors restricting adoption of sustainable agricultural practices in a smallholder agro-ecosystem: a case study of Potshini community, upper Thukela region, South Africa. 2011.
11. Armah RN, Al-Hassan RM, Kuwornu JK, Osei-Owusu Y. What influences farmers' choice of indigenous adaptation strategies for agrobiodiversity loss in Northern Ghana? Br J Appl Sci Technol. 2013;3(4):11–62.
12. Nkegbe PK, Shankar B. Adoption intensity of soil and water conservation practices by smallholders: evidence from Northern Ghana. Bio-based Appl Econ. 2014;3(2):159–74.
13. Ghana Statistical Service. 2010 Population and Housing Census. Summary Report of Final Results. Accra. 2012.
14. Dinye R. Irrigated agriculture and poverty reduction in Kassena Nankana district in the upper-east region, Ghana. J Sci Technol (Ghana). 2013;33(2):59–72.
15. Dinye RD, Ayitio J. Irrigated agricultural production and poverty reduction in Northen Ghana: a case study of the Tono Irrigation Scheme in the Kassena Nankana District. Int J Water Resour Environ Eng. 2013;5(2):119–33.
16. Agyedu GO, Donkoh F, Obeng S. Teach yourself research methods. Kumasi: Ghana; 2013. p. 94–5.
17. Winkelmann R. Counting on count data models. Bonn: IZA World of Labor; 2015.
18. Ramirez OA, Shultz SD. Poisson count models to explain the adoption of agricultural and natural resource management technologies by small farmers in Central American countries. J Agric Appl Econ. 2000;32(01):21–33.
19. Greene WH. Econometric analysis. India: Pearson Education; 2003.
20. Katchova A. Count data models. 2013. https://sites.google.com/site/econometricsacademy/econometrics-models/count-data-models. Accessed 6 Jun 2017.
21. Williams R. Models for count outcomes. 2015. https://www3.nd.edu/~rwilliam/stats3/CountModels.pdf. Accessed 6 Sep 2015.
22. Greene W. Functional forms for the negative binomial model for count data. Econ Lett. 2008;99(3):585–90.
23. Raghu PT, Manaloor V. Factors influencing adoption of farm management practices in agro-biodiversity hotspots of India: an analysis using Negative Binomial Count Data model. J Nat Resour Dev. 2014;04:46–53. https://doi.org/10.5027/jnrd.v4i0.07.
24. Garming H, Waibel H. Do farmers adopt IPM for health reasons? The case of Nicaraguan vegetable growers. Paper presented at the Proceedings of the Tropentag Conference Utilisation of Diversity in Land Use Systems: Sustainable and Organic Approaches to Meet Human Needs. 2007; p 9–11.

## 12

Effect of climate-smart agricultural practices on household food security in smallholder production systems: micro-level evidence from Kenya

Bright Masakha Wekesa[*], Oscar Ingasia Ayuya and Job Kibiwot Lagat

Abstract

Background: Climate change in Sub-Saharan Africa has had a negative impact on agricultural production leading to food insecurity. Climate-smart agricultural (CSA) practices have the potential to reverse this trend because of its triple potential benefits of improved productivity and high income, reduction or removal of greenhouse gases and improved household food security. Hence, we empirically find the determinants of choice and the effect of CSAs on household food security among smallholder farmers in Kenya.

Methods: Primary data were collected in Teso North Sub-county, Busia County of Kenya, among smallholder farmers. CSA practices used by farmers were grouped by principal component analysis and linked to food security by multinomial endogenous switching regression model.

Results: With the application of principal component analysis, we clustered the CSA practices into 4 components: crop management, field management, farm risk reduction and soil management practices. We find that the greatest effect of CSA adoption by smallholder farmers on food security is when they use a larger package that contains all the four categories of practices. Adopters of this package were 56.83% more food secure in terms of HFCS and 25.44% in terms of HDDS. This package mitigates upon the impacts of climate change as well as enhancing nutrient availability in the soils for higher productivity. Further, adoption of this package was positively influenced by gender of the household head, farm size and value of productive farm assets.

Conclusions: CSAs have the potential to alleviate food insecurity among smallholder farmers if used in combinations and to a larger extend. To enhance adoption, land fragmentation should be discouraged through civic education and provision of alternative income-generating activities for farmers to benefit when practiced on relatively bigger land. Farmers should be sensitized on the need to invest in farm productive assets in order to absorb the risks of climate change while enhancing adoption of CSA practices.

Keywords: Climate-smart agricultural practices, Food security, Climate change, Smallholder farmers, Multinomial endogenous switching regression analysis

*Correspondence: wekesabright@gmail.com
Department of Agricultural Economics and Agribusiness Management,
Egerton University, P.O. Box 536-20115, Egerton, Kenya

Introduction

Climate change is a threat to food security systems and one of the biggest challenges in the twenty-first century [1]. The ability to contain the pace of climate change by keeping temperature rise within 2 °C threshold is now curtailed, and the global population will have to deal with its consequences [2]. This is in the context that agricultural production systems are expected to produce food for the global population that is projected to be 9.1 billion people by 2050 and above 10 billion by the year 2100 [3]. According to [4], agricultural systems should be transformed to increase the productive capacity and stability in the wake of climate change. Climate change has already caused significant impacts on water resources, human health and food security [1]. The steady rise in temperature and irregular rainfall patterns affect agricultural production with the attendant decline in crop and livestock production.

In Sub-Saharan Africa (SSA), poverty reduction and food security improvement are among the many challenges that governments face. These governments constantly face a trade-off between food production which generates significant amounts of green house gas (GHG) and mitigation of climate change which requires reduction in some agricultural activities [3]. For instance, ruminant production contributes a significant amount of methane gas to the atmosphere, yet it is an important exercise to meet the food demand and income for farmers [1]. Addressing these antagonistic objectives has proved challenging. Attention in the literature has mostly focused on the low and stagnant returns from African agriculture [3, 5]. Moreover, many ecosystem services, including nutrient cycling, nitrogen fixation, soil regeneration and biological control of pests and weeds, are under threat in African food production systems and have serious implications on smallholder sustainable food security [6–8]. SSA continues to significantly face declining fallow periods, with inadequate investment in sustainable intensification and veering off from diversification in favour of mono-cropping in otherwise traditionally complex farming systems [6]. The result of this trend is food insecurity brought by the low agricultural production, especially under the conditions of climate change.

Climate change in Kenya is quite evident indicated by a continuous rise in temperature [9]. Generally, irregular rainfall patterns continue to be experienced with intense downpours causing floods in many parts which appear in cycles with severe droughts. Specifically, both day and night temperatures have significantly been on a rising trend since the 1960s. For instance, the night temperature (minimum) has risen by 0.7–2.0 °C and the day temperature (maximum) by 0.2–1.3 °C, depending on the season and the region [10, 11]. Further, these unprecedented changes in climate have accompanied losses that have already been experienced in the country [10]. For instance, evidence indicates that between 1999 and 2000 droughts in Kenya caused damages equivalent to 2.4% of gross domestic product (GDP) [9]. The report further indicates that the projected annual cost of climate change impacts will be in the tune of USD 1–3 billion by the year 2030 [9].

Majority of smallholder farmers in Kenya depend on agriculture for survival [12]. Building their adaptive capacity and resilience to climate change is key to enable them protect their livelihoods and ensuring their food security. The ability to cope with the impacts of weather shocks and natural disasters brought by the effects of climate change depends largely on the household's resilience, or its capacity to absorb the impact of, and recover from, a shock [13]. One way of combatting the effects of climate change is through climate-smart agricultural (CSA) practices [1, 11, 14, 15]. Promoters of CSA adoption seek to sustainably increase agricultural productivity and incomes by building resilience through adapting to changes in climate and reducing and/or removing GHGs emissions relative to conventional practices [1]. Strengthening Adaptation and Resilience to Climate Change in Kenya Plus (StARCK+) Programme identifies poverty, weak institutions and under-investment in key sectors as the main factors which stifle Kenya's ability to cope with climate change.

Climate change is a serious threat to local food production and family well-being resulting in malnutrition, hunger and persistent poverty in many regions of Kenya [16]. Despite the multiple benefits of CSAs and the deliberate efforts by the government and development partners to encourage farmers to invest in them, there is still a lack of evidence on farmers' incentives, conditioning factors that hinder or accelerate usage and impact of CSAs on food security status. Thus, an improved understanding of farmers' adoption behaviour and the potential welfare effects in terms of food security is important in informing the strategies policy makers and other development partners could champion in enhancing usage and effectiveness of CSA practices in smallholder production systems.

Based on the foregoing, the objectives of this study are twofold. We first seek to determine the factors that influence the choice of CSA practices in smallholder production systems. Secondly, we explore the effect of the CSA practices on household food security. To achieve these objectives, we use a micro-level data set of smallholder farmers in Kenya. This paper contributes to the literature as follows. First, we group the CSA practices based on usage by farmers in a principal component analysis (PCA). This departs from use of the conventional groups used by earlier researchers [8, 17, 18] which could potentially present difficulties, especially where few or even one strategy represents the entire group leading to a weak attribution of the impacts of such groups. Secondly, we also evaluate the influence of farmer perception on soil conditions and past experiences with climate-related shocks on adoption of climate-smart agricultural practices. Lastly, we link smallholder farmer's usage of CSA practices with household food security status to provide micro-level evidence. A multinomial endogenous switching treatment effects approach is used to control for selection bias while determining the impact of CSAs on food security. This is demonstrated using data from a cross-sectional survey of rural smallholders who participate in agricultural production amidst the challenges of climate change.

Methodology
Study area
This study was conducted in Teso North Sub-county, Busia County in Kenya. The area was selected for study because of its high potential for food production in the entire Busia County which is attributed to its better soils but under threat of massive soil degradation. It lies on the Northern part of Busia County and has six wards (Malaba Central, Malaba South, Malaba North, Ang'urai South, Ang'urai North, and Ang'urai East) and covers an area of 261 km^2 with a population of 117,947 [16]. The Sub-county has two main rivers Malakisi and Malaba on the northern part. The dry season with scattered rains falls from December to February. The Sub-county receives an annual rainfall of between 760 mm and 2000 mm. [16] indicates that 50% of the rainfall falls during the long rain season which is at its peak between late March and late May, while 25% falls during the short rains between August and October. The annual mean maximum and minimum temperatures range between 26 and 30 and 14 and 22 °C, respectively.

The Sub-county has experienced environmental degradation including loss of quality and quantity of natural biodiversity, soil erosion and flooding which poses a threat to its food production potential. As stated in the county's integrated development plan, varying rainfall patterns have affected both land preparation and good production leading to lower yields [16]. There is also a remarkable decline in water volumes in rivers, wells, pans, and springs with the average distance to watering point averaging at 1.5 km.

The data used for this study were obtained from a farm household survey carried out between May and July 2016 by well-trained enumerators. The sample for this study was drawn from smallholder farmers in Teso North Sub-county. Multistage sampling procedure was employed to select respondents, whereby in stage one, Teso North

Sub-county was purposively selected based on its high food production potential in the entire Busia County. In stage two, three wards (Malaba South, Malaba North and Ang'urai South) were randomly selected from the six wards in Teso North Sub-county. Finally, in the last stage, simple random sampling was used to select 384 farmers for the interview from a source list acquired from the office of County Director of Agriculture using a pretested interview schedule. The interview schedule was administered through face-to-face interviews by well-trained enumerators.

Analytical framework
First, CSA practices used in Teso North were identified and grouped into heterogeneous principal clusters by the use of principal component analysis. The components were rotated using orthogonal rotation (varimax method) [19, 20] so that smaller number of highly correlated practices would be put under each component for easy interpretation and generalization about a group. The result of the rotation was 4 principal components from a possible 14 extracted with eigenvalues > 1 following the [21] criterion. Principal component analysis was useful in reducing the dimensionality of data without loss of much information. This was important as it allowed determination of the relationship between practices based on usage and subsequent analysis by fitting the groups into the model and reaching conclusions. The approach is superior to the use of conventional grouping of practices which would make it difficult to conclude about a group in cases where few practices could represent the entire group.

The practices were grouped using principal component analysis with iteration and varimax rotation in the model represented as shown below:

$$Y_1 = a_{11}x_{12} + a_{12}x_2 + \cdots + a_{1n}x_n$$
$$Y_j = a_{j1}x_{j1} + a_{j2}x_2 + \cdots + a_{jn}x_n \tag{1}$$

where $Y_1,...Y_j$ = principal components which are uncorrelated, $a_1 - a_n$ = correlation coefficient, $X_1,...X_j$ = factors influencing choice of a particular strategy. The CSA practices identified and grouped through a principal component analysis are presented in Table 1. Selection of these practices prior to the field study was guided by the successful CSA practices established by a previous study done by Forum for Agricultural Research in Africa in the region [7].

After grouping the CSA practices, multinomial endogenous switching regression (MNLESR) model was then used to model the determinants of choice and effect of CSA practices on food security of smallholder farmers.

Table 1 Climate-smart agricultural practices identified to be actively used by farmers

S. No.	CSA practices
1	Use of improved crop varieties
2	Use of legumes in crop rotation
3	Use of cover crops
4	Changing planting dates
5	Efficient use of inorganic fertilizers
6	Use of terraces
7	Planting trees on crop land
8	Use of live barriers
9	Diversified crop and animal breeds
10	Irrigation
11	Use of improved livestock breeds
12	Use of organic fertilizers
13	Planting crops on tree land
14	Use of mulching

Food security status of the respondents was measured using Household Food Consumption Score (HFCS) and Household Dietary Diversity Scores which are measures of dietary diversity and quality.

In the first stage, farm households were assumed to face a choice of 7 mutually exclusive combinations/packages for responses to changes in mean temperature and rainfall (climate change). In the second stage, MNLESR econometric model was used to investigate the effect of different CSA practices on food security status.

Stage 1: Multinomial adoption selection model

At this stage, multinomial logit was used to determine the determinants of choice of CSA packages. Farmers were assumed to maximize their food security status, Y_i by comparing the revenue provided by 7(M) alternative CSA strategies. The requirement for farmer i to choose any strategy, j over other alternatives M is that $Y_{ij} > Y_{iM}$ $M \neq j$, that is, j provides higher expected food security than any other strategy. Y_{ij}^* is a latent variable that represents the expected food security level which is influenced by the observed household, plot characteristics, climate shocks and unobserved features expressed as follows:

$$Y_{ij}^* = X_i \beta_j + \varepsilon_{ij}. \tag{2}$$

X_i captures the observed exogenous variables (household and plot characteristics), while the error term ε_{ij} captures unobserved characteristics. The covariate vector X_i is assumed to be uncorrelated with the idiosyncratic unobserved stochastic component ε_{ij}, that is, $E(\varepsilon_{ij}|X_i) = 0$, whereby error terms ε_{ij} are assumed to be identically Gumbel distributed and independent, that is, under the

independent irrelevant alternatives (IIA) hypothesis. The selection model (2) leads to a multinomial logit model [22] where the probability of choosing strategy $j(p_{ij})$ is:

$$p_{ij} = p(\varepsilon_{ij} < 0|X_i) = \frac{\exp(X_i \beta_j)}{\sum_{M=1}^{j} \exp(X_i \beta_M)}. \tag{3}$$

Stage 2: Multinomial endogenous switching regression model

Here, endogenous switching regression (ESR) was used to investigate the impact of each response packages on food security by applying [23] selection bias correction model. Farm households face a total of 7 regimes with regime $j = 1$ being the reference category (non-responsive). The food security status equation for each possible regime is defined as:

$$
\begin{aligned}
&\text{Regime 1} \quad Q_{i1} = z_i \alpha_1 + \mu_{i1} \text{ if } i = 1 \\
&\quad \vdots \qquad\qquad \vdots \\
&\text{Regime } j \quad Q_{ij} = z_i \alpha_j + \mu_{ij} \quad \text{if } i = j
\end{aligned}
\tag{4}
$$

From the above equation, Q_{ij}'s represent the food security status, Z_i represents a set of exogenous variables (that is, household, plot, location characteristics, institutional variables and climate shocks), and the ith farmer in regime j and the error terms μ_{ij}'s are distributed with $E(\mu_{ij}|x, z) = 0$ and $\text{var}(\mu_{ij}|x,z) = \sigma_j^2$. Q_{ij} is observed if, and only if, CSA strategy j is used, which occurs when $Y_{ij}^* > \max_{M \neq 1}(Y_{im})$; if the error terms in (3) and (4) are not independent, OLS estimates for Eq. (4) were biased. A consistent estimation of α_j requires inclusion of the selection correction terms of the alternative choices in Eq. (3). MNLESR assumes the following linearity assumption: $E(\mu_{ij}|\varepsilon_{i1} \ldots \varepsilon_{ij}) = \sigma_j \sum_{m \neq j}^{j} r_j(\varepsilon_{im} - E(\varepsilon_{im}))$. By construction, the correlation between the error terms in (3) and (4) was zero.

Using the above assumption, Eq. (3) can be expressed as follows:

$$
\begin{aligned}
&\text{Regime 1} \quad Q_{i1} = z_i \alpha_1 + \sigma_1 \lambda_1 + \omega_{i1} \text{ if } i = 1 \\
&\quad \vdots \qquad\qquad \vdots \\
&\text{Regime } j \quad Q_{ij} = z_i \alpha_j + \sigma_j \lambda_j + \omega_{ij} \quad \text{if } i = j
\end{aligned}
\tag{5}
$$

σ_j is the covariance between ε's and μ's, while λ_j is the inverse Mills ratio computed from the estimated probabilities in Eq. (5) as follows:

$$\lambda_j = \sum_{m \neq j}^{j} \rho_j \left[\frac{p_{im} In(p_{im})}{1 - p_{im}} + In(p_{ij}) \right]. \tag{6}$$

ρ in the above equation represents the correlation coefficient of ε's and μ's, while ω_{ij} are error terms with an expected value of zero. In the multinomial choice setting expressed earlier, there were $j - 1$ selection correction terms, one for each alternative CSA practice. The standard errors in Eq. (5) were bootstrapped to account for the heteroskedasticity arising from the generated regressors given by λ_j.

Estimation of average treatment effects

At this point, a counterfactual analysis was performed to examine average treatment effects (ATT) by comparing the expected outcomes of adopters with and without adoption of a particular CSA strategy. ATT in the actual and counterfactual scenarios were determined as follows [8, 17]:

Food security status with adoption/usage

$$E(Q_{i2}|i = 2) = z_i\alpha_2 + \sigma_2\lambda_2 \qquad (7a)$$

$$E(Q_{ij}|i = j) = z_i\alpha_j + \sigma_j\lambda_j. \qquad (7b)$$

Food security status without adoption (counterfactual)

$$E(Q_{i1}|i = 2) = z_i\alpha_1 + \sigma_1\lambda_2 \qquad (8a)$$

$$E(Q_{i1}|i = j) = z_i\alpha_1 + \sigma_1\lambda_j. \qquad (8b)$$

ATT can be defined as the difference between (7a) and (8a) which is given by:

$$\begin{aligned} ATT &= E(Q_{i2}|i = 2) - E(Q_{i1}|i = 2) \\ &= z_i(\alpha_2\alpha_1) + \lambda_2(\rho_2 - \rho_1). \end{aligned} \qquad (9)$$

The right-hand side indicates the expected change in adopters' mean food security status, if adopters' characteristics had the same return as non-adopters, for instance, if adopters had the same characteristics as non-adopters, while λ_j is the selection term that captured all potential effects of difference in unobserved variables.

Variables used in econometric analysis are presented in Table 2 and were derived from review of past studies [7, 8: 14, 24: 17].

Measurement of food security

To measure food security status of the farm households, Household Food Consumption Score (HFCS) and Household Dietary Diversity Scores were used as proxies for food security of farmers. These tools were developed by WFP and are commonly used as proxies for access to food [25]. HFCS is a weighted score based on dietary diversity, food frequency and the nutritional importance of food groups consumed. The HFCS of a household is calculated by multiplying the frequency of foods consumed within 7 days with the weighting of each food group. The weighting of food groups was determined by WFP according to the nutrition density of the food group

Table 2 Variables used in econometric analysis

Variable	Description	Measurement	Mean	SD
FOODSEC	Food security status of the household	Food consumption score	63.22	19.24
		Household Dietary Diversity Score	6.73	1.65
AGE	Age in years of the household head	Continuous	46.51	14.69
GENDER	Gender of the household head	Dummy = 1 if male 0 = female	0.77	–
EDUC	Years of education of the household head	Discrete	10.00	4.45
H/SIZE	# of household size	Discrete	6.87	2.61
OFF-FARM	Participation in off-farm employment	Dummy = 1 if yes 0 = otherwise	0.44	–
ASSETS	Value of productive farm assets	Continuous	62,965.81	63,951.31
LAND	Owned farm size in acres	Continuous	2.54	1.57
TERRAIN	Terrain of the land	1 = sloppy 0 = otherwise	0.52	–
S/FERTILITY	Level of soil fertility	1 = poor 2 = medium 3 = fertile	1.70	–
EROSION	Severity of soil erosion	1 = severe 2 = moderate 3 = low	2.06	–
FLOOD	If household experienced floods in the last 5 years	Dummy = 1 yes 0 = otherwise	0.39	–
RAINS	If the household experienced insufficient rains in the last 5 years	Dummy = 1 yes 0 = otherwise	0.71	–
H/STRMS	If the farm household experienced hailstorms in the last 5 years	Dummy = 1 yes 0 = otherwise	0.63	–
DISTNCE	Walking time in minutes to the input and output market	Continuous	52.36	37.45
EXTN	Number of annual contacts with extension agents	Discrete	5.50	3.70
GRPMSHIP	If the household head is a member of a farmer-related group or association	Dummy = 1 if a member 0 = otherwise	0.66	–
CREDIT	Whether household received credit	Dummy = 1 if yes 0 = otherwise	0.60	–

[26, 27]. Appendix 1 presents the various food components used to determine the HFCS. HDDS is similar to HFCS with slight differences in the components of the various food clusters. While HFCS takes into account food items consumed within 7 days, the HDDS takes into account food items consumed within the last 24 h. Appendix 2 shows food group and weights for determination of HDDS. The two indicators measure food diversity which is strongly correlated with dietary quality and adequacy [28]. While recording the food items, foods taken during ceremonies and major occasions were skipped to reduce the bias that would have arisen in capturing such meals. Thus, for both the indicators such days were dropped.

Results and discussion
Principal component analysis (PCA)
Table 3 contains principal components (PCs) and the coefficients of linear combinations called loadings. A visual inspection of Table 3 reveals that the four PCs explained 74.19% of total variability in the data set. The results presented in Table 3 present a good fit, indicating that the PCA results highly explained the data. The first component explained 35.65% variance and is correlated with changing crop varieties, use of legumes in crop rotation, use of cover crops, changing planting dates and efficient use of inorganic fertilizer all with positive effects (factor loadings). Thus, this component was named crop management practices.

Principal components 2, 3 and 4 accounted for 20.12, 11.08 and 7.35% variances, respectively. This means that the first four components have more importance in explaining the variance in data set. The second PC was associated with use of organic manure, planting of food crops on tree land (as part of agroforestry) and use of mulching all with positive loadings too. The third PC contained crops and livestock diversification and use of improved livestock breeds both with highly negative loadings and use of irrigation with positive loadings. Finally, the last PC was associated with use of planting trees on crop land and use of live barriers with high positive effects (loadings) and use of terraces with a high negative effect.

The communality column shows the total amount of variance of each variable retained in the four components. MacCallum et al. [29] noted that all items in PCs should have communalities of over 0.60 or an average communality of 0.7 for small sample sizes precisely below 50 to justify performing a PCA analysis. With the sample size of 384, the communalities presented in Table 3 meet the minimum criteria as they contribute more than 60% variance in the PCs. For the interpretation of the PCs, variables with high factor loadings and high communalities were considered from the varimax rotation [19, 30].

Table 4 presents the descriptive statistics of composition of each component (climate-smart strategies). The most commonly used component was of crop management practices with 96.09% of farmers using at least a unit of this component. This component comprised of practices such as: use of improved crop varieties, use of

Table 3 Loadings of the four components for CSA compositions

Strategies	Comp1	Comp2	Comp3	Comp4	Communality
Changing crop varieties	0.5467	− 0.3965	0.2579	− 0.2853	0.6040
Use of legumes in crop rotation	0.6431	− 0.3903	0.2574	− 0.2224	0.6894
Use of cover crops	0.6257	− 0.3138	− 0.2292	− 0.1559	0.6344
Changing planting dates	0.5223	− 0.3779	0.3280	− 0.2981	0.6121
Crop and livestock diversification	0.3910	0.3482	− 0.4904	0.3216	0.6180
Use of organic manure	0.2550	0.6522	− 0.3156	− 0.3036	0.5086
Efficient use of inorganic fertilizer	0.5537	0.2032	0.3940	− 0.3311	0.6127
Use of terraces	0.2485	0.3343	− 0.3243	− 0.6249	0.6691
Irrigation	0.3816	0.3986	0.4546	0.2423	0.6283
Trees on crop land	0.2459	− 0.3013	− 0.4518	0.6024	0.7183
Food crops on tree land	0.3202	0.6198	0.3715	0.3424	0.7419
Use of live barriers	0.3190	− 0.3308	− 0.3845	0.5146	0.6238
Mulching	0.2811	0.5512	0.3483	0.3819	0.6500
Use of improved livestock breeds	0.2510	0.3794	− 0.7011	− 0.1492	0.7207
Eigenvalues	4.9160	2.8161	1.5505	1.0287	
Eigenvalues % contribution	35.6543	20.1153	11.0751	7.3479	
Cumulative %	35.6543	55.7696	66.8447	74.1926	

Table 4 List of climate-smart strategies

Group	Percentage of users	Components
Crop management practices (C)	96.09%	Use of improved crop varieties
		Use of legumes in crop rotation
		Use of cover crops
		Changing planting dates
		Efficient use of inorganic fertilizers
General field management practices (F)	81.51%	Use of terraces
		Planting trees on crop land
		Use of live barriers
Farm risk reduction practices (R)	39.84%	Diversified crop and animal breeds
		Irrigation
		Use of improved livestock breeds
Soil conservation practices (S)	22.92%	Use of organic fertilizers
		Planting food crops on tree land
		Use of mulching

legumes in crop rotation, use of cover crops, changing planting dates and efficient use of inorganic fertilizers. The second most used component was of general field management practices for soil erosion control used by 81.51% of farmers. This component entailed of use of terraces and contour bunds, planting trees on cropland and use of live barriers.

Farm risk reduction measures were only used by 39.84% of farmers. The practices in this component included: crop and livestock diversification, irrigation and use of improved livestock breeds. Finally, the least used component comprised of specific soil conservation practices which included: use of organic manure, planting crops on tree land and application of mulching. This component was used by 22.92% of farmers.

Econometric results

The determinants of choice of CSA packages are given followed by their impact on food security. CSA practices can be adopted in a wide range of different combinations, and this has implication on household's food security status. Given the set of available packages, understanding what drives an individual to select specific packages is important for policy direction.

Table 5 presents different packages (combinations), whereby 7 out of 16 possible combinations/packages were used by farmers. Few farmers (3.6%) were non-users/non-adopters of any CSA package. About 2.6% of farmers used package $C_1F_0R_1S_0$. This package comprised

of crop management practices and farm risk reduction measures only. Another 4.4% used package $C_1F_0R_1S_1$ that had crop management, farm risk reduction measures and soil management practices. Further, 7.0% of farmers used package $C_1F_1R_0S_1$ that contained crop management, field management and soil conservation practices. Another 8.3% of farmers used package $C_1F_0R_0S_0$ that contained only crop management practices. Approximately 12% of farmers used package $C_1F_1R_1S_1$ with all the four groups of CSA strategies. About 21% used package $C_1F_1R_1S_0$ that contained crop management, general field management for soil erosion control and farm risk reduction practices only).

The largest share of farmers (41.1%) used a package $C_1F_1R_0S_0$ that had crop management and general field management for soil erosion control. This reveals the efforts of many subsistence farmers to achieve food production despite the challenges of land degradation caused by soil erosion. This observation is similar to the findings of [7] which suggested that farmers in the region executed such responsive strategies for survival amidst challenges of climate change. A keen look at Table 5 reveals that all users of CSA practices (96.4% of all farmers) used packages that included at least a crop management practice. This observation demonstrates the need of most farmers to meet their basic crop production for food generation.

Determinants of choice of specific CSA packages

This section describes the factors that influence choice of CSA packages and then followed by quantification of the effect of using packages on food security status of farmers in the last stage. This was achieved using the multinomial endogenous switching regression (MNLESR) model which is a two-stage regression analysis model. The first stage of the MNLESR is the multinomial logit model which determines factors that influence the choice of CSA packages. This is an important stage as it guides on the necessary interventions to improve the adoption of CSA packages. In the second stage, the impact of usage of CSA packages on household food security was determined. The marginal effects from the MNL model that measured the expected change in the probability of a particular choice being made with respect to a unit change in an independent variable are reported in Table 6.

Non-use of all practices ($C_0F_0R_0S_0$) was the base category compared to other seven packages (refer to Table 5 for the packages) used by farmers. The results show seven sets of parameter estimates, one for each mutually exclusive combination of strategies. The Wald test that all regression coefficients are jointly equal to zero is rejected [x^2 (119) = 445.52; $p = 0.000$]. Thus, the results show that

Table 5 Specification of CSA strategy combinations to form the packages

Choice (j)	Binary quadruplicate	C = crop management		F = field management		R = risk reduction		S = specific soil management		Frequency	Percentage
		C_0	C_1	F_0	F_1	R_0	R_1	S_0	S_1		
1	$C_0F_0R_0S_0$	✓		✓		✓		✓		14.0	3.60
2	$C_0F_0R_0S_1$	✓		✓		✓			✓	0.00	0.00
3	$C_0F_0R_1S_1$	✓		✓			✓		✓	0.00	0.00
4	$C_0F_1R_1S_1$	✓			✓		✓		✓	0.00	0.00
5	$C_1F_1R_1S_1$		✓		✓		✓		✓	45.0	11.7
6	$C_1F_1R_1S_0$		✓		✓		✓	✓		82.0	21.1
7	$C_1F_1R_0S_0$		✓		✓	✓		✓		157	41.1
8	$C_1F_0R_0S_0$		✓	✓		✓		✓		32.0	8.30
9	$C_0F_1R_0S_1$	✓			✓	✓			✓	0.00	0.00
10	$C_1F_0R_1S_0$		✓	✓			✓	✓		10.0	2.60
11	$C_1F_0R_0S_1$		✓	✓		✓			✓	0.00	0.00
12	$C_0F_1R_0S_0$	✓			✓	✓		✓		0.00	0.00
13	$C_0F_1R_1S_0$	✓			✓		✓	✓		0.00	0.00
14	$C_0F_0R_1S_0$	✓		✓			✓	✓		0.00	0.00
15	$C_1F_0R_1S_1$		✓	✓			✓		✓	17.0	4.40
16	$C_1F_1R_0S_1$		✓		✓	✓			✓	27.0	7.00
Total										384	100

The binary quadruplicate represents the possible CSA packages. Each element in the quadruplicate is a binary variable for a CSA combination: crop management (C), general field management for soil erosion control, farm risk reduction (R) and soil management practices (S). Subscript 1 = adoption and 0 = otherwise

the estimated coefficients differ substantially across the alternative packages.

Age of the household head was negatively associated with usage of $C_1F_0R_0S_0$ and positively associated with usage of $C_1F_1R_0S_1$ at 10% and 5% significant levels, respectively. Increase in age of the household head by one year reduced the likelihood of using package $C_1F_0R_0S_0$ by 0.19%, while increased the likelihood of using $C_1F_1R_0S_1$ by 0.16%. This indicates that as age increases, farmers shift from smaller packages to larger ones. Older farmers may be more experienced with regard to production technologies and may have accumulated more physical and social capital thus to afford larger and better packages. Contrary, [31] noted that old age had a negative relationship to adopting climate change adaptation strategies, explaining that agriculture is a labour-intensive venture which requires healthy, risk-bearing and energetic farmers. Again, older farmers may not be aware of recent innovations.

With regard to gender of the household head, male-headed households were 2.7% more likely to use package $C_1F_1R_1S_1$ that contains crop management practices, field management, farm risk reduction practices only at 5% significant level relative to $C_0F_0R_0S_0$ (non-use of any CSA practices) compared to females. Women generally face constraints in terms of accessing resources and time. This may explain the negative relationship with usage of

CSA practices in this study. FARA [7] reported that gender remains a significant barrier to the adoption of CSAs by women, stemming largely from customary gender roles. They further stated in the report that women have less access than men to resources such as land, inputs, credit, education and extension services, all of which may be important to support transitions to CSA. Land ownership systems also present more entrenched barriers to female-led households. Land tenure systems in Western Kenya, for example, require women who want to adopt CSA to obtain permission from male relatives, thus derailing them [32].

Years of education of the household head negatively influenced usage of $C_1F_1R_0S_0$ which contains crop and field management practices only. One more year of education reduced the probability of using this package by 2% at 5% significance level. It could be that educated farmers opted out of this package since it does not offer risk reduction measures which could safeguard their investment against prevailing risks of climate change. This category of farmers avoided taking the risk of using this package with increase in their years of education. Similarly, [33] argues that higher levels of education tend to build the innovativeness and ability to assess risks by farmers for proper farm adjustments.

There was a positive and significant relationship between the value of productive farm assets (a proxy of

Table 6 Marginal effects estimates for the determinants CSA packages by MNL

Variables	$C_1F_0R_0S_0$ dy/dx	$C_1F_0R_1S_0$ dy/dx	$C_1F_0R_1S_1$ dy/dx	$C_1F_1R_0S_0$ dy/dx	$C_1F_1R_0S_1$ dy/dx	$C_1F_1R_1S_0$ dy/dx	$C_1F_1R_1S_1$ dy/dx
Socio-economic factors							
Age of HH	−0.0019*	0.0007	0.0015	−0.0028	0.0016**	0.0016	0.0000
Gender of HH	−0.0434	0.0045	0.0340	−0.0284	−0.0047	−0.0047	0.0271**
Years of education of HH	0.0013	0.0015	0.0033	−0.0204**	0.0019	0.0019	0.0000
Household size	0.0075	−0.0006	−0.0020	−0.0239	0.0056	0.0056	0.0004
Participation in off-farm employment	−0.0251	0.0022	0.0341	−0.0518	−0.0168	−0.0168	0.0012
Natural log of farm assets	0.0032	0.0009	−0.0045	−0.0722***	0.0014	0.0014***	0.0307*
Farm size	−0.0378***	−0.0104	−0.0268**	−0.0171	0.0110*	0.0110***	0.0013**
Farm characteristics							
Perception on terrain of land	−0.0007	0.0057	−0.0123	0.0995	−0.0177	−0.0177	0.0011
Perception Severity of soil erosion	−0.0107	−0.0342**	0.0198	−0.0452	−0.0451**	−0.0451***	0.0007
Perception of soil fertility	−0.0064	−0.0002	0.0105	0.1871***	−0.0128	−0.0128***	0.0007
Bad incidences							
Frequent floods	0.0284	−0.0266	−0.0204	0.0205	0.0330*	0.0330	0.0003
Hailstorms	0.0268	0.0052*	−0.0053	−0.0126	0.0193	0.0193	0.0006
Insufficient rains	−0.0023	0.0008	−0.0169	0.0628	−0.0311	−0.0311	0.0004
Institutional factors							
Walking time from farm to market	0.0002	−0.0003	−0.0005*	0.0011	−0.0007**	−0.0007**	0.0001
Membership to a farmer group	0.0279	0.0148	−0.0126	0.1888**	0.0221	0.0221**	0.0000
Contacts with extension agents	−0.0046	0.0021	0.0073	−0.0296***	0.0046	0.0046**	0.0002
Access to credit	−0.0461*	−0.0044	−0.0083	−0.1571**	0.0028	0.0028***	0.0001
Number of observations = 375; Wald χ^2 (119) = 445.52; p = 0.000							

$C_0F_0R_0S_0$ is the reference base category in the MNL; HH is household head

***Significant at 1% level

**Significant at 5% level

*Significant at 10% level

wealth) and usage of CSAs. Resource-endowed farmers (those with greater value of productive farm assets) were more likely to use larger packages $C_1F_1R_1S_0$ and $C_1F_1R_1S_1$ as opposed to non-use of any package. Precisely, the probability of using these packages increased by 0.14% and 3.07%, respectively, for resource-endowed farmers. It is likely that wealthier farmers have the capacity to use CSA practices, particularly expensive ones like use of improved livestock breeds and crop varieties available in these packages. Further, these assets enhance ability to absorb the risks associated with failure and the time it takes before realizing meaningful effects of using CSAs. This is consistent with [34] who noted that lack of productive assets limits the ability to adopt climate-smart practices that require huge resource allocation. Ochieng et al. [35] as well notes that wealthier households have higher capacity to invest in such measures to improve crop production. However, on the other hand the probability of using $C_1F_1R_0S_0$ reduced by 7.2% with increase in farm assets perhaps due to lack of risk reduction measures in this package.

Farm size owned positively influenced the use of packages $C_1F_0R_1S_1$, $C_1F_1R_0S_1$, $C_1F_1R_1S_0$ and $C_1F_1R_1S_1$ and negatively associated with the use of package $C_1F_0R_0S_0$. This implies that an increase in size of land by 1 acre (0.40 ha) increased the probability of using packages $C_1F_0R_1S_1$, $C_1F_1R_0S_1$, $C_1F_1R_1S_0$ and $C_1F_1R_1S_1$ by 2.7%, 1.1%, 1.1% and 0.13%, respectively, while reduced the probability of using package $C_1F_0R_0S_0$ by 3.8%. It follows therefore that farmers with larger farm size had the capacity to use larger packages as opposed to non-usage of any package. Availability of land provides opportunity for farmers to experiment these important technologies, thus influencing usage of the large packages. This result is consistent with the result of [36] who stated that bigger farm size accrues benefits of economies of scale to farmers and also provide a means of diversifying production. Users of package $C_1F_0R_0S_0$ which contains crop management practices only were less likely to use the package with increase in their farm sizes. The possible explanation could be that these farmers chose to rent out their increasing farms for other users rather than farming since this small package may not offer meaningful production in the circumstances of harsh weather. Renting in farmers may not be motivated to implement long-term packages, thus reducing the usage of CSA practices on these particular farms.

The perception of severity of soil erosion by farmers was negatively associated with the use of the following packages: $C_1F_0R_1S_0$, $C_1F_1R_0S_1$ and $C_1F_1R_1S_0$. The probability of using these packages reduced by 3.4%, 4.5% and 4.5%, respectively, for the farmers who regarded their plots as severely eroded. It appears that farmers were highly motivated to implement CSA practices on less severely eroded farms and vice versa. In essence, these farmers were not quite responsive to countering the effects of severe soil erosion but were rather discouraged by severe soil erosion in implementing CSA technologies. Contrary, [37] noted a positive relationship with adoption of many soil conservation practices with the argument that farmers were responsive to soil degradation effects brought by soil erosion.

The perception of farmers towards soil fertility of the farm had a positive and significant influence on the usage of $C_1F_1R_0S_0$ and a negative influence on the usage of $C_1F_1R_1S_0$. The likelihood of using packages $C_1F_1R_0S_0$ and $C_1F_1R_1S_0$ increased by 18.7% and reduced by 1.3%, respectively, for farmers who regarded their farms as being more fertile. This implies that farmers who regarded their farms as being more fertile were more likely to use a small package $C_1F_1R_0S_0$ as opposed to non-use of any package. This is a lean package without significant soil fertility replenishing avenues. But those who regarded their farms as being less fertile implemented a larger package $C_1F_1R_1S_0$ that contains more soil nutrient enriching practices in the risk reduction component. Manda et al. [18] argues that the propensity to adopt sustainable agricultural practices such as improved maize is expected to be greater on plots with fertile soils, because most improved maize varieties require the application of expensive inorganic fertilizers.

Factors related to past experiences with extreme weather conditions by farmers also influenced choice of CSA packages. For instance, farmers who experienced frequent floods in the past were more likely to use package $C_1F_1R_0S_1$. The probability of using this package increased by 3.3% for the farmers who experienced frequent floods in the recent past. It is likely that these farmers were keener to the flood-related shocks, thus implementing a responsive strategy to curb it with proper field and soil management to abate soil degradation. Contrary, [24] noted that adoption of improved climate change adaptation technologies such as crop rotation and drought-resistant seeds is negatively and significantly influenced by harsh conditions brought by flooding such as waterlogging and frost stress.

Past experience with hailstorms was also positively associated with the use of package $C_1F_0R_1S_0$. It was revealed that the likelihood of using this package increased by 0.52% for farmers who had experienced frequent hailstorms in the recent past. Similarly, these farmers could be implementing a responsive strategy that included farm risk reduction through diversified production means. Previous study by [38] had a contrary result where frequent hailstorms were the greatest source of production risks related to climate change that discouraged adoption of production techniques posing a threat to yield stability in rural Amhara Ethiopia.

Distance (measured by walking time) to the input and output market negatively influenced usage of CSA practices. An increase in time taken to reach the market by 1 min reduced the probability of using packages $C_1F_0R_1S_1$, $C_1F_1R_0S_1$ and $C_1F_1R_1S_0$ by 0.05, 0.07 and 0.07%, respectively. Longer distance to the market for such larger packages increases the transaction costs involved in input purchase and output sale. Teklewold et al. [8] noted that apart from affecting the access to the market, distance can also affect the accessibility of new technologies, information and credit institutions, thus having a negative relationship.

Group membership had a positive and significant influence on the usage of packages: $C_1F_1R_0S_0$ and $C_1F_1R_1S_0$. Rather than not using any package, belonging to a farmer group increased the probability of using these two packages by 18.8% and 2.2%, respectively. Farmer groups are important channels through which extension agents and other farmer service providers (like insurance) use to access farmers. Further, field management practices like construction of terraces could be possibly achieved in mobilized labour in groups. Again, through group networks, members get to exchange ideas, handle farm demonstrations and also get connections to dissemination of important research findings. Ward and Pede [39] notes that learning from the experiences of peers increases the probability of technology adoption due to the fact that farmers trust more practical experiences demonstrated by their peers since they share much in common including shared labour.

The number of contacts with extension service providers positively influenced the use of $C_1F_1R_1S_0$ and negatively influenced the use of $C_1F_1R_0S_0$. One more annual contact with extension agents increased the probability of using $C_1F_1R_1S_0$ by 0.46% but reduced the probability of using $C_1F_1R_0S_0$ by 3.0%. This suggests that extension service played a crucial role in implementation of larger packages by farmers. It further suggests that the information disseminated had inclusion of a climate change dimension that promoted the use of the larger package. However, on the other hand, reduction in probability of using $C_1F_1R_0S_0$ suggests that the goal of promoting CSA technologies by extension service agents had mixed effects. It appears that farmers who used package $C_1F_1R_0S_0$ with only crop and field management practices were sceptical about the veracity of the information and its ability to improve their production, thus opting not to use any package. This is consistent with the findings of a study in Zambia by [14] which indicated that extension agents were involved in a lot of activities that include delivering inputs and administering credit; hence, farmers may question their skills impacting on their trust and eventual decline in implementation.

Access to credit had a positive and significant influence on the use of $C_1F_1R_1S_0$ but a negative influence on the use of $C_1F_0R_0S_0$ and $C_1F_1R_0S_0$. The results indicate that farmers who received credit in the previous farming season were 0.28% more likely to use $C_1F_1R_1S_0$. Credit access enables farmers to meet costs involved in implementing CSA technologies, especially including expensive ones like use of improved livestock breeds and irrigation present in this large package. Similarly, [40] explain that credit constraints negatively influence investment in improved seed and inorganic fertilizers, suggesting that credit-constrained households are less likely to adopt CSA technologies that require cash outlays. Access to credit reduced the probability of using packages $C_1F_0R_0S_0$ and $C_1F_1R_0S_0$ by 4.6% and 15.7%, respectively. A negative influence of credit access to usage of $C_1F_0R_0S_0$ and $C_1F_1R_0S_0$ may suggest that these farmers diverted credit to fund non-farming expenses like school fees and medical, thus opting not to use any package.

Average adoption treatment effects for the CSA packages

After determining the drivers of choice of CSA packages in the first stage, treatment effects were determined in the second stage to find the effect of usage of the packages on household food security. The ordinary least squares regression of Household Food Consumption Scores (HFCS) and Household Diversity Scores (HDDS) of the households were estimated for each combination of CSA practices, taking care of the selection bias correction terms from the first stage. At this stage, treatment effects which are the most important part of this stage were reported.

Appendices 1 and 2 present the food categories for HFCS and HDDS. For interpretation, HFCS were preferred to HDDS as the latter only capture meals taken within 24 h which may not include occasional meals taken on particular days like market days within a week. However, HDDS were used for sensitivity analysis. It is also important to note that the two scores were strongly correlated (0.97) as indicated in Table 7.

Table 7 presents the average adoption effects in terms of HFCS and HDDS under actual and counterfactual conditions. In Table 7, X_1 represents the treated group (adopters) and X_2 represents untreated (non-adopters), β_1 represents treated characteristics (adoption state) and β_2 untreated characteristics (non-adoption state). The level effect is the difference in food security status as a result of usage of the specified package. The impact is as a result of the difference between treated with treatment characteristics and the untreated with untreated characteristics $(\beta_1 X_1) - (\beta_2 X_2)$. Except users of $C_1F_0R_1S_0$, $C_1F_1R_1S_0$ and $C_1F_1R_1S_1$, all the rest using other packages would be better off in the counterfactual scenarios (non-usage) suggesting availability of other better options. All packages that included farm risk reduction practices apart from $C_1F_0R_1S_1$ had a positive impact on the welfare of farmers. This implies that farmers need to manage their farm risks to be assured of improved food security in the uncertain events of climate change.

For larger packages ($C_1F_1R_0S_1$, $C_1F_1R_1S_0$ and $C_1F_1R_1S_1$), all users were more food secure compared to their counterparts who did not use CSAs in the actual scenarios. Based on these results, a complete package with crop management practices, field management practices and farm risk reduction practices and soil management ($C_1F_1R_1S_1$) had the greatest overall effect of 30.14 and 1.72 scores on the welfare of farmers estimated using both HFCS and HDDS, respectively. This implies that farmers who used this package were 56.83% and 25.44% more food secure compared to their counterparts who chose not to use any CSA practice. Thus, farmers may be more food secure if they use climate-smart technologies within this package. This package is quite comprehensive as it addresses a wider spectrum of both field and soil conditions while also mitigating upon soil degradation for production stability. In general terms, the overall result is that non-usage of this ($C_1F_1R_1S_1$) package would be irrational as farmers will be better off in terms of food security if they use this package as it addresses a wide range of climate change challenges.

Conclusions and policy implications

The results indicate that adoption rate of CSAs was still low with many farmers implementing low capital requirement practices. This may be attributed to smallholder agriculture that is resource constrained. Crop management practices were the most dominant perhaps due to their low-cost implications. This suggests the need for farmer empowerment to progressively move towards more capital-intensive practices.

A larger package which comprised of crop management, field management, risk reduction practices and

Table 7 Impact of use and non-use of CSA packages on food security estimated using HFCS of farmers by ESR

Package		HFCS			HDDS		
		Treated characteristics (β_1)	Untreated characteristics (β_2)	Impact/returns	Treated characteristics (β_1)	Untreated characteristics (β_2)	Impact/returns
$C_1F_0R_0S_0$	Treated (X_1)	49.14 (1.92)	49.52 (0.96)	−0.38	5.31 (0.21)	6.06 (0.12)	−0.25
	Untreated (X_2)	52.35 (2.23)	65.07 (0.80)	−12.72	5.68 (0.019)	6.89 (0.07)	−1.21
	Level effects	−3.21	−15.54***	−15.93	−0.37*	−0.83***	−1.58
$C_1F_0R_1S_0$	Treated	65.75 (7.24)	56.52 (2.25)	9.23	7.20 (0.55)	6.36 (0.18)	0.84
	Untreated	63.29 (3.68)	63.65 (0.78)	−0.36	6.69 (0.31)	6.74 (0.07)	−0.05
	Level effects	2.46	−7.13***	2.1	0.51	−0.38**	0.46
$C_1F_0R_1S_1$	Treated	61.09 (3.37)	80.84 (2.72)	−19.75	6.56 (0.30)	6.63 (0.10)	0.07
	Untreated	57.40 (2.63)	63.82 (0.80)	−6.42	6.25 (0.23)	6.76 (0.06)	−0.51
	Level effects	3.69	17.02***	−2.73	0.32	−0.13	−0.20
$C_1F_1R_0S_0$	Treated	55.77 (1.09)	65.81 (1.01)	−10.04	6.14 (0.09)	7.04 (0.09)	−0.90
	Untreated	59.44 (0.96)	69.11 (0.93)	−9.67	6.29 (0.09)	7.18 (0.09)	−0.89
	Level effects	−3.67***	−3.30***	−13.34	−0.15	−0.14	−1.04
$C_1F_1R_0S_1$	Treated	63.89 (2.18)	69.99 (0.80)	−6.10	6.70 (0.23)	7.52 (0.09)	−0.82
	Untreated	63.59 (1.94)	63.69 (0.83)	−0.10	6.76 (0.07)	6.75 (0.14)	0.01
	Level effects	0.30	6.30***	0.20	−0.05	0.76***	−0.05
$C_1F_1R_1S_0$	Treated	74.70 (1.03)	62.72 (0.83)	11.98	7.66 (0.10)	6.35 (0.09)	1.31
	Untreated	75.75 (1.20)	60.64 (0.89)	15.11	7.90 (0.11)	6.51 (0.08)	1.39
	Level effects	−1.05	2.08*	27.09	−0.25**	−0.16*	1.15
$C_1F_1R_1S_1$	Treated	83.92 (1.01)	68.04 (0.82)	15.88	8.48 (0.11)	7.06 (0.10)	1.42
	Untreated	79.09 (1.23)	53.51 (0.82)	15.58	8.19 (0.12)	6.76 (0.07)	1.43
	Level effects	4.83***	4.53***	30.41	0.29**	0.31***	1.72
Pairwise correlation							
	HDDS	**HFC**					
HDDS	1						
HFC	0.9652***	1					

Standard errors are in parentheses. *C* crop management, *F* field management, *R* risk reduction, *S* specific soil management

specific soil management practices ($C_1F_1R_1S_1$) had the highest impact on food security. This package is quite comprehensive as it addresses a wider spectrum of both field and soil conditions while also mitigating upon soil degradation for production stability. Thus, for farmers to benefit more from CSAs, they need to incorporate all CSAs as much as possible. Findings were that the likelihood of usage of this package was positively influenced by gender, farm size and farm assets. Its usage was more likely on larger pieces of self-owned plots, for male-headed households with more farm assets. Thus, CSAs have the potential to alleviate food insecurity among smallholder farmers if used in combinations and to a larger extend.

Farmers should then be encouraged to incorporate larger CSAs packages which comprise at least a member in each of the four categories: crop management, field management, risk reduction practices and specific soil management practices, to have a higher effect on food security status. This could be through first sensitization on the need to invest in productive farm assets to enable them absorb risks associated with climate change at the same time enhancing their ability to uptake important CSAs. The sensitization could be done in groups by extension service providers. Secondly, land fragmentation should also be discouraged through civic education and engagement in alternative income-generating activities by farmers to benefit more from CSAs when practiced on relatively bigger portions of land.

Authors' contributions
This research was conceived and developed by all the three authors listed. All authors read and approved the final manuscript and unanimously agreed to publish it in this journal. All authors read and approved the final manuscript.

Acknowledgements
We would like to acknowledge the financial support from African Economic Research Consortium (AERC) who funded this study. Special thanks go to all smallholder farmers who responded to our questions and the enumerators for the valuable effort during data collection.

Competing interests
The authors declare that they have no competing interests.

Funding
This research was fully funded by African Economic Research Consortium (AERC) based in Nairobi, Kenya.

Appendix 1
See Table 8.

Table 8 Food groups for HFCS bY WFP

Food item	Food group	Weight
Rice	Cereals and tubers	2
Wheat/other cereals		
Potato (including sweet potatoes)		
Pulses/beans/nuts	Pulses	3
Milk/milk products	Milk	4
Meat and fish	Meat and fish	4
Poultry		
Eggs		
Fish and sea food (fresh/dried)		
Dark green vegetables—leafy	Vegetables	1
Other vegetables		
Sugar/honey	Sugars	0.5
Fruits	Fruits	1
Oil	Fats and oils	0.5
Spices, tea, coffee, salt, fish power, small amounts of milk for tea	Condiments	0

The maximum FCS has a value of 112 which would be achieved if a household ate each food group every day during the last 7 days. The total scores are then compared with pre-established thresholds. Poor food consumption 0–2, borderline food consumption 28.5–42 and acceptable food consumption > 42

Appendix 2
See Table 9.

Table 9 Food groups for HDDS

Food groups	Score
Cereals	1
White tubers and roots	1
Vegetables	1
Fruits	1
Meat	1
Eggs	1
Fish and other sea food	1
Legumes, nuts and seeds	1
Milk and milk products	1
Oils and fats	1
Sweets	1
Spices, condiments and beverages	1

Dietary diversity scores are calculated by summing the number of food groups consumed in the household or by the individual respondent over the 24-h recall period out of a maximum of 12 per day

References
1. Food and Agriculture Organization. Multiple dimensions of food security. Rome: The State of Food Insecurity in the World; 2013.
2. IPCC. Climate Change 2014: Impacts adaptation and vulnerability, part B regional aspects, Working Group II Contribution to the Fifth Assessment Report of the Intergovernmental Panel on Climate Change, Cambridge University Press, New York; 2014.
3. World Bank. Climate smart agriculture: increased productivity and food security, enhancing resilience and reduced carbon emissions for sustainable development, opportunities and challenges for a converging agenda: country examples. Washington, DC: The World Bank; 2011.
4. Branca G, McCarthy N, Lipper L, Jolejole MC. Climate smart agriculture: a synthesis of empirical evidence of food security and mitigation benefits from improved cropland management. FAO Working Paper No. LAC3/10. Rome; 2011.
5. Bluffstone R, Kohlin G, editors. Agricultural investment and productivity: Building Sustainability in East Africa. UK: Routledge; 2011.
6. Lee DR. Agricultural sustainability and technology adoption: issues and policies for developing countries. Am J Agric Econ. 2005;87(1):325–34.
7. FARA. State of knowledge on CSA in Africa: Case Studies from Ethiopia, Kenya and Uganda. Accra, Ghana; 2015.
8. Teklewold H, Kassie M, Shiferaw B, Kohlin G. Cropping system diversification, conservation tillage and modern seed adoption in Ethiopia: impacts on household income, agrochemical use and demand for labor. J Agric Econ. 2013;64(3):597–623.
9. GOK. National Climate Change Response Strategy. Ministry of Environment and Mineral Resources. Nairobi; 2010.
10. FICCF. A review of climate smart agriculture in the non-ASAL areas of Kenya: finance innovation and climate change fund. Nairobi; 2014.
11. Kabubo-Mariara J, Kabara M. Climate change and food security in Kenya. Environment for Development. Discussion Paper Series No. 15-05; 2015.
12. Ochieng J, Kirimi L, Makau J. Adapting to climate variability and change in rural Kenya: farmer perceptions, strategies and climate trends. Nat Resour Forum. 2016;41(4):195–208.
13. Wineman A, Mason NM, Ochieng J, Kirimi L. Weather extremes and household Welfare in Rural Kenya. Food Secur. 2017;9(2):281–300.
14. Food and Agriculture Organization. Climate smart agriculture: policies, practices and financing for food security. Rome: Adaptation and Mitigation; 2010.
15. Arslan A, McCarthy N, Lipper L, Asfaw S, Catteneo A. Adoption and intensity of adoption of conservation farming practices in Zambia. Agric Ecosyst Environ. 2014;187:72–86.

16. Lukano LS. Busia County integrated development plan, 2013–2017. County Government of Busia: Ministry of Planning and County Development; 2013.

17. Di Falcao S, Veronesi M. On adaptation to climate change and risk exposure in the Nile Basin of Ethiopia. IED Working Paper, 15. ETH, Zurich; 2011.

18. Manda J, Alene AD, Gardebroek C, Kassie M, Tembo G. Adoption and impacts of sustainable agricultural practices on maize yields and incomes: evidence from rural Zambia. J Agric Econ. 2015;67(1):1–24.

19. Goswami R, Biswas MS, Basu D. Validation of participatory farming situation identification: a case of rainfed rice cultivation in selected area of West Bengal, India. Indian J Tradit Knowl. 2012;11:471–9.

20. Chatterjee S, Goswami R, Bandyopadhyay P. Methodology of identification and characterization of farming systems in irrigated agriculture: case study in West Bengal State of India. J Agric Sci Technol. 2015;17:1127–40.

21. Kaiser HF. The varimax criterion for analytic rotation in factor analysis. Psychometrika. 1958;23:187–200.

22. McFadden D. Conditional logit analysis of qualitative choice behaviour. In: Zarembka, P. Frontiers in Econometrics. Academic Press, London, pp. 105–142; 1973.

23. Bourguinon F, Fournier M, Gurgand M. Selection bias corrections based on the multinomial model: Monte Carlo comparisons. J Econ Surv. 2007;21:174–205.

24. Kassie M, Zikhali P, Pender J, Kohlin G. The economics of sustainable land management practices in the Ethiopian highlands. J Agric Econ. 2010;61:605–27.

25. World Food Program. Food consumption analysis: calculation and use of the food consumption score in food security analysis. Rome: Italy; 2009.

26. Bickel G, Nord M, Price C, Hamilton W, Cook J. Guide to Measuring Household Food Security: U.S. Department of Agriculture, Food and Nutrition Service, USDA Guide 2000. Alexandria VA; 2000.

27. Jayawardena R, Byrne NM, Soares MJ, Katulanda P, Yadav B, Hills AP. High dietary diversity is associated with obesity in Sri Lankan adults: an evaluation of three dietary scores. J Agric Food Secur. 2013;13:314. https://doi.org/10.1186/1471-2458-13-314.

28. Arimond M, Wiesmann D, Becquey E, Carriquiry A, Daniels MC, Deitchler M, Fanou-Fogny M, Joseph ML, Kennedy G, Martin-Prevel Y, Torheim LE. Simple food group diversity indicators predict micronutrient adequacy of women's diets in 5 diverse, resource-poor settings. J Nutr. 2010;140(11):2059S–69S.

29. MacCallum RC, Widaman KF, Preacher KJ, Hong S. Sample size in factor analysis: the role of model error. Multivar Behav Res. 2001;36(4):611–37.

30. Lorenzo-Seva U. How to report the percentage of explained common variance in exploratory factor analysis. Technical Report. Department of Psychology, Universitat Rovirai Virgili, Tarragona; 2013.

31. Ali A, Erenstein O. Assessing farmer use of climate change adaptation practices and impacts on food security and poverty in Pakistan. Clim Risk Manag. 2017;16:183–94.

32. Silici L. Conservation agriculture and sustainable crop intensification in Lesotho. Integr Crop Manag. 2010;10:978–1052.

33. Gido EO, Sibiko KW, Ayuya OI, Mwangi JK. Demand for agricultural extension services among small-scale maize farmers: micro-level. J Agric Educ Ext. 2015;21(2):177–92.

34. Teklewold H, Mekonnen A, Köhlin G, Di Falco S. Does adoption of multiple climate smart practices improve farmers' climate resilience? Empirical evidence from the Nile Basin of Ethiopia. Environment for Development, Discussion Paper Series EfD DP 16-21; 2016.

35. Ochieng J, Kirimi L, Mathenge M. Effects of climate variability and change on agricultural production: the case of small scale farmers in Kenya. NJAS-Wagenigen J Life Sci. 2016;2016(77):71–8.

36. Belay A, Recha JW, Woldeamanuel T, Morton JF. Smallholder farmers' adaptation to climate change and determinants of their adaptation decisions in the Central Rift Valley of Ethiopia. J Agric Food Secur. 2017;6(24):1–12.

37. Haghjou M, Hayati B, Choleki DM. Identification of factors affecting adoption of soil conservation practices by some rain fed farmers in Iran. J Agric Sci Technol. 2014;16:957–67.

38. Gebeyehu MG. The impact of technology adoption on agricultural productivity and production risk in Ethiopia: Evidence from Rural Amhara Household Survey. Open Access Libr J. 2016;3:2369–84.

39. Ward PS, Pede VO. Capturing social network effects in technology adoption: the spatial diffusion of hybrid rice in Bangladesh. Aust J Agric Resour Econ. 2014;59:225–41.

40. Shiferaw T, Dargo F, Osman A. Agropastoralist evaluations of integrated sorghum crop management packages in Eastern Ethiopia. Adv Crop Sci Technol. 2015;3(5):2329–8863.

Barriers to and determinants of the choice of crop management strategies to combat climate change in Dejen District, Nile Basin of Ethiopia

Zerihun Yohannes Amare[1*], Johnson O. Ayoade[2], Ibidun O. Adelekan[2] and Menberu Teshome Zeleke[3]

Abstract

Background: Climate change without adaptation is projected to impact strongly the livelihoods of the rural communities. Adaptation to climate change is crucial for least developed country like Ethiopia due to high population and dependency on agriculture. Hence, this study was initiated to examine the barriers to and determinants of the choice of crop management strategies to combat climate change. The Intergovernmental Panel on Climate Change concepts of climate change response provided the framework. Stratified and snowball sampling techniques were employed to select a sample of 398 households. The household survey was employed to collect data on current adaptation strategies. Logistic regression was used to analyse the determinants of the choice of adaptation strategies. Logistic regression analyses were carried out at $p \leq 0.05$.

Results: Small farmland size, agro-ecology, farmland location, financial constraints, and lack of skills were the major barriers to adoption of crop management strategies. Age, farming experience, income, family size, government experts' extension services, agro-ecology setting, and crop failure history of households significantly affect the choice of most of the crop management strategies.

Conclusions: Socio-economic and institutional factors determined rural communities' ability and willingness to choose effective adaptation strategies. Policy priority should be given based on agro-ecology and households demand of policy intervention such as providing extension services and subsidizing the least adopted strategies due to financial constraints.

Keywords: Adaptation, Climate change impact, Crop management practices, Nile Basin of Ethiopia

Background

The warming trends observed over the past few decades continued in 2014. World Meteorological Organization (WMO) has ranked as nominally the warmest year since modern instrumental measurements began in the mid-1800s [1]. The global average near-surface temperature for 2014 was comparable to the warmest years in the 165-year instrumental record. In 2014, the global average temperature was 0.57 ± 0.09 °C above the 1961–1990 average of 14 °C. It was 0.08 °C above the average anomaly of 0.50 °C for the past 10 years (2005–2014) [1].

According to Food and Agricultural Organization [2], due to climate change and variability almost one billion people experienced hunger in 2010 globally. This implies the most marginalized people cannot access enough of the primary macronutrients. Perhaps, other billions are thought to suffer from hidden hunger, in which essential micronutrients are missing from their diet, with consequent risks of physical and mental impairment [3]. The majority (85%) of the Ethiopian population is dependent on agriculture. As a result, agriculture will continue to be

*Correspondence: Zerihun.yohannes19@gmail.com
[1] Institute of Disaster Risk Management and Food Security Studies, Bahir Dar University, BahirDar, Ethiopia
Full list of author information is available at the end of the article

the most important sector in its need to adapt to climate change.

There are many reasons and convincing arguments for a more comprehensive consideration of adaptation as a response measure to climate change. Firstly, given the amount of past greenhouse gas emissions and the inertia of the climate system, we are already bound to some level of climate change, which can no longer be prevented even by the most ambitious emission reductions [4]. Second, the effect of emission reductions takes several decades to fully manifest, whereas most adaptation measures have more immediate and sustainable benefits [5].

Third, adaptations can be effectively implemented on a local or regional scale such that its efficiency is less dependent on the actions of others, whereas mitigation of climate change requires international cooperation. Fourth, most adaptations to climate change also reduce the risks associated with current climate variability, which is a significant hazard in many world regions. According to Gbetibouo [6], there are two adaptation assessment approaches, namely top-down and bottom-up assessment approaches. The top-down approach starts with climate change scenarios, and estimates impact through scenario analysis, based on which possible adaptation practices are identified. Most of the top-down adaptations represent possible or potential measures, rather than those that have been used [6].

Most studies, e.g. [7–9], carried out in Ethiopia and Africa using top-down approach predicted the impact of climate change on the agricultural sector with adverse effects on crop yields. The bottom-up approach takes a vulnerability perspective where adaptation strategies are considered more as a process involving the socio-economic, and policy environments, and elements of decision-making [6]. In line with this notion, Schröter et al. [10] argue that in choosing adaptation options to climate change and developing policies to implement these possibilities the affected community should actively participate. This study adopts the bottom-up approach that seeks to identify actual adaptation strategies at the local level and the factors that determine the choice of crop management strategies in Dejen District, Nile Basin of Ethiopia.

Methods

The study area

The study district is located in west-central Ethiopia (Fig. 1) at a road distance of 335 km south of the regional state capital, Bahir Dar City, and 230 km northwest of the capital city of Ethiopia, Addis Ababa, in the Amhara Regional State at the edge of the canyon of the Blue Nile. The district lies between longitude 38°6′E and 38°10′E, and between latitude 10°7′N 10°11′N, with an elevation of 1071 and 3000 m above sea level (m.a.s.l). The average temperature and total annual rainfall of the district range between 20 and 24 °C and 800 and 1200 mm, respectively. Dejen District is located in the valley of the Blue Nile which is highly undulated topography and frequent susceptibility to climate-related problems such as erratic rainfall, crop pests, livestock diseases, and malaria outbreaks. In the past 8 years (2009–2016) due to climate change impacts, the district loses 50,555 quintals of crops, which has the potential to feed 27,701 individuals. The study district is categorized into three agro-ecological zones, 41% *Dega* (highland), 31% *Woinadega* (midland), and 28% *Kolla* (lowland) [11, 12].

Research design and sampling procedure

The study employed cross-sectional research design with both quantitative and qualitative research methods. This study used a multi-stage sampling technique to select the agro-ecology, *Kebeles* (the lower administrative unit next to district), and households. At the first stage, Dejen District of the Nile Basin was selected purposely due to its highly undulated topography and frequent susceptibility to extreme events and representativeness of the three agro-ecological zones such as highland, midland, and lowland. In the second stage, six *Kebeles* (two from each agro-ecological setting) were selected purposely based on the above-listed district selection criteria.

Climate change affects the rural communities differently in different agro-ecological zones. As a result, communities' knowledge and skill to adapt to the climate change impacts varies from different agro-ecological settings. In the third stage, stratified sampling was employed to select households. Under the stratified sampling, the population was divided into male- and female-headed households, and then the sample was selected from each male- and female-headed household to constitute a representative sample.

The sample size was determined proportionately (Table 1). In Ethiopia context, female-headed households are those who do not have husband due to either being divorced, widowed, or separated. In Ethiopia, in some of the rural communities, disclosing of the marital status of older females is culturally not feeling comfort them. Thus, to get female-headed households, snowball sampling was employed. Based on Yamane [13] at the 95% confidence interval and 5%, level of precision, 398 households were selected at the six *Kebeles* of the district (Table 1).

Data sources and data collection methods

The primary and secondary data sources were quantitative and qualitative in nature. The study used two main data sources to analyse the barriers and determinants of the adoption of crop management strategies to climate

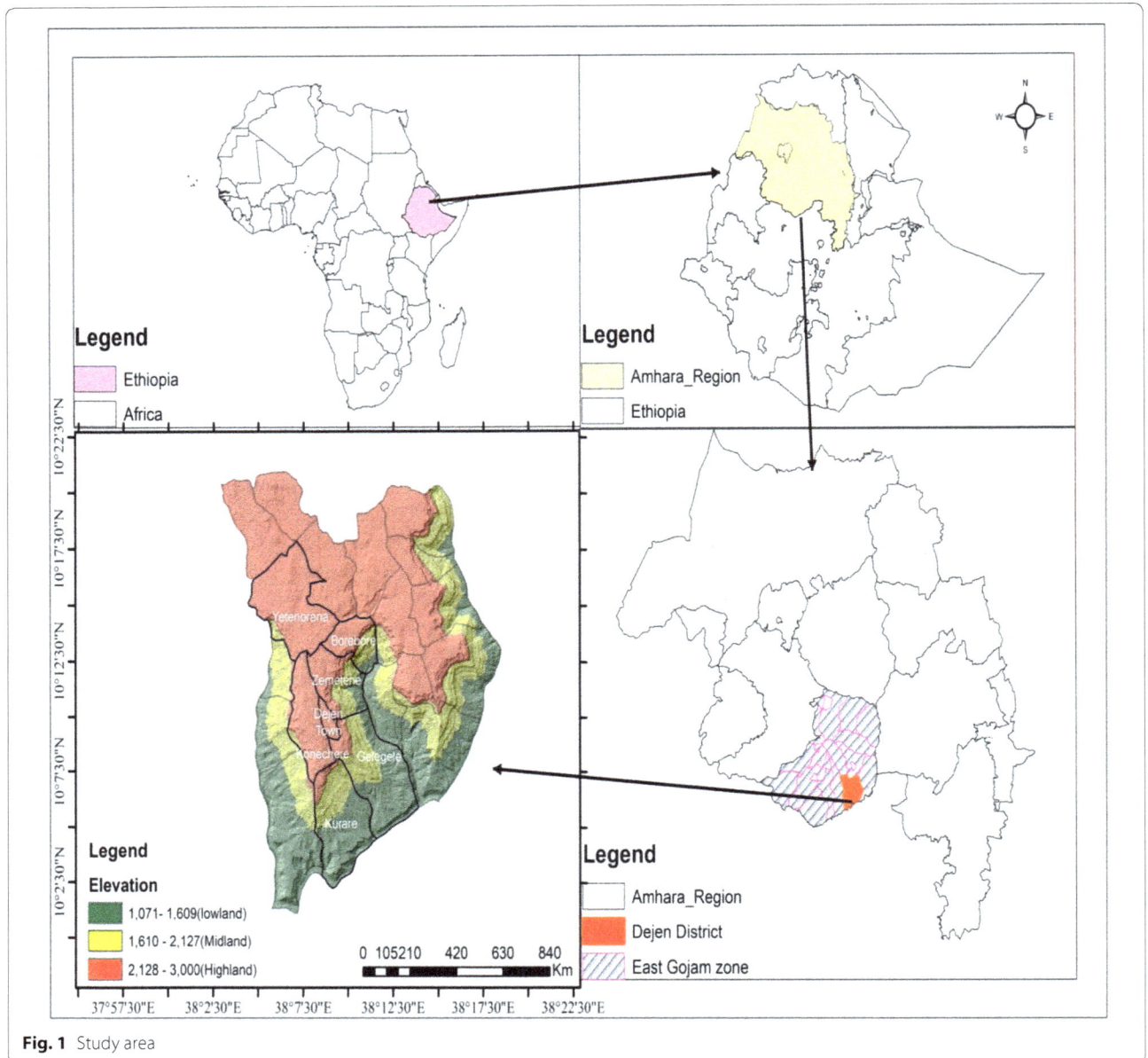

Fig. 1 Study area

change impacts. The main data source was the socio-economic data collected through household survey.

Household survey

The household survey was used to collect quantitative data on households' current adaptation strategies, the barriers to adapt, and factors that affect rural communities' choice of crop management strategies. The survey questions were prepared in English language. It was translated into local language (*Amharic*) during data collection and then encoded into SPSS in *English*

language for data processing and analysis. The household survey was initially pretested to check the appropriateness and validity of the questions from the three agro-ecological zones. Pretested *Kebeles* and participant households were not involved in the actual survey.

After pretesting, ambiguous words were rephrased and inappropriate questions were modified. The experience of the author in the study area played a paramount role in choosing the data collectors who have been working for many years in the rural community in the field of agriculture, environment, and land administration. The data were collected by the trained data

Table 1 Distribution of sample size based on the total study population. (*Source*: computed based on [14])

Sample *Kebeles*	Number of household heads			% of study population	Share of *Kebeles* from 398 HHs
	Male	Female	Total		
Koncher (highland)	995	225	1220	16.5	66
Yetnora (highland)	1388	767	2155	29	115
Zemeten (midland)	652	280	932	12.6	50
Borebor (midland)	644	281	925	12.5	50
Kurar (lowland)	726	363	1089	14.7	59
Gelgele (lowland)	732	345	1077	14.6	58
	5137	2261	7398	100	398

HHs Households

collectors under close supervision of the author in the period March to October 2016.

Methods of data analysis

The descriptive statistics such as percentage and frequencies were used to summarize and categorize the information gathered from households. The binary logistic regression was used to analyse determinants of the choice of crop management strategies.

Before the data collection, a multinomial logistic (MNL) model was proposed based on aspects of the literature. However, in the course of this study, the surveyed households chose more than one adaptation strategies simultaneously. As a result, the use of multinomial

logistic regression was inappropriate. To fix such problems, the possible remedy suggested by Bryan et al. [15] was to combine similar measures into single categories. However, such grouping into self-defined categories may lead to miss interpretation [15, 16]. Besides, the set of explanatory variables affecting the households' decision was also expected to be different for different adaptation strategies. Therefore, this study employed binary logistic regression technique to analyse the determinants of the choice of crop management strategies. Analysis of the logistic regression of binary response model can be defined as:

$$y = \frac{1}{e^{-[\beta 0 + \beta 1(x1) + \beta 2(x2) + \beta n(xn)]}} e^{\beta 0 + \beta 1(x1) + \beta 2(x2) + \beta 3(x3) + \beta n(xn)},$$
(1)

where y is response variable; $y = 1$, outcome is present (adopt); $y = 0$, outcome is not present (not adopt); β_0 is constant.

$\beta_1 + \beta_2 + \beta_3 + \cdots \ \beta_n$ are regression coefficients that explain the change in the log odds for each unit change in x. $x_1 + x_2 + \cdots x_n =$ set of predictor variables included in the model (Table 2). e^{β} represents the change in the odds of the outcome (multiplicatively) by increasing x by 1 unit. The current crop management adaptation strategies in the study area, such as using crop diversification, improved seeds, changing planting date, and replanting failed crops, were used as dependent variables. The choices of these adaptation strategies were determined by socio-economic and demographic characteristics, institutional factors, and agro-ecological settings (Table 2). Based on Agresti [17], logistic regression method can be used when the dependent variable Y is dichotomous. Y

Table 2 Description of explanatory variables that affect households' choice of crop management strategies. *Source*: Author (2016)

Explanatory variables	Description
Sex	Dummy, 1 = [a]Male, 0 = Female
Age	Discrete (years), [a]18–35, 36–55, > 55
Education	Discrete (years), [a]cannot read and write, primary school and above)
Farming experience	Continuous (years), [a]10, 10–20, 21–30, > 30
Income	Continuous (ETB), [a]< 10,000, 10,001–30,000, 30,001–50,000, > 50,000
Family size	Discrete (number), [a]< 4, and > 4
Access to weather information	Dichotomous, 1 = [a]Yes, 0 = No
Access to farmer to farmer extension services	Dichotomous, 1 = [a]Yes, 0 = No
Access to government experts' extension services	Dichotomous, 1 = [a]Yes, 0 = No
Agro-ecology settings	Dummy, 1 = [a]highland; 2 = midland; 3 = lowland
Farmland size	Continuous (hectare), [a]< 1.2, and > 1.2
Crop failure history of households	Dichotomous, 1 = [a]Yes, 0 = No

[a] Taken as reference (base for analysis), ETB is Ethiopian currency ($1 = 22.3ETB) [18]

is coded as 1 when the outcome is present and coded 0 when the outcome is not present.

Test of goodness of fit and multicollinearity

The fitness of the logistic regression model to the data was measured by using the SPSS classification table (crosstabs) and the Hosmer–Lemeshow test. Besides, collinearity among predictor variables was checked using multicollinearity statistics. The Hosmer–Lemeshow test used 95% confidence interval (CI) and asymptotically follows a X^2 distribution.

Empty cells or small frequency was checked by doing crosstabs between categorical predictor variables and the outcome variables. When the cell has very few cases, the model becomes unstable. Based on Kothari [19] and SPSS [20], the Hosmer–Lemeshow statistics indicate a poor fit if the significance value (p) is less than 0.05 (Table 3).

Techniques to remedy poor fit of the model were by: (1) using re-categorized (for example educational levels were categorized as cannot read and write and primary school level and above), (2) dropping the least theoretically important explanatory variables that contribute to the model a poor fit to the data. For example, explanatory variables such as family size and crop failure history of households were excluded from entering and competing in the model in the improved seed and crop diversification adaptation options, respectively.

In the logistic regression model, the Exp (B) is the "Odds ratio" which explains the effect of the independent variable (X_n) on the dependent variable. The beta coefficient (β) is the estimated logit coefficient which is the rate of change in the Y (the dependent variables) as X (independent variable) changes. When the beta coefficient (β) is negative, it shows that the dependent and independent variables have an inverse relationship, and when it has a positive coefficient, there is a positive relationship. Odds ratio = 1 indicates the same probability of an event occurring between the two situations. Odds ratio > 1 probability of an event occurring with a unit increase in the independent variable is higher than the reference/base variable. Odds ratio < 1 probability of an event occurring with a unit increase in the

Table 3 Hosmer and Lemeshow goodness-of-fit test results of logistic regression model. (*Source*: Computed based on household survey data, March–October (2016))

Dependent variables	Chi-square	df	P value (> 0.05)
Crop diversification	7.763	8	0.457
Improved seeds	10.78	8	0.214
Changing planting date	9.239	8	0.323
Replanting failed crops	2.416	8	0.966

df degree of freedom

independent variable is lower than the reference/base variable (Table 2).

Multicollinearity was assessed by examining tolerance and variance inflation factors (VIF). The variance inflation factor (VIF) quantifies how much the variance is inflated. The tolerance is the percentage of the variance in a given predictor that cannot be explained by the other predictors. When VIF > 5, X (the explanatory variable) is highly correlated with the other explanatory variables [20, 21]. Mathematically, variance inflation factor (VIF) can be expressed by:

$\text{VIF} = \frac{1}{1-R^2_j}$, where R^2_j is the coefficient of determination of a linear regression model that uses X_i as the response variable and all other X variables as the explanatory variables. Tolerance is the reciprocal of VIF (i.e. tolerance $= \frac{1}{\text{VIF}}$). A tolerance of less than 0.20 and/or a variance inflation factor (VIF) greater than 5 and above indicates a multicollinearity problem. However, in this study, the results of the variance inflation factor indicated that there was no multicollinearity problem.

Results

Barriers to crop management strategies

Implementation of crop management strategies used by the rural communities varied among households. The study identified a number of constraints faced by the households to adopt crop management strategies to combat climate change impact.

Crop diversification

The major constraints identified by respondents to not to adopt crop diversification were: small land size (57%) followed by soil fertility decrement (25.3%), shortage of money to buy some expensive crop varieties (1.3%), shortage of labour (2.5%) to implement some labour-intensive farming practices, and they prefer the easiest crop type, lack of skill (3.8%) to sow different crops and stick to one types of crop, topography of farmland location (2.5%) which permits only some crop types to grow, and the remaining (2.5%) do not/have small land size. The majority (58%) of rural communities have less than 1.2 hectares. As a result, small land size/no land at all consequences is barrier to the choice of crop management strategies. On the other hand, rural communities who are located in the lowlands of the undulated topography encountered farm soil fertility decrement. Due to this reason, farmers were obliged to leave their farmland (fallowing) for some periods instead of diversifying different crops as an adaptation strategy to climate change.

Improved seeds

The study communities adopt improved seeds (84.4%). However, the remaining households did not use improved seeds for one or another reason: financial constraints (42.6%), lack of skills (18%) on how to use, compatibility problems with their farmland (16.4), small/no land at all (13.1%), and lack of information (9.8%). Rural communities need all the support they can get to fight the adverse impacts of climate change and extreme weather events. Improved seed varieties developed by research institutes offer higher yields and stronger resistance to challenges related to climate change such as drought. Improved seed tolerates weeds and other climate change-related diseases.

Changing planting dates

Planting dates are growing season during which the rainfall and temperature allow plants to grow. Among interviewed households, 78.9% used changing planting dates, whilst others do not use this method. Among those who did not use changing planting dates, 87.6% of the respondents attributed lack of skills as a barrier to adaptation methods, whilst 12.3% have not/small land size.

Replanting damaged crops

Weather events such as flooding, hailstorms, disease outbreaks can damage previously planted crops in all or a portion of farm fields. This requires technical assistance for decision-making in replanting. The majority (92.5%) of respondents replant their failed crops. However, among those who did not use, the majority (63%) of households indicated lack of skills about future weather forecast and economic return of the replanting crops, suitability of land and cropping season (7.4%) and small land size/no land (29.6%) contributes for not using replanting damaged crops.

Determinants of the choice of crop Management strategies

The determinant factors of the choice of adaptation strategies are presented in Table 2. Analyses were carried out at $p \leq 0.05$.

Changing crop management practice is one of the adaptation practices to climate change impacts. For this study using crop diversification, improved seeds, changing planting date, and replanting failed crops were selected in the context of the study sites. The applications of these strategies have been determined by a number of socio-economic, biophysical, and institutional factors (Table 4).

Age of the household head

Crop diversification is considered as an important adaptation strategy to combat climate change impacts. The logistic regression model results indicated that adult-headed households have a significant ($p = 0.010$) effect on adopting crop diversification (Table 4). This means adult household heads (age 36–55) are 3.506 times more likely to use crop diversification than young-headed households (age 18–35). Adult-headed households (36–55) have a significant ($p = 0.010$) effect on adopting improved seeds ($p = 0.011$). The beta coefficient ($+1.415$) shows positive relationships in explaining adopting improved seeds. This indicates there is an increase in the log of odds using improved seeds by 1.415 in adult-headed households (HHH1). Exp (B) of 4.115 indicates that adult-headed households are 4.115 times more likely to use improved seeds than young- and old-headed households. Old-headed households have a significant ($p = 0.038$) effect on adopting changing planting date (age > 55 years) (HHH2); Exp (B) of 5.985 indicates 5.985 times more likely to adopt changing planting date than (age 18–35). This indicates that, as age increases, the probability of adopting changing planting date as adaptation strategy increased.

Adult- and old-headed households have a significant ($p = 0.005$; 0.014) power in explaining replanting failed crops. Age shows positive relationship (Beta = HHH1, $+1.689$ and HHH2, $+3.470$). This indicates that there is an increase in the log of odds by 1.689 and 3.470. Adult-headed households (HHH1) Exp (B) of 5.416 indicate 5.4416 times more likely, and old-headed households (HHH2) Exp (B) of 32.143 times much more likely to use replanting their failed crops than young-headed households (age 18–35 years). The possible explanation is that age of household head increases the possibility of pursuing replanting failed crops as climate change adaptation strategy.

Farming experience

Farmers in the range of farming experience 10–20 years (HHH1) have a significant effect on adopting improved seeds ($p = 0.000$). The beta coefficient indicates positive relationships in adopting improved seeds. This implies there is an increase in the log of odds by 2.319 in using improved seeds. The EXP (B) of 10.166 indicates that households having farming experience of 10–20 (HHH2) are 10.166 times more likely to adopt improved seeds than households having farming experience of fewer than 10 years (HHH), 21–30 years (HHH2), and > 30 years (HHH3).

The farming experience of 21–30 years and > 30 years has a significant effect ($p = 0.007$ and 0.018) on adopting changing planting date. Farming experience of 21–30 years (HHH2, Beta = -1.567) and > 30 years (HHH3, Beta = -1.659) indicates that there is a decrease in the log of odds by 1.567 and 1.659 (inverse relationships). The Exp (B) of 0.209 and 0.190 indicates farming

Table 4 Determinants of households' choice of crop management strategies to climate change impact. (*Source*: Computed from household survey, March–October (2016))

Explanatory variables	Crop diversification		Improved seed		Changing planting dates		Replanting failed crops	
	p	Exp (B)	p	Exp (B)	p	Exp (B)	p	Exp (B)
Sex_HHH (1)	.170	1.850	.983	.991	.055	.376	.104	.332
Age_HHH	.035		.039		.102		.008	
Age_HHH (1)	.010*	3.506	.011*	4.115	.092	2.360	.005*	5.416
Age_HHH (2)	.110	3.662	.111	3.886	.038*	5.985	.014*	32.143
Edu_HHH (1)	.731	1.149	.226	1.690	.725	.867	.456	1.513
Farm_exp_HHH	.233		.004		.046		.603	
Farm_exp_HHH (1)	.108	2.384	.000*	10.166	.120	.428	.261	2.246
Farm_exp_HHH (2)	.884	.917	.238	2.100	.007*	.209	.496	1.796
Farm_exp_HHH (3)	.533	.636	.078	3.943	.018*	.190	880	.864
Income_HHs	.000		.013		.097		.697	
Income_HHs (1)	.000*	8.481	.012*	3.408	.639	1.378	.558	1.490
Income_HHs (2)	.000*	17.510	.029*	3.632	.051	4.163	.241	2.701
Income_HHs (3)	.000*	18.539	.002*	9.064	.281	2.275	.444	1.998
Family_size_HHs (1)	.820	.902	N/C	N/C	.416	1.490	.002*	.101
Weather_info_HHH (1)	.364	.668	.182	.526	.102	.533	.774	.848
Farmer to farmer extension (1)	.809	1.118	.424	1.455	.658	.806	.093	3.286
Government experts extension (1)	.003*	.271	.635	.815	.124	.520	.064	.290
Agro-ecol._HHs	.005		.000		.000		.022	
Agro-ecol._HHs (1)	.036*	4.082	.055	5.446	.002*	5.412	.200	2.922
Agro-ecol._HHs (2)	.093	.496	.002*	.218	.000*	145.815	.006*	11.247
Farmlandsize (1)	.001*	4.286	.865	.931	.000*	.150	.02*	4.570
Crop failure	N/C	N/C	.205	1.656	.044*	.345	.211	0.146
Constant	.121	.245	.305	.347	.002	23.795	.894	1.196

N/C not computed, *HHH* household head, *HHs* households

*Significant at 0.05

experience of (HHH2) only 0.209 times and (HHH3) only 0.190 times (much less) likely to adopt changing planting date.

Income

Income has a positive and significant ($p = 0.000$) effect on adopting crop diversification. Exp (B) of income (10,001–30,000) 8.481 times, income (30,001–50,000) 17.510 times, and income (>50,000) 18.539 times more likely to diversify crops than low-income groups (< 10,000). Income of households has a positive (Beta, + 1.226, + 1.290, and + 2.204) and significant ($p = 0.012$, 0.029, and 0.002) effect on adopting improved seeds. This indicates there is an increase in the log of odds by 1.226, 1.290, and 2.204 in adopting improved seeds as an adaptation strategy. The EXP (B) of income (HHs1) 3.408 times, income (HHs2) 3.632 times, and income (HHs3) 9.064 times is more likely to adopt improved seeds than low-income households (< 10,000).

Family size

Family size has a significant ($p = 0.002$) effect on adopting replanting failed crops. Family size has an inverse (Beta $= -2.297$) in the log of odds by 2.297. The Exp (B) of 0.101 indicates households having family size > 4, only 0.101 times (much less) likely to replant failed crops. The inverse of Exp (B) of 0.101 indicates small family sizes (< 4) are 9.91 times (much more) likely to replant failed crops than large family sizes.

Access to government experts' extension services

Formal extension services from government experts have a significant ($p = 0.003$) effect on adopting crop diversification to combat climate change impacts. The beta coefficient shows an inverse relation (-1.306) in adopting adaptation strategies. This indicates households who did not get extension service; there is a decrease in the log of odd in diversifying crops by 1.306. The Exp (B) of 0.271 indicates that households who did not get formal extension services were only 0.271 times (i.e. much less) likely to diversify crops than households who have got extension service during 12 months of the year.

Agro-ecology settings

Significant variation in the adoption of crop diversification was observed across agro-ecological zones (midland, $p = 0.036$). For example, higher crop diversification was identified in the midland than highland and lowland agro-ecological settings. The Exp (B) of 4.082 indicates households who live in the midland 4.082 times more likely to use crop diversification than households reside in highland. The Exp (B) of 0.496 indicates the lowland households are only 0.496 times much less likely diversifying crops than highlands. This means highland resident households are 2.016 times more likely to diversify crops than lowland households (i.e. invert, $1/0.496 = 2.016$).

The lowland agro-ecology with an inverse beta value (-1.552) has a significant effect ($p = 0.002$) on adopting improved seeds. This indicates a decrease (1.522) in the log of odds on adopting improved seeds in the lowland agro-ecology zones. The EXP (B) of 0.218 indicates the lowland households are only 0.218 times (much less) likely to use improved seeds than highland households. This indicates that highland households are 4.587 times (much more) likely to use improved seeds than lowland resident households (inverse of $1/0.218 = 4.587$).

The midland and lowland agro-ecologies have a significant effect on changing cropping date ($p = 0.002$ and $p = 0.000$), respectively. The coefficient of beta ($+1.689$ midland (HHs1) and $+4.982$ lowland (HHs2)) indicates there is an increase in the log of odds by 1.689 and 4.982 on adopting changing planting date. The Exp (B) of 5.412 indicates that households who reside in the midland agro-ecology are 5.412 times much more likely to change their cropping date than highland households.

The mid- and lowland agro-ecologies have a significant ($p = 0.006$; 0.020) effect on adopting replanting failed crops. There is an increase (Beta $= +1.072$ midland (HHs2) and $+2.420$ lowland (HHs2)) in the log of odds by 1.072 and 2.420. The Exp (B) of 2.922 indicates households who live in the midland agro-ecologies are 2.922 times more likely to use replanting than highland households. The Exp (B) of 11.247 indicates the lowland households are 11.247 times (much more) likely to replant their failed crops.

Farmland size

Farmland is the most significant ($p = 0.001$) factor to diversify crops in the study communities. The beta coefficient of households having farm size of > 1.2 hectare has a positive relation to diversify crops. This implies there is an increase in the log of odds in diversifying crops by 1.455. The Exp (B) of 4.286 means households

having farmland size > 1.2 hectares are 4.286 times more likely to diversify crops than households with < 1.2 hectare farmland. Farmland size (> 1.2 hectares) of households has an inverse (beta, -1.898) and significant ($p = 0.000$) effect on adopting changing planting date. This indicates a decrease (inverse relationships) in the log of odds by 1.898 in changing planting date. The Exp (B) of 0.150 indicates households having large farmland size (> 1.2 hectares) are only 0.150 times changed their planting date. Households having small land size (< 1.2 hectare) are 6.76 times more likely to use changing planting date. Farmland size has a positive relationship (Beta, $+1.519$) with no significant effect on adopting replanting failed crops. There is an increase in the log of odds by 1.519. The Exp (B) of 4.570 indicates households having > 1.2-hectare land are 4.2570 times more likely to replant failed crops than households with < 1.2 hectares.

Crop failure history of households

Crop failure has a significant ($p = 0.044$) effect on adopting changing planting date. Households who never faced crop failure in the past 10 years have an inverse relationship (Beta $= -1.064$) in employing changing planting date. The Exp (B) of 0.345 indicates households who never faced crop failure in the past 10 years are only 0.345 times adopted changing planting date. The invert of Exp (B) of 0.345 is 2.8986, which indicates households who faced crop failure are 2.8986 times more likely to change planting date. This implies most farmers learn only when they faced problems.

Discussion

The rural communities of Dejen District adopt crop management strategies to combat climate change impacts. However, the key barriers identified in the study district were shortage of money, lack of access to information, and small land size. Previous studies (e.g. [22–24]) stated that financial barriers are one of the barriers that restrict implementation of adaptation strategies. This implies every form of adaptation requires some direct or indirect costs. For instance, the use of improved varieties of crops has been reported as one of the key adaptation strategies for farmers in Dejen District, Nile Basin of Ethiopia, where this study confirmed. In the context of this study, improved seeds include high yielding varieties, drought tolerant, short maturing, pest- and disease-resistant species either induced or indigenous. When improved seed varieties are available, their price may be prohibiting making it difficult for many rural households to access. Thus, framers have often sought to use their own saved seeds. One of the possible causes of financial barriers in the study area could be due to lack of credit facilities to rural communities.

Access to information on weather and climate change is an important tool that can be used to enhance the adaptation and implementation of adaptation strategies by rural communities of the study area. Access to information is particularly important for Africa [25] and Ethiopia in particular, where there are few climate projections due to lack of appropriate climate data. This is crucially important, considering that most farming systems in Dejen District depend on rain-fed agricultural systems. Hence, lack of appropriate climate information could be crucial for rural communities' food security.

Age of household heads significantly determined crop diversification, improved seeds and changing planting date, and replanting failed crops. Crop diversification and replanting of the failed crops require more energy and experience. Thus, adult household heads are more matured and active in sowing different crops than old and young household heads. The possible reason for the positive and significant association is due to the fact that age is the proxy indicator that may likely to endow the farmers with the requisite experience that enables them to make a better decision in the choice of crop management strategies. This is in line with studies by Deressa et al. [26] who found that an increase in age of household head does mean an increase in farming experience which would increase rural communities' local knowledge to respond to hazards resulted in climate change and variability.

Farming experience is one of the significant variables that affect the rural communities' choice of adaptation strategies. Farming experience is a proxy indicator of age. Like crop diversification, the middle age household heads have ability and willingness to adopt improved seeds to adjust climate change impacts. This implies as one become more experienced in farming, the probability of one to use improved seeds increases more than a farmer with less farming experience.

On the other hand, farming experience has an inverse relationship with changing planting date. The reason for an inverse relationship might be that experienced farmers will have access to irrigation and water harvesting for their agricultural activities and plant their seeds without changing the planting date. This implies a farmer with more experience would know when climate variability is occurring in the area and which method of adaptation strategies works well in that specific agro-ecology zone. As expected, income is positively and significantly associated with the household decision to pursue crop diversification and improved seeds. This means crop diversification and purchasing of improved varieties of seeds require money. This implies the rate of using crop diversification and improved seeds is increased as income of households increased.

Family size is negatively and significantly associated with the households' decision to pursue replanting failed crops. Households who have large family size are supposed to have an opportunity in pursuing various adaptation options to combat impacts of clime change and variability. This argument is raised by previous studies [26, 27] who argued that large family size is associated with higher labour endowment which would enable a household to accomplish various agricultural tasks. The possible reason for an inverse relationship might be due to the fact that community's expectation of the benefits of using adaptation strategy. In this regard, Barungi and Maonga [28] based on the rational choice theory; argue that the behaviour of human beings is motivated by the possibility of gaining benefit. The possible explanation could be households who have large family size have the possibility to engage in off-farm activities, and they will ignore the failed crops to replant. Therefore, rural communities are rationale consumers of new technologies, and they will only adopt technology as they foresee it will result in increased productivity.

Access to government extension services has a negative and significant association with the likelihood of choosing crop diversification to combat climate change impacts. This result is in contrary with previous studies [29, 30] who noted that farmers who obtain agricultural extension services through extension workers are more likely informed about the climatic situation and the responses followed. The contributing factors for this inverse relationship could be barriers to adopting crop diversification such as inadequate extension services, constraints of money, labour, skills, and farmland locations.

Households who live in the midland and highland agro-ecologies have a significant and positive effect on adoption of crop diversification. This is because the suitability of highland agro-ecology to sow different types of crops and access to government extension services due to proximity to the administration. For instance, in this study finding, the midland agro-ecology has got more access to extension services (77%) than the lowland agro-ecology (47%) communities by the government extension experts in the past cropping season.

The lowland agro-ecology resident households have a negative and significant effect on adoption of improved seed varieties. The possible explanation is that lowland households did not use improved seeds because of suitability problem of the lowland agro-ecology and topography to use improved seeds to their farmland. This was confirmed by households report on the barriers to adopt improved seeds as crop management strategy. On the other hand, the lowland agro-ecology has a positive and significant effect on pursuing changing planting dates to combat climate change impacts. This is because lowland agro-ecologies are characterized by erratic rainfall and other extreme events that lead households to change their planting date. The midland and lowland agro-ecologies have a significant effect on employing replanting failed crops as crop management strategy to combat climate change impacts. This is due to the fact that the midland and lowland households are characterized by climate variability such as erratic rainfall than the highland agro-ecology households. The exposure of climatic variability gave them more experience in adopting replanting their failed crops than highland households.

As expected, farm size has a significant and positive effect on adopting crop diversification to combat climate change impacts. Households with larger farm sizes were more likely to diversify their crops. On the other hand, larger farmland size has a negative and significant effect on using changing planting dates as crop management strategies. This means households having small land size (< 1.2 hectares) are more likely use changing planting dates. This reminds us "a hunter who has only one arrow does not shoot with careless aim". This implies households who have small land size took care of their farmland and changed their planting dates when there is a change in weather conditions. Even if farmland size has no significant effect on changing failed crops, it shows a positive effect on using changing failed crops. The possible explanation could be the more farmland plot they have, the more is the probability of having failed croplands that could lead them to replant their failed crops.

Conclusion and policy recommendations

The study communities have tremendous ideas to adapt for current and future climate change impacts with a strong motivation to move out of poverty. However, the mere willingness to adopt climate change adaptation strategies was not enough. Their ability to adopt is constrained by many internal and external factors. Rural communities who did not employ adaptation strategies gave many reasons for their failure to adopt. These includes poor or no access to water sources, limited knowledge, and skill, shortage of labour, lack of and/or shortage of farmland, lack of money, lack of information, lack of agricultural extension services, and other institutional factors.

The most significant determinant factors of the choice of crop management strategies were age, farming experience, income, agro-ecology setting, and farmland size. Agro-ecology setting has a significant effect on all adaptation strategies. Due to the soil characteristics, the lowland agro-ecology zones were not suitable for adopting improved seeds. However, the government bodies in the office of agriculture did not realize the problems. This implies, in the process of diffusion of adaptation strategies, climate change adaptation process should require close collaboration and active participation of climate change researchers, decision-makers, policy analysts, the community, and partners. Government policies should strengthen the current adaptation strategies practised by rural community households and support the adoption of crop management strategies. Besides, the less adopted crop management strategies due to financial constraints should be subsidized by government and aid organizations. This study contributes to the academic discourse on climate change impact adaptations by providing empirical evidence to deepen understanding of the barriers and determinants that confronts rural communities in their attempt to implement adaptation strategies to manage the negative impacts of climate change and variability.

Abbreviations

df: degree of freedom; HHs: households; HHHs: household heads; NC: not computed; m.a.s.l: metres above sea level; VIF: variance inflation factor.

Authors' contributions

ZYA designed the data collection tools, conducted fieldwork and analysis, and developed the manuscript. JOA contributed in commenting the data collection tools, recommending data analysis methods, reviewed and made editorial comments on the draft manuscript. IOA contributed in commenting the data collection tools, data analysis methods, reviewed and made editorial comments on the draft manuscript. MTZ contributed to developing the data collection tools, commenting data analysis methods, reviewed, and made editorial comments on the draft manuscript. All authors read and approved the final manuscript.

Author details

Institute of Disaster Risk Management and Food Security Studies, Bahir Dar University, BahirDar, Ethiopia. [2] Department of Geography, Faculty of the Social Sciences, University of Ibadan, Ibadan, Nigeria. [3] Department of Geography and Environmental Studies, Debre Tabor University, Debre Tabor, Ethiopia.

Acknowledgements

This study was made possible by the financial support of Pan-African University (PAU), a continental initiative of the African Union Commission (AU), Addis Ababa, Ethiopia. Further, authors would like to thank the data collectors and field assistants for their effective coordination, support, and time spent in organizing and conducting successful household interviews. Special thanks to the Dejen District rural communities who willingly volunteered the information in this study.

Competing interests

The authors declare they have no competing interests.

Funding

This study was sponsored by the Pan-African University (PAU), a continental initiative of the African Union Commission (AU), Addis Ababa, Ethiopia.

References
1. WMO. Statement on the status of the global environment, vol. 1152. Geneva: Japan Meteorological Agency, in Cooperation with the World Meteorological Organization; 2015. p. 250.
2. FAO. Climate change and food safety: a review. Food Res Int. 2010;43(7):1745–65.
3. Foresight U. The future of food and farming. Final project report. London: The Government Office for Science; 2011.
4. Füssel H-M, Klein RJ. Climate change vulnerability assessments: an evolution of conceptual thinking. Clim Change. 2006;75(3):301–29.
5. Rahman MI-U. Climate change: a theoretical review. Interdiscip Descr Complex Syst. 2013;11(1):1–13.
6. Gbetibouo GA. Understanding farmers' perceptions and adaptations to climate change and variability: the case of the Limpopo Basin, South Africa, vol. 849. Washington: The International Food Policy Research Institute; 2009.
7. Segele ZT, Lamb PJ. Characterization and variability of Kiremt rainy season over Ethiopia. Meteorol Atmos Phys. 2005;89(1):153–80.
8. NMA. Climate change national adaptation programme of action (Napa) of Ethiopia. National Meteorological Services Agency (NMA), Ministry of Water Resources, Federal Democratic Republic of Ethiopia, Addis Ababa; 2007.
9. You GJ-Y, Ringler C. Hydro-economic modeling of climate change impacts in Ethiopia. International Food Policy Research Institute (IFPRI); 2010.
10. Schröter D, Polsky C, Patt AG. Assessing vulnerabilities to the effects of global change: an eight step approach. Mitig Adapt Strat Glob Change. 2005;10(4):573–95.
11. DDARDO. Dejen District Agricultural and Rural Development Office (DDRDO), Annual Report, East Gojjam zone, Dejen, Ethiopia; 2016.
12. DDEPO. Dejen District Environmental Protection Office (DDEPO), East Gojjam zone, Dejen, Ethiopia; 2016.
13. Yamane T. Statistics: an introductory analysis. New York: Harper and Row; 1967.
14. DDFEDO. Dejen District Finance and Economic Development Office (DDFEDO) Population Projection (DDFED, 2014); 2014.
15. Bryan E, et al. Adapting agriculture to climate change in Kenya: household strategies and determinants. J Environ Manag. 2013;114:26–35.
16. Abid M, et al. Farmers' perceptions of and adaptation strategies to climate change and their determinants: the case of Punjab province, Pakistan. Earth Syst Dyn. 2015;6(1):225.
17. Agresti A. An introduction to categorical data analysis. 2nd ed. Hoboken: Wiley; 2007.
18. NBE National Bank of Ethiopia. Commercial Banks' Exchange rate; 2016.
19. Kothari CR. Research methodology: methods and techniques. Chennai: New Age International; 2004.
20. SPSS. Statistical Package for the Social Sciences (SPSS). Version20.
21. Kothari GG. Research methodology. 3rd ed. New Delhi: New Age International Publishers; 2014.
22. Bryan E, et al. Adaptation to climate change in Ethiopia and South Africa: options and constraints. Environ Sci Policy. 2009;12(4):413–26.
23. Kithiia J. Climate change risk responses in East African cities: need, barriers and opportunities. Curr Opin Environ Sustain. 2011;3(3):176–80.
24. Peterson C. Fast-growing groundnuts keep Ghana's farmers afloat amid climate shifts. Retrieved July, 2013, vol. 16, p. 2013.
25. IPCC. Climate change: climate change impacts, adaptation and vulnerability. The fourth assessment report of the Intergovernmental Panel on Climate Change. Geneva, Switzerland; 2007.
26. Deressa TT, et al. Determinants of farmers' choice of adaptation methods to climate change in the Nile Basin of Ethiopia. Glob Environ Change. 2009;19(2):248–55.
27. Menberu TZ, Aberra Y. Determinants of the adoption of land management strategies against climate change in Northwest Ethiopia. Ethiop Renaiss J Soc Sci Humanit. 2014;1:93–118.
28. Barungi M, Maonga BB. Adoption of soil management technologies by smallholder farmers in central and southern Malawi. J Sustain Dev Afr. 2011;13(3):28–38.
29. Maddison D. The perception of and adaptation to climate change in Africa, vol. 4308. Washington: World Bank Publications; 2007.
30. Nhemachena C, Hassan R. Micro-level analysis of farmers adaption to climate change in Southern Africa. Washington: The International Food Policy Research Institute; 2007.

Effect of planting time on growth, yield components, seed yield and quality of onion (*Allium cepa* L.) at Tehuledere district, northeastern Ethiopia

Maria Tesfaye[1], Derbew Belew[2], Yigzaw Dessalegn[3] and Getachew Shumye[4*]

Abstract

Background: Onion (*Allium cepa* L.) is member of the family Alliaceae and the most widely grown herbaceous biennial vegetable crop. Quality planting material is one of the major inputs to successful vegetable production. However, it is one of the major constraints in Ethiopia. Northeastern Ethiopia has suitable agro-climatic condition for onion seed production. However, onion seed production packages, including its appropriate planting time, are not yet determined. Evidences on effects of the different planting time on quality and yield level are not well explored. Therefore, this experiment was conducted at Jari small-scale irrigation scheme from September 2015 to April 2016 to determine an appropriate planting time for a better plant growth, yield components, seed yield and quality of Adama red onion variety.

Methods: The experiment was laid out in randomized complete block design with three replications. Treatments were nine planting dates: September 1st, September 16th, October 1st, October 16th, October 31st, November 15th, November 30th, December 15th and December 30th. Data were collected on growth, yield components, seed yield and quality parameters and analyzed using SAS version 9.2 statistical software.

Results: Analysis of variance revealed that plant height, number of leaves per plant, number of scapes per plant, scape diameter, scape height, days to 50% flowering and maturity, umbel diameter, number of seeds per umbel, 1000-seed weight, seed yield and germination percentage were significantly influenced by planting time. The highest seed yield (1032.7 kg/ha) and the highest germination percentage (94.3%) were recorded from bulbs planted early (September 1st). On the other hand, the lowest seed yield (29.7 kg/ha) and germination percentage (15.3%) recorded from bulb planted late (December). The correlation values explain the apparent association of the planting time parameters with each other and clearly indicated the magnitude and directions of the association and relationships.

Conclusion: The September 1st is recommended as appropriate planting time for onion seed production at Jari, northeastern Ethiopia.

Keywords: Onion, Planting date, Seed yield, Seed quality

Background

Onion (*Allium cepa* L.) is one of the most important vegetable crops grown in Ethiopia. It ranks first among

Allium species grown in Ethiopia both in area coverage and total production [1]. Its area coverage was 24,357.7 ha, and total annual production was 219,735.3 tons with average productivity of 9 t/ha during 2013/14 cropping season [1]. The area coverage of onion is steadily increasing mainly due to its high profitability, ease of production, and the expansion of irrigation infrastructure in different parts of the country [2]. Likewise, the

*Correspondence: aytenew2001@gmail.com
[4] Department of Plant Science, College of Agriculture, Wollo University, P.O. Box 1145, Dessie, Ethiopia
Full list of author information is available at the end of the article

demand for quality onion seed is steadily increasing over time [3]. However, seed supply is inadequate, its price is increasing, and onion seed available in the market is poor in quality [2]. Onion seeds lose viability within a year, and hence, they are poor in maintaining quality. Owing to these challenges, onion seed production gradually started by smallholder farmers in different parts of the country.

Onion seed production is influenced by many factors, among which varieties, bulb size, soil, climate, spacing, fertilizer application and date of planting are important. Cool weather with ample moisture supply is required for flower stalk initiation. Then, drier conditions with good sunshine are required for pollination, seed maturity, harvesting and processing [2]. High temperature during flowering results in flower abortion and subsequently results in lower seed yield. On the other hand, very low temperature, foggy weather and rainfall during flowering time affect the movement of honey bees and the pollination process. Rainfall during harvesting time adversely affects the quality of onion seed. Therefore, selection of appropriate planting date to align the time of maturity during dry season is crucial for onion seed production in a given locality.

The effect of planting time on onion seed production and its significant effects on both productivity and quality was studied and reported by several researchers in different parts of the world. The effect of different planting dates on onion seed production in Bangladesh was done, and October 30th was recommended as the best planting date [4]. Similarly, November 15th identified as the best planting date for onion seed production on Borga region of Bangladesh with a total rain fall of 351 mm [5] and Giza region of Egypt with mean minimum and maximum temperature of 15.60 and 29.8 °C, respectively [6]. In Iran, the best planting dates for onion seed production was from September 22nd to October 6th and from October 21st to November 5th for onion varieties Texas Early Grano 502 and Germez Iranshah, respectively, with mean minimum and maximum temperature of 21.48 and 9.09 °C, respectively [7].

Although Ethiopia has very diverse agro-ecology, studies on the effect of planting time on onion seed production are limited. Studies in Central Ethiopia showed that onion seed production is best if mother bulbs are planted in September and October which helped to align the time of flowering on the months of January and February, during cooler and drier months [2]. However, in Kobo October 25th was recommended as the best planting time for onion seed production with mean annual rainfall of 668 mm and mean minimum and maximum temperature of 15 and 31 °C, respectively [8]. These findings depict the importance of identifying appropriate planting date for onion seed production in each locality and variety.

Therefore, the present study was conducted to identify the appropriate planting time for onion seed production and associated selected quality indicators in Tehuledere district, northeastern Ethiopia.

Methods

Description of the study area

The experiment was conducted at Jari small-scale irrigation scheme, Tehuledere district in northeastern Ethiopia during the period of September 2015 to April 2016. Geographically the experimental site is located at 11°14′N latitude and 39°40′E longitude and at an elevation of 1700 m above sea level (masl). The soil of the experimental site is sandy loam in texture, and its mean annual rainfall is 1204.6 mm [9]. The mean maximum and minimum temperature during the growing season was 28.0 and 6.0 °C, respectively. Monthly temperature and rainfall data of the trial site during the experiment period are given in Table 1.

Experimental materials and design

The experiment was conducted under irrigation from September 2015 to April 2016. Onion cultivar 'Adama Red' was used as experimental material for this study. Recommended bulb size with a bulb diameter ranging from 4.1 to 5 cm and free from insect, disease and mechanical injury was selected and used for the study. Treatments of the experiment were nine planting dates such as September 1st, September 16th, October 1st, October 16th, October 31st, November 15th, November 30th, December 15th and December 30th.

The experiment was laid out using randomized complete block design (RCBD) with three replications. The size of each experimental plot was 2.1 m wide and 3 m long. Each plot had six rows. The spacing between irrigation furrows, planting rows and plants within a row was 50, 30 and 20, respectively. A spacing of 1 and 1.3 m was maintained between plots and blocks. Oxen plowing

Table 1 Monthly temperature (°C) and rain fall (mm) at Jari northeastern Ethiopia during the trial period. *Source*: Kombolcha Meteorological Directorate (2016)

Months	Temperature			Rain fall
	Maximum	Minimum	Mean	
September	28.4	8.4	18.4	66.43
October	27.2	5.2	16.2	20.8
November	25.8	2.6	14.2	75.2
December	25.8	3.4	14.6	22.4
January	27.4	7.2	17.3	39.2
February	31.0	7.6	19.3	29
March	31.2	6.8	19.0	98.3
April	28.8	11.6	20.2	97.6

method and manual harrowing were used for land preparation and fine seedbed preparation, respectively. About one third of the growing portion of mother bulbs was cut for easy and quick sprouting of growing buds. The lower portion with disk-like stem and roots was dusted with ash to prevent decay due to fungal infection and used for planting. Mother bulbs were planted by hand. The recommended fertilizer rates of urea 100 and 200 kg/ha DAP [2] were applied. DAP was applied at the time of planting, and urea was applied in two split doses of equal amounts, at planting and 45 days after planting. The field was irrigated 3 days after planting to facilitate for easy germination of bulbs. Then, it was irrigated every 7 days until full flowering and then at every 10-day interval followed by 10–15-day interval near maturity [2]. Weeding was done from 15 days after planting up to harvesting within 7–14-day interval at each growing phases.

Data collection and analysis

The central four rows of each plot and ten plants from each plot were used for data collection. Data were collected for different plant characters such as plant height (cm), number of leaves per plant, number of scapes per plant, scape height (cm), scape diameter (at the base), umbel diameter, number of seeds per umbel, seed yield per plant (g) and thousand seed weight (g). On the other hand, data on days 50% flowering and maturity and seed yield per hectare (kg) were collected on net plot size basis. Seed germination (%) test was conducted at Dessie seed laboratory, and seed germination percentage was calculated [10]. Data collected on various parameters were subjected to analysis of variance (ANOVA) using SAS version 9.2 software [11]. Differences among treatment means were compared using the least significant difference test (LSD) at 5% probability level.

Results

Growth and phonological parameters

Both plant height and number of leaves per plant were significantly ($p \leq 0.01$) affected by planting date (Table 2). Plant height ranged from 61.8 to 82.6 cm with an average of 74.1 cm. The maximum plant height (82.6 cm) was recorded from bulbs planted on September 1st, and the lowest plant height (61.8 cm) was recorded from bulbs planted on December 30th (Table 2). Number of leaves per plant ranged from 15.4 to 54.4 with an average of 31.6. The maximum number of leaves per plant (54.4) was recorded from plants planted in September, while the minimum number of leaves per plant (15.4) was recorded from bulbs planted in December.

Planting time highly significantly ($p \leq 0.01$) influenced scape number per plant, scape height and scape diameter (Table 2). The number of scapes per plant ranged from 2.7 to 6.3 with an average of 4.4. The height and diameter of scapes ranged from 49.9 to 72.5 cm and from 0.58 to 1.48 cm, respectively. The maximum number of scapes per plant, scape height and scape diameter was recorded from bulbs planted early such as on September 1st. On the other hand, bulb planted late such as on December 15th and 30th had short and minimum number of scapes per plant with small scapes diameter.

Seed yield and quality parameters

Planting time was significantly influenced onion seed yield and seed yield-related parameters such as umbel diameter, number of seeds per umbel, 1000-seed weight, seed yield per plant and seed yield per hectare (Table 3). Bulb planted in September and October had significantly higher umbel diameter, number of seeds per umbel, 1000-seed weight, seed yield per plant and seed yield per hectare compared to those planted in November and December.

Table 2 Effect of planting time on growth and phonological parameters of onion at Jari northeastern Ethiopia

Planting time	Plant height (cm)	No. of leaves/plant	No. of scape/plant	Scape diameter (cm)	Scape height (cm)	Days to 50% flowering	Days to 50% maturity
Sept 1st	82.6^a	46.7^a	6.3^e	1.48^a	72.5^a	79.3^{abc}	130.2^b
Sept 16th	80.9^a	54.4^a	5.0^{bc}	1.25^b	70.4^a	80.7^{ab}	132.8^a
Oct 1st	80.1^{ab}	46.5^a	4.7^{cd}	0.88^e	69.8^a	77.3^{cd}	129.0^b
Oct 16th	77.7^{ab}	25.3^b	5.9^{ab}	0.99^d	64.7^a	58.7^e	129.6^b
Oct 31st	72.5^{abcd}	27.3^b	4.0^{de}	1.12^c	69.5^a	68.0^d	124.0^d
Nov 15th	69.5^{bcd}	31.9^b	3.5^{ef}	0.72^f	64.7^a	60.7^e	126.7^c
Nov 30th	76.2^{abc}	14.6^c	4.8^{cd}	0.72^f	55.3^b	81.0^{ab}	120.3^e
Dec 15th	65.3^{cd}	15.4^c	2.7^f	0.61^g	53.4^b	73.2^{cd}	120.0^e
Dec 30th	61.8^d	22.5^{bc}	2.7^f	0.58^g	49.9^b	84.3^a	110.1^f
Mean	74.1	31.6	4.4	0.93	63.4	73.7	124.8
LSD (5%)	11.4	9.8	0.96	0.09	9.2	6.7	2.1
CV (%)	8.9	17.9	12.63	5.96	8.4	5.2	0.95

Means followed by the same letter within a column are not significantly different at 5% probability level

Table 3 Effect of planting dates on onion seed yield and quality parameters at Jari northeastern Ethiopia

Planting time	Umbel diameter (cm)	No. of seed/umbel	1000-seed weight (g)	Seed yield/plant (g)	Seed yield kg/ha	Germination percentage
Sept 1st	5.8[a]	533.3[a]	3.6[a]	10.0[a]	1032.7[a]	94.3[a]
Sept 16th	5.3[ab]	395.2[b]	3.6[a]	5.5[b]	652.3[b]	71.0[b]
Oct 1st	5.2[ab]	351.9[b]	2.9[b]	4.9[bc]	691.5[b]	50.3[c]
Oct 16th	4.7[bc]	218.4[c]	2.9[b]	4.4[c]	552.5[c]	50.7[c]
Oct 31st	4.4[bc]	107.5[d]	2.4[cd]	0.8[e]	102.2[e]	25.0[e]
Nov 15th	5.0[abc]	60.2[e]	2.2[d]	0.4[e]	37.4[e]	18.0[f]
Nov 30th	4.8[bc]	153.0[d]	2.7[bc]	2.1[d]	280.8[d]	39.0[d]
Dec 15th	4.3[c]	59.4[e]	2.4[cd]	0.5[e]	57.4[e]	15.3[f]
Dec 30th	3.3[d]	47.4[e]	2.3[cd]	0.3[e]	29.7[e]	21.7[e]
Mean	4.8	214.0	2.8	3.2	381.8	42.8
LSD (5%)	0.9	45.88	0.4	0.6	73.7	6.1
CV (%)	10.9	12.38	8.9	10.9	11.2	8.3

Means followed by the same letter within a column are not significantly different at 5% probability level

Planting time significantly influenced the umbel diameter, which was ranged from 3.3 to 5.8 cm with a mean of 4.8 cm. The highest umbel diameter (5.8 cm) was recorded from bulbs planted on September 1st but was not significantly different from those planted on September 16th, October 1st and November 15th. On the other hand, the smallest umbel diameter (3.3 cm) was recorded from bulbs planted on December 30th.

Planting time significantly influenced ($p \leq 0.01$) number of seeds per umbel which ranged from 47.4 to 533.3 seeds per umbel with an average of 214.0. In the earlier studies, number of seeds per umbel (256.6–515.3) and (93.0–299.9) were reported in Ethiopia (Kobo area) and Bangladesh, respectively [5, 8]. However, in the present study, the highest number of seeds per umbel (533.3) was recorded from September 1st planting date. On the other hand, the minimum number of seeds per umbel (47.4) was recorded from December 30th planting date, but it was statistically on par with December 15th and November 15th plantings dates.

Planting time significantly influenced ($p \leq 0.01$) thousand seed weight (Table 3). Similarly, the significant effect of planting dates on thousand seed weight was reported in other studies [6, 8]. Thousand seed weight ranged from 2.2 to 3.6 g with an average of 2.8 g. The maximum thousand seed weight (3.6 g) was recorded from early planting time such as September 1st and September 16th. On the other hand, the minimum thousand seed weight (2.2 g) was recorded from bulbs planted on November 15th and it was statistically on par with those planted on December 15th, December 30th and October 31st.

Seed yield per plant was significantly influenced by planting time (Table 3). The average seed yield per plant ranged from 0.3 to 10.0 g with an average of 3.2 g. The

highest seed yield per plant (10.0 g) was obtained from bulbs planted on September 1st. On the other hand, the lowest seed yield per plant (0.25 g) was recorded from bulbs planted on December 30th but not significantly different from those planted on December 15th, November 15th and October 31st.

Planting dates significantly influenced seed yield per hectare and ranged from 29.7 to 1032.7 kg/ha with an average of 381.8 kg/ha (Table 3). The maximum seed yield (1032.7 kg/ha) was obtained from September 1st planting. On the other hand, the least seed yield (29.7 kg/ha) was recorded from December 30th planting, but the seed yield was on par with December 15th, October 31st and November 15th plantings. Therefore, early planting dates are suitable for higher onion seed yield at Jari irrigation scheme.

The germination percentage of onion seeds produced significantly influenced by planting time (Table 3). The percentage of germination ranged from 15.3 to 94.3% with overall average of 42.8%. The highest germination percentage was recorded from seeds harvested from bulbs planted on September 1st. On the other hand, low germination percentage was recorded from seeds harvested from those planted on December 15th, but there was no significant difference with those planted on November 15th.

Correlation analysis among planting time and different parameters

Correlation coefficient (r) values computed to determine the relationships between and within the planting time and parameter are depicted in Table 4. The correlation values explain the apparent association of the planting time parameters with each other and clearly indicated the magnitude and directions of the association and relationships.

Plant height ($r = -0.915$), number of leaves/plant ($r = -0.858$), number of scape/plant ($r = -0.834$), diameter of scape ($r = -0.905$), scape height ($r = -0.929$), days to 50% maturity ($r = -0.901$), umbel diameter ($r = -0.861$), number of seed/umbel ($r = -0.916$), thousand seed weight ($r = -0.855$), seed yield/plant ($r = -0.881$), seed yield/hectare ($r = -0.895$) and germination percentage ($r = -0.876$) were negatively but strongly correlated with planting time (Table 4).

Discussion
Growth and phonological parameters
Bulb planted in September showed vigorous vegetative growth (plant height and number of leaves) compared to those planted in December. The increase in plant height could mainly be due to early planting which might have provided plants with relatively cooler period compared to the latter eight plantings. The cooler period stimulates cytokine and gibberellins' accumulation, modifying the hormonal balance and leading the plant to increase the plant development and responsible for elongation of flower stalk [12]. The taller plant height provides more photosynthetic capacity to the plant than shorter height with more number of leaves compared to those planted in November. This could be attributed to the increase in the vegetative growth of the onion plant through the effect of planting time, and the cooler time was important for the synthesis of different growth component of

onion stem and seed. This good foliage indicates higher growth, development and productivity of plant. Similarly, in other study, in Kobo area, the significant effect of planting dates was recorded, where bulb planted in October had maximum plant height compared to those planted in November [8]. Likewise, the significant effect of planting dates on plant height and number of leaves per plant was also reported [7]. In this study, bulbs planted in September were longer and with more number of leaves compared to those planted in November. On the other hand, other study indicated that bulb planted in November was longer and with more number of leaves per plant compared to those planted in October [5]. This could be attributed to climatic variation among the study sites.

The maximum number of scapes per plant, scape height and scape diameter was recorded from bulbs planted early such as on September 1st. These showed that early planting resulted in vigorous plants. In agreement with this, the maximum number of scapes per plant, scape length and scape diameter was recorded from early planting dates [6, 8].

Concerning flowering date, the bulb planted in December 30th took maximum number of days for flowering but no significant difference with bulb planted in September and November 30th. This might be attributed to the coincidence of growth stage of the crop and occurrence of cold weather to induce bolting. Similarly, other report indicated that planting time had marked influence on the number of days required for 50% flowering [8]. In

Table 4 Correlation analysis of planting time among growth, yield components, seed yield and quality of onion at Jari northeastern Ethiopia

	PT	PTH	NL	NS	DS	SH	FL	DM	UD	NSU	TSW	SYPP	SYPH	GP
PT	1	−0.915*	−0.858*	−0.834*	−0.905*	−0.929*	0.034	−0.901*	−0.861*	−0.916*	−0.855*	−0.881*	−0.895*	−0.876*
PTH		1	0.684*	0.920*	0.786*	0.798*	0.039	0.850*	0.875*	0.879*	0.846*	0.845*	0.890*	0.860*
NL			1	0.506	0.718*	0.808*	0.188	0.724*	0.686*	0.816*	0.746*	0.716*	0.735*	0.731*
NS				1	0.782*	0.679*	−0.072	0.738*	0.754*	0.816*	0.804*	0.863*	0.877*	0.871*
DS					1	0.848*	0.035	0.730*	0.708*	0.835*	0.821*	0.828*	0.788*	0.854*
SH						1	−0.234	0.887*	0.794*	0.726*	0.621	0.669*	0.685*	0.656
FL							1	−0.339	−0.074	0.306	0.337	0.226	0.217	0.293
DM								1	0.885*	0.714*	0.688*	0.676*	0.710*	0.663
UD									1	0.803*	0.730*	0.776*	0.781*	0.754*
NSU										1	0.948**	0.976**	0.980**	0.973**
TSW											1	0.925*	0.923*	0.966**
SYPP												1	0.986**	0.981**
SYPH													1	0.966**
GP														1

PT planting time, *PH* plant height, *NL* number of leaves/plant, *NS* number of scape/plant, *DS* diameter of scape, *SH* scape height, *FL* days to 50% flowering, *DM* days to 50% maturity, *UD* umbel diameter, *NSU* number of seed/umbel, *TSW* thousand seed weight, *SYPP* seed yield/plant, *SYPH* seed yield per hectare, *GP* germination percentage

*, **Significant at 5 and 1% probability levels, respectively. The decimal numbers without any asterisk are nonsignificant ($p > 0.05$)

the present study, the bulb planted in September 16th took maximum number of days to mature although bulbs planted in December 30th required minimum number of days to reach maturity. This indicates that bulb planted in relatively hot season mature early condition matured early compared to those which were planted in relatively cold season [4].

Seed yield and quality parameters

Bulb planted in September had significantly higher umbel diameter but not significant different with planted in October 1st and November 15th. In the same way, September 1st had significantly different with other planting time for number of seeds per umbel, seed yield per plant and seed yield per hectare compared to other planting dates.

Similarly, umbel diameter was recorded with a range of 3.0–6.9 cm in Bangladesh [5] and from 4.8 to 6.0 cm in kobo, Ethiopia [8]. The highest umbel diameter recorded from bulb planted in October compared to in November [8] and in November compared to those planted in December and January [6]. This shows that early planting increases umbel diameter compared to late planting. On the other hand, in Bangladesh the maximum umbel diameter was recorded from onion bulbs planted in November compared to those planted in October [5]. This difference might be attributed to the climatic variability among the study sites.

The number of seeds per umbel ranged from 47.4 to 533.3 with the highest number of seeds per umbel (533.3) was recorded from September 1st planting date. The variation in number of seeds per umbel might be due to flower abortion, indicating the time when high temperature caused such even, lack of efficient pollinators of all the flowers in the umbel, shortage of nutrition which caused high competition and death of the weak florets in the umbel [8]. So, selection of appropriate time in a given location is crucial in onion seed production.

Thousand seed weight ranged from 2.2 to 3.6 g with the maximum thousand seed weight (3.6 g) was recorded from early planting times (September 1st and September 16th), the minimum thousand seed weight (2.2 g) was recorded from onion bulbs planted on November 15th, and it was statistically on par with those planted on December 15th, December 30th and October 31st. This might be attributed to climatic condition prevailing during the seed filling stage. Therefore, early planting results in well-filled seeds compared to late plantings. In addition, the seed filling period of late planting dates was significantly shorter than early planting time. However, in Bangladesh, heavy 1000-seed weights from bulbs planted in November compared to those planted in October [5].

The average seed yield per plant ranged from 0.3 to 10.0 g with the highest seed yield per plant (10.0 g) was

obtained from bulbs planted on September 1st. Similarly, the effects of planting dates on seed yield per plant were significant [4–6]. Therefore, early planting time resulted in higher seed yield per plant. Cool temperature for flower development in early planting time and subsequent favorable temperature could have increased the final seed yield per plant on early planting dates. High atmospheric temperature causes early maturity of bulbs before attaining sufficient growth of plant, thereby resulting in low seed yield in onion. The difference in seed yield per plant might be due to the number of scapes per plant, number of seed per umbel and cool temperature for flower development in early planting, and subsequent favorable temperature could have increased the final seed yield per plant. Similarly, in Kobo area, higher seed yield per plant was recorded from bulb planted in October compared to those planted in November [8].

The maximum seed yield per hectare (1032.7 kg/ha) was obtained from September 1st planting, while the least seed yield (29.7 kg/ha) was recorded from December 30th planting which is not significantly different with December 15th, October 31st and November 15th plantings. Similarly, in Bangladesh higher seed yield was recorded from bulbs planted in October compared to those planted in November [4]. In late planting, it might be resulted in reduced cycle and less yield, because the plants received stimulus for bulb development before reaching full vegetative development. Therefore, early planting times are suitable for higher onion seed yield at Jari irrigation scheme.

In the same way, the germination percentages were ranged from 44 to 84% and from 77.1 to 97.6% in Bangladesh and Ethiopia (Kobo area), respectively [5, 8]. The highest germination percentage was recorded from seeds harvested from bulbs planted on September 1st. On the other hand, low germination percentage was recorded from seeds harvested from those planted on December 15th, but there was no significant difference with those planted on November 15th. The reason for increasing the percentage of seed germination in early planting may be due to the highest seed size and seed weight. Therefore, early planting is suitable to produce high-quality onion seed at Jari irrigation scheme. Similarly, early planting date is recommended to produce high-quality seed at Kobo area [8].

Correlation analysis among planting time and different parameters

With increase in time of planting (from September 1st to December 30th), there was increase in temperature. Onion requires ample moisture with cool environment. So the increase in temperature influences flower stalks development, flowering and seed maturation.

Plant height ($r = -0.915$) and number of leaves/plant ($r = -0.858$) were negatively but strongly correlated with planting time (Table 4). Both plant height and number of leaves/plant could mainly important for photosynthesis and pollination. This implied that the cooler time was important for the synthesis of different growth components of onion stem and seed. This good foliage indicates higher growth, development and productivity of plant.

Umbel diameter ($r = -0.861$), number of seed/umbel ($r = -0.916$), thousand seed weight ($r = -0.855$), seed yield/hectare ($r = -0.895$) and germination percentage ($r = -0.876$) were negatively but strongly correlated (Table 4). For that reason, it is possible to say that umbel size of the onion plant is one of the major characters highly demanded for flower and seed production and positive relation with seed yield. The result indicated that the above-mentioned parameters can be increased by early planting in which the plant can accumulate high seed yield and quality due to the extended vegetative growth of the plant. Similarly, parameters can be increased by extending crop cycle by early planting [13].

The tallest scape height was associated with the highest yields and the greatest number of seeds per plant. The taller scape height might have provided more photo-assimilates to the plant causing the weight of each seed to be greater than the weight of seed from plants with short scape. The highest seed yield (1032.7 kg/ha) as well as the highest germination percentage (94.3%) was recorded from onion planted on September 1st. On the other hand, the lowest seed yield (29.7 kg/ha) and germination percentage (15.3%) recorded from onion planted in December. It was evident that plants sown in September 1st exhibited superior performance both in growth, yield and quality characters. This may be because there was enough time for the early planted onion plants to complete both their growth and developmental stages, which in turn enhanced the production and partitioning of photo-assimilates, thus leading to an increase in growth and yield characters. Plants sown late (December 30th) had no adequate time to complete their life cycle fully.

Conclusion

The finding showed significant differences among the different planting dates with regard to growth, yield and quality parameters, viz. number of leaves/plant, number of scape/plant, scape diameter, days to 50% flowering, days to maturity, number of seed per umbel, thousand seed weight and seed yield/plant. The highest seed yield/

plant, seed yield per hectare and germination percentage were obtained from early planting date, while significantly the lowest values were recorded from late planting. Therefore, based on the finding of the current study, early planting (September 1st) can be used for high yield and better quality of onion seed. Several growth, yield component and quality parameters were negatively but strongly correlated with planting time of bulbs. Hence, September 1st was identified and recommended as the optimum planting time for seed production of "Adama red" at Jari irrigation scheme, Tehuledere district of northeastern Ethiopia.

Authors' contributions
MT initiated the research, wrote the research proposal, conducted the research, did data entry and analysis and wrote the manuscript. DB and YD involved in methodology, writing, reviewing and editing of research proposal and manuscript; and GS did data entry and analysis and writing the manuscript. All authors read and approved the final manuscript.

Author details
Department of Plant Science, College of Agriculture, Mekdela Amba University, P.O. Box 32, Tuluawlia, Ethiopia. ² College of Agriculture and Veterinary Medicine, Jimma University, P.O. Box 378, Jimma, Ethiopia. ³ International Livestock Research Institute, P.O. Box 527, Bahir Dar, Ethiopia. ⁴ Department of Plant Science, College of Agriculture, Wollo University, P.O. Box 1145, Dessie, Ethiopia.

Acknowledgements
We would like to acknowledge ILRI-_IVES project funded by Global Affairs of Canada for the financial support to conduct this study.

Competing interests
The authors declare that they have no competing interests.

Funding
The research funded by ILRI-LIVES project which funded by Global Affairs of Canada. The funders had no role in study design, data collection and analysis, decision to publish or preparation of the manuscript.

References
1. Central Statistical Agency (CSA). Agricultural sample survey report on area and production of major crops for the period 2013/14 cropping season. Vol. 1, statistical bulletin 532, Addis Abeba, Ethiopia, 2014.
2. Olani N, Fikre M. Onion seed production techniques: a manual for extension agents and seed producers. Addis Abeba: FAO; 2010.
3. Amsalu A, Afari-Sefa V, Bezabih E, Fekadu FD, Tesfaye B, Milkessa T. Analysis of vegetable seed systems and implications for vegetable development in the humid tropics of Ethiopia. Int J Agric For. 2014;4(4):325–37.
4. Ud-Deen MM. Effect of mother bulb size and planting time on growth, bulb and seed yield of onion, Bangladesh. J Agric Res. 2008;33:531–7.
5. Mollah MRA, Ali MA, Ahmad M, Hassan MK, Alam MJ. Effect of planting dates on the yield and quality of true seeds on onion. Int J Appl Sci Biotechnol. 2015;3(1):67–72.
6. El-Helaly MA, Karam SS. Influence of planting date on the production and quality of onion seeds. J Hortic Sci Ornam Plants. 2012;4(3):275–9.
7. Mehri S, Forodi BR, Kashi AK. Influence of planting date on some morphological characteristic and seed production in onion (*Allium cepa* L.) cultivars. Agric Sci Dev. 2015;4(2):19–21.

8. Teshome A, Derbew B, Sentayehu A, Yehenew G. Effects of planting time and bulb size on onion (*Allium cepa* L.) seed yield and quality at Kobo Woreda, Northern Ethiopia. Int J Agric Res. 2014. https://doi.org/10.3923/ijar.2014.

9. Jari Agricultural Sub-center. Socio-economical profile of Tehuledere district of Jari; 2015 (**Unpublished**).

10. International Seed Testing Association (ISTA). International rules for seed testing. International Seed Testing. Seed Sci Technol. 1985;13:299–355.

11. Statistical Analysis Software (SAS). SAS/STAT version 9.2 user's guide. Cary: SAS Institute Inc.; 2008.

12. Rakhimbaev IR, Ol'Shaskaya RV. Dynamics of endogenous gibberellins during transition of garlic bulbs from dormancy to active growth. Fisiologye Rasteii. 1976;23:76–9.

13. Bewuketu H. Effect of planting date on tuber yield and quality of potato (*solanum tuberosum* L.) varieties at Anderacha district, South-western Ethiopia. Int J Res Agric Sci. 2012;2(6):2348–3997.

Factors affecting adoption of upland rice in Tselemti district, northern Ethiopia

Hadush Hagos[1]*, Eric Ndemo[2] and Jemal Yosuf[2]

Abstract

Background: Rice cultivation is a new practice to Tselemti district of Tigray region, Ethiopia. Adoption of rice technologies is very slow in spite of its potential in the area. This research intended to identify factors affecting adoption of rice technologies.

Methods: A multistage sampling technique was employed to select 150 sample households for this study. Descriptive statistics and inferential statistics were employed to see mean and percentage differences between adopter and non-adopter categories. Besides, binary logistic regression model was employed to identify the factors affecting adoption of rice technology.

Results: Result of the descriptive and inferential analysis showed that adopters had better farm size, livestock holding, farm income, labor availability, education level, perception on rice yield, access to credit service, contacts with extension agents, participation in off-farm activities, participation in training and field days as compared to non-adopters. Moreover, the binary logistic regression model result showed that the level of education, perception on rice yield, access to credit service, participation in off-farm activities, participation on field day and participation in training were found to positively and significantly influence the adoption decision of rice technology at 1%, 5% and 10% significant level. However, market distance influences rice technology adoption negatively and significantly at 10% significant level.

Conclusions: The variables education, rice yield, access to credit, off-farm activities, market distance, participation on field day and training determine the farmers' continued adoption decision behavior of rice technology. Therefore, the adoption of rice technology should be sustained by paying attention and moving along with those variables which influenced the adoption significantly.

Keywords: Adoption, Binary logistic model, Upland rice, Tselemti district

Background

Agriculture is the mainstay of the Ethiopian economy. Although the transformation toward a more manufacturing and industrially oriented economy is well underway, the agriculture sector continues to be the most dominant aspect of the Ethiopian economy, accounting for nearly 46% of gross domestic product (GDP), 73% of employment, and nearly 80% of foreign export earnings [1]. The

bulk of agricultural GDP for the period 1960–2009 had come from cultivation of crops (90%) and the remaining (10%) from livestock production [8].

Rice has become a commodity of strategic significance in Ethiopia for domestic consumption as well as export market for economic development [3]. Besides, rice is among the target commodities that have received due emphasis, by the government, for the promotion of agricultural production. The crop is considered as the "millennium crop" for Ethiopia as it is expected to contribute for food security improvement. Rice production has brought a significant change in the livelihood of farmers and created job opportunities for a number of citizens in different areas of the country. As a result, the demand

*Correspondence: hadat2009@yahoo.com
[1] Socioeconomic and Research Extension Directorate, Mekelle Agricultural Research Center, P.O. box 492, Mekelle, Ethiopia
Full list of author information is available at the end of the article

for improved rice technologies is increasing from time to time from different stakeholders [7].

Even if rice cultivation was not practiced in Tselemti district, it is confirmed that this crop can find appropriate ecological niches since 2007. The governmental and non-governmental organizations have promoted improved rice varieties in order to create awareness and encourage the use of the improved varieties in the district. Demonstration and promotion have also been undertaken on the use of different improved crop management technologies such as plowing, planting, fertilizing, weeding, harvesting, post-harvest handling, and these practices are being adopted by the rice growers. The recent trends in the area and production of rice along with its high compatibility in the traditional consumption habits show that rice is becoming one of the staple foods and important for ensuring household food security in the district.

Meanwhile, the district food security and nutrition of the people depend on the amount and stability of their farm output and income. Poor choice of crops to be grown and its management system accounts for the low productivity of the district. Moreover, crops grown in the swampy soils of the district give almost no yield before rice was introduced to the district. Since it was new to the district, farmers did not fully accept the rice crop to grow on its appropriate niche. The wide range and diversified problems associated with the rice sector need to be addressed in order to make it competitive. Otherwise, the poor adoption of rice technologies by farmers would eventually lead to high cost of production with corresponding low yield. Consequently, the potential area for rice production would remain unproductive due to little motivation of farmers to grow rice.

Although rice production has increased during the last decade in Ethiopia, the country's production capacity is far below the national requirement. Disseminating improved varieties and other modern inputs to rice farmers is very important to reduce food insecurity and the rate of rice importation. Adoption of improved rice production technologies should lead to significant yield increase in rice production. As an attempt to address these problems, research institutes introduced varieties accompanied by other management practices that will produce higher yield in order to boost food security. Despite all these efforts, research findings still indicate that rural farmers in most cases find it difficult to obtain improved rice production inputs that are suitable to their local conditions [12, 13]. Reasonable proportions of the farmers are aware of the potential of rice crop production, but they have not adopted them.

The importance of farmers' adoption of new agricultural technology has long been of interest to agricultural economists, extensionists and rural development policy makers. It is believed that an effective way to increase productivity and enhance peoples' livelihoods is broad-based adoption of new farming technologies [6]. Rice area coverage in Ethiopia is still low compared to its potential [7] despite several agricultural policies and programs aimed at increasing productivities of rice implemented by the government to ensure improved food security. However, it is one thing to adopt and continue the use of the newly introduced rice crop technology; it is another thing to reject the technology adoption which is an important component that needs to be addressed in technology adoption decision process. Therefore, study on the different socioeconomic, institutional and psychological factors that influence the adoption of rice technology is useful for technology development and design of policies and strategies that foster adoption of rice technologies to cope up the current food insecurity of the district. Hence, this study was designed to identify the factors that affect adoption of rice technology in the study area to fulfill the existing knowledge gap.

Methods
Study site description
The study was conducted in Tselemti district of Tigray National Regional state of Ethiopia. It is found 1172 km far from the capital city Addis Ababa and geographically located $13°48'$N latitude and $38°15'$E longitude. It is bordered with Asgede Tsimbla, Welkait, Tanqua Abrgelle districts and Amhara region to the north, west, east and south, respectively. The district covers an altitude ranging from 800 to 2870 m above sea level. The mean annual temperature of the area is 16 °C (November–January) and 38 °C (February–May) minimum and maximum, respectively. Some of the major crops grown in the area include sorghum, finger millet, maize, chickpea and sesame. A map indicating the study location is presented in Fig. 1.

Sampling procedures and data collection
Multistage sampling technique was employed to select the sample respondents. First, Tselemti district was purposefully selected due to its rice production potential in northwestern zone of Tigray. Secondly, from a total of 23 rural kebeles of the district, only 11 kebeles were considered for the study due to their rice production potential. The rest 12 kebeles were agro-ecologically not suitable and not experienced in rice production. From the 11 kebeles, three were randomly selected based on the relative similarity of the kebeles in the adoption of rice technology. A kebele is the smallest administrative unit in Ethiopia. Thirdly, two categories of farmers were included in the household sample; farmers who were producing rice for three or more years (adopters) and those not producing rice but have been exposed to

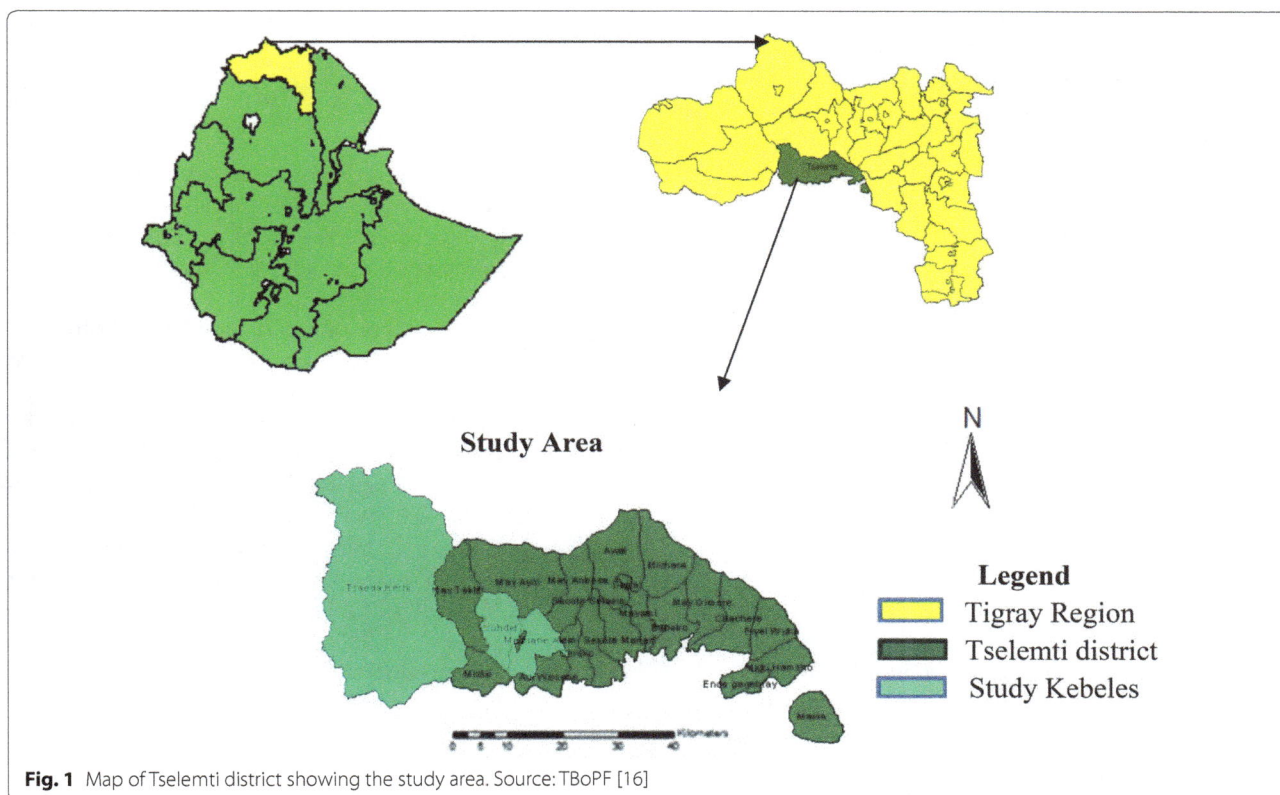

Fig. 1 Map of Tselemti district showing the study area. Source: TBoPF [16]

rice technology (non-adopters). There was a difference in rice adoption status of each sampled kebeles. Some of the kebeles were better in rice acceptance and utilization though there is a similarity in rice production potential of the area. The sample size of each category was determined by using proportional to size of the adopters and non-adopter farmers to take a representative sample for the study. Finally, of 150 households 62 adopters and 88 non-adopters were drawn with proportional to the sample size of adoption status from the three kebeles. In addition, focus group discussions (FGD) in each kebeles were conducted to supplement the data collected through individual interview.

The data collected from both primary and secondary data sources were qualitative and quantitative in nature. Semi-structured interview schedule and checklist were used as data collection tools. Prior to the interviews, the semi-structured questionnaire was pretested to control validity, and modifications were made to enhance its utility in addressing the relevant issues. The household survey was conducted in January 2015 by four trained junior researchers who administered the questionnaire to the household head or any other senior member whenever the household head was not present at the time of the interview. Moreover, Secondary data were also collected from published and unpublished documents like reports of office of agriculture and rural development, documents, research publications, journals and internet browsing.

Method of data analysis

Descriptive, inferential, and econometric model was employed to analyze the data. Descriptive statistical analysis methods were employed to discuss the result of survey using frequency, percentages, mean, and standard deviation. The Chi-square tests and t test were used to see the presence of statistically significant differences and the systematic association between those who adopt and those who do not in terms of the hypothesized variables.

The logit econometric model was applied for analyzing factors influencing the adoption of rice technology. Logit model was used to determine the relative influence of various explanatory variables on the dependent variable. This model was chosen because it has an advantage that it reveals the relative influence on the probability of adoption of the technology and can predicate the probability on the extent of adoption in a proper way.

The dependent variable in this case is dummy variable which takes the value of 1 for adopter farmers or 0 otherwise. Logit model which helps to test the determinants of adoption can mathematically be specified as follows:

$$P_i = E(Y = 1|X_i) = \beta_0 + \beta_i X_i \tag{1}$$

where $Y=1$ means a given farmer participates in production.

X_i is a vector of independent variables.

β_o is the constant and β_i, $i=1, 2...n$ are the coefficients of the independent variables to be estimated.

$$P_i = E(Y = 1|X_i) = \frac{1}{1 + e^{-(\beta_0 + \beta_i X_i)}}$$

$$P_i = \frac{1}{1 + e^{-z_i}} = \frac{e^z}{1 + e^z}$$

where $Z_i = \beta_0 + \beta_i X_i$.

If P_i is the probability of being adopter, then $(1 - P_i)$, the probability of being non-adopter of improved rice variety is $1 - P_i = \frac{1}{1+e^{z_i}}$. Therefore, we can write this equation as $\frac{p_i}{1-p_i} = \frac{1+e^{z_i}}{1+e^{-z_i}} = e^{z_i}$. Hence, $\frac{p_i}{1-p_i}$ is the odds ratio in favor of adopters.

In other words, it is the ratio of the probability that a given farmer participate in production to the probability that the farmer will not participate in production. Then, if we take the natural logarithm of equations (e) we obtain

$$L_i = \mathrm{Ln}\left[\frac{p(i)}{1 - p(i)}\right] = \ln\left[e^{\beta_0} + \sum_{i=1}^{m} \beta_I \chi_i\right] = Z_{(i)}$$

If the disturbance term a is taken into account, the logit model becomes

$$L_i = Z(i) = \beta_0 + \beta_1 X_1 + \beta_2 X_2 + \beta_3 X_{3...} + \beta_i X_i + U_i \tag{2}$$

Therefore, L_i, which is the log of the odds ratio, is called logit or logit model [2]. Hence, the above logit model was employed to estimate the effect of the hypothesized explanatory variables on the adoption decision of farmers to use improved rice variety.

To avoid the problem of multicollinearity, both continuous and dummy variables were checked prior to executing the logit model. Different methods are often suggested to detect the existence of multicollinearity problem. Among them, variance inflation factors (VIF) technique was employed to detect multicollinearity in continuous explanatory variables [2] and the contingency coefficient (CC) for dummy variables. This analysis was carried using SPSS version 16. The variables used in logit model are presented in Table 1.

Result and discussion
Effects of explanatory variables on adoption of improved upland rice variety

The descriptive statistics of the selected variables of the sample households examined in the study are presented in Tables 2 and 3. As shown from the Table 2, t value was computed for all continuous variables and it was found statistically significant for farm size, annual farm income, livestock holding, and labor availability in adult equivalent at 1% and 5% level of significance. This implies that there was a significant difference in all these variables between the two categories (adopters and non-adopters).

Table 1 Description of explanatory variables and their measurement

Variables	Type	Measurement
Sex of household head	Dummy	1 male, 0 female
Age of households head	Continuous	Age in years
Level of education	Categorical	1. Illiterate, 2. Read and write, 3. Primary school, 4. Secondary school and 5. College
Farm size	Continuous	Size of land under cultivation (ha)
Annual farm income	Continuous	Annual farm income earned (Ethiopian Birr)
Access to credit	Dummy	1 if the farmers has access to credit, 0 otherwise
Livestock ownership	Continuous	Number of livestock owned in TLU
Labor availability	Continuous	Family members in adult equivalent
Market distance	Continuous	Distance to market in Kilometer
Off-farm income	Dummy	1 engagement in off-farm income activities, 0 otherwise
Perception on rice yield	Dummy	1 if the farmer perceives rice has better yield than other crops, 0 otherwise
Contact with extension agents	Dummy	1 if the farmer has contact with extension agents, 0 otherwise
Participation in field day	Dummy	1 if the farmer has participated in field days, 0 otherwise
Participation in training	Dummy	1 if the farmer has participated in training, 0 otherwise

The *Chi-square* test was computed for the categorical and dummy variables and it was found statistically significant for educational level, contact with extension agents, credit access, off-farm income, perception on rice yield, participation on field days and participation on training at 1% and 5% significant level in these variables between the two categories.

Rice is an important component of household food intake and income in the surveyed rice grower households. Rice technology was introduced in 2007 in the study area, by Maitsebri Agricultural Research Center, which was first conducted as an observation trail of two varieties. Two nationally released rice varieties (NERICA-3 and NERICA-4) were identified and recommended to the area following the adaptation trial conducted. In addition to this, the research center released one NERICA variety (NERICA-13) nationally in 2014. NERICA rice varieties had been selected in the area for its high yielding, short maturation periods, tolerance to a biotic stresses, tolerance to lodging and high response to mineral fertilization. Owing to its success, the NERICA varieties were covered more than 95% of the area covered with rice crop. Besides, the farmers were utilized recommended inorganic fertilizer (DAP and Urea) and weed management activities.

Currently, almost all of the rice growers are food sufficient and some of them have an excess rice grain available round the year. From the total rice grower sample

Table 2 Descriptive analysis results of the continues variables

Variables	Mean of adoption categories			*t* value	*p* value
	Adopter	Non-adopter	Total		
Age of the household head	44.05	43.77	43.91	0.137	0.891
Farm size	1.17	0.89	1.02	3.232	0.002***
Livestock holding	10.65	7.14	8.89	4.089	0.000***
Market distance	9.55	9.85	9.70	− 0.325	0.746
Annual farm income	13,541.40	8115.30	10,828.35	3.689	0.000***
Labor availability	5.19	4.67	4.93	2.013	0.046**

At 5% and * at 1% probability level

Table 3 Descriptive analysis results of the dummy and categorical variables

Variable	Description	Percentage between adoption categories			χ^2 value	*p* value
		Adopter	Non-adopter	Total		
Sex of household head	Male	93.6	89.8	91.3	0.655	0.418
	Female	6.4	10.2	8.7		
Education level	Illiterate	37.1	73.9	58.7	22.874	0.000***
	Read and write	14.5	6.8	10		
	Primary school	32.3	17	23.3		
	Secondary school	16.1	2.3	8		
Contact with extension agents	Yes	98.4	88.6	92.7	5.089	0.024**
	No	1.6	11.4	7.3		
Credit access	Yes	66.1	40.9	51.3	9.261	0.002***
	No	33.9	59.1	48.7		
Off-farm income activities	Yes	37.1	12.5	22.7	12.554	0.000***
	No	62.9	87.5	77.3		
Perception on rice yield	Yes	91.9	53.4	69.3	25.392	0.000***
	No	8.1	46.6	30.7		
Participation on field days	Yes	72.6	26.1	45.3	31.660	0.000***
	No	27.4	73.9	54.7		
Participation on training	Yes	66.1	13.6	35.3	43.865	0.000***
	No	33.9	86.4	64.7		

At 5% and * at 1% probability level

respondents, 48% were facing to food shortage before they started to grow rice. It was due to lack of appropriate crop to their land, limited fertilizer utilization, management problem and water logging. Furthermore, the variety development and agronomic practice have been found promising to enhance production and productivity so as to improve the livelihood of small scale farmers in the region.

Factors determining the adoption of upland rice technology

The main purpose of this study was to explore the important factors that influence farmers' decisions to adopt upland rice technology. The goodness-of-fit measures were employed to check and validate that the model fits the data well. The *Chi-square* goodness-of-fit test statistics of the model show that the model fits the data with significance at the 1% significance level. This shows that the independent variables were relevant in explaining the farmers' decision to adopt upland rice technologies. The model predication result also shows that about 88% of the overall sample cases were correctly predicted by the model.

Out of 14 explanatory variables included in the model, seven were found to be significant in influencing the farmers' decision to adopt or not to adopt rice technologies. The remaining seven variables: sex, age, farm size, labor availability, livestock holding, farm income and contact with extension were not significant (Table 4).

That implies they do not determine the farmers' continued adoption decision behavior of rice technology.

Level of education

As expected education level of the household head had a positive and significant relationship with the probability of adoption of rice technology. The odds ratio in favor of adopting improved rice technology, other factors kept constant, increases by a factor of 2.256 for the farmer whom assumed household heads become literate than that who did not. This could be due to the fact that relatively educate farmers have more access to information and they become aware of new technology, and this awareness enhances the adoption of technologies. This result is consistent with finding of Umeh and Chukwu [20], Tiamiyu et al. [17], Rahman and Bulbul [14] and Leake and Adam [5] which suggested that the more educated the farmer was, the more likely to adopt new technology. On the contrary, the negative influence of education was also observed in other study by Tura et al. [19] which justifies households headed by literates were relatively less likely to adopt improved maize varieties in Central Ethiopia. This was due to the fact that the relatively more educated household heads were youngsters and the land ownership among the youth was minimal.

Perception on rice yield

Farmers' perception about the yield of rice was found to—influence adoption of rice technology positively and

Table 4 The maximum likelihood estimation of the binary logit model

No.	Variable	B value	SE	Wald	Sig.	Odds ratio
1	Age of the household head	0.035	0.027	1.632	0.201	1.035
2	Sex of the household head	−0.513	1.029	0.249	0.618	0.599
3	Level of education	0.814	0.333	5.980	0.014**	2.256
4	Farm size	0.166	0.602	0.076	0.783	1.181
5	Livestock holding	0.032	0.069	0.211	0.646	1.032
6	Perception on rice yield	2.468	0.750	10.822	0.001***	11.797
7	Off-farm income	1.113	0.643	2.997	0.083*	3.044
8	Access to credit	0.981	0.515	3.627	0.057*	2.666
9	Market distance	−0.084	0.050	2.838	0.092*	0.919
10	Contact with extension agent	0.407	1.313	0.096	0.757	1.502
11	Participation on field days	1.755	0.585	8.996	0.003***	5.784
12	Participation on training	1.287	0.569	5.127	0.024**	3.624
13	Annual farm income	0.000	0.0001	1.580	0.209	1.000
14	Labor availability	0.087	0.201	0.189	0.664	1.091
15	Constant	−8.035	2.397	11.239	0.001***	0.000
χ^2 value			100.541***			
−2 log likelihood			102.873			
Overall model predictions			88%			

***, ** and * significant at 1%, 5% and 10% significant level, respectively

significantly at less than 1% significance level. Farmers' perception of the superiority on yield of rice crop compared to other crops in the area creates interest to adopt rice technology. The odds ratio, other things kept constant, implies that the probability of adopting rice technology increases by a factor of 11.797 as the perception of yield benefit relative to other crops increases. A similar result was reported by Langyintuo and Mungoma [4] and Wen-chi et al. [21] that imply yield potential plays a crucial role in choosing a given variety. Therefore, the probability of adoption of improved technologies would increase once a farmer perceives that the yield potential of the given crop was higher than that of the existing one.

Off-farm activities

The engagement in an off-farm income activities was found to have positive and significant influence on the adoption of upland rice. The odds ratio, other things kept constant, implies that the probability of adopting rice technology increases by a factor of 3.044 as the farmer's engagement in off-farm source of income increase by one unit. This could be associated with the farmers opportunity of using money from off-farm activities for purchasing of inputs that enable them to adopt rice technology. This was consistent with the findings of Tura et al. [19] and Olalekan and Simeon [11] that shows the possibility of using money from off-farm activities for purchasing of inputs necessary to continue growing improved varieties.

Access to credit service

As the model result indicates, the variable access to credit had positively and significantly influenced the likelihood of adoption of rice technology. From this result, it can be stated that those farmers who have access to formal credit, from agricultural office and cooperatives were more probable to adopt rice technology than those who have no access to formal credit. The odds ratio indicated in the model with regard to credit implies that, another thing being held constant, the odds ratio in favor of adopting rice technology increases by a factor of 2.666 as farmers get access to credit. Earlier studies by Ogutu and Obare [10] and Leake and Adam [5] also indicated that credit availability significantly affects the adoption of improved technologies and the quantity of inputs farmers apply.

Market distance

The variable market distance affects the adoption decision of farmers negatively and significantly at 10% level of significance. The result of the odds ratio indicated that as the market distance increases, the logs of odds ratio in favor of households' adoption of rice technology decrease by 0.919. This implies that farmers who are distant from the input and output market have less likelihood to adopt the improved upland rice technology. The study by Solomon et al. [15], Yemane [22] and Olalekan and Simeon [11] were consistent with this result; that shows distance from the nearest market affects adoption of improved agricultural technology negatively and significantly. It also indicates that the shorter the distance from the household to the nearest market, the higher the probability of adopting new technology.

Rice crop was matched with the farmer's needs and field realities of the area. Farmers do not operate their rice farm on a large scale in isolation, but within the wider market system. The farmers were able to produce surplus paddy rice in their farms. However, most of the rice producer farmers respond that there were market problems in their nearest place. The major rice marketing difficulties were related to non-availability of market, a small number of market actors, low quality product that can meet consumer demand, and absence of rice polisher. As a result, a major portion of the total production was consumed and very little sold in the local market for consumers and as a seed formally and informally.

Participation on field days

It was found that exposure to information in relation to attending field days had positively and significantly influenced the probability of adoption of rice technology at less than 1% significant level. When farmers practically observe a new practice they could weigh the advantage and disadvantages of the new technology. This can facilitate adoption and helps them to implement the new technology properly in their own situation. Another thing held constant, the odds ratio of the variable attending field days implies that as farmers' exposure to agricultural information through field day increases, the odds ratio in favor of adopting improved rice technology increases by a factor of 5.784. This result goes along with the study done by Mustapha et al. [9] and Yemane [22] that revealed exposure to information due to participation in field days had a positive and significant influence on the probability of adoption of improved upland rice variety.

Training participation

Training was important for acquiring information, knowledge and developing abilities or attitudes, which result in greater competence in the performance of a work. It was found that participation in training had positively and significantly influenced the probability of adoption of rice technology at less than 5% significant level. The more the opportunity to participate in rice technology related training programs had increased farmers' adoption of rice technologies. This implies that if the households participated in training, the logs of odds ratio in favor of households' adoption of rice technology increased by a factor of 3.624. A study by Tsado et al. [18] showed the responses of the participants on the training and adoption of the improved rice package had most impacted their lives. The farmers have received training and experience sharing on rice variety development, agronomic practices, post-harvest handling and utilization on different food recipes. The wereda office of agriculture and rural development, Agricultural Research and NGOs were the main rice training and experience sharing providers in the area.

Conclusion and policy implication

Descriptive statistical analysis results show that adopters of rice technologies were better on education level, access to farmland, family labor force, livestock ownership, earning annual farm income and perception of rice yield. In addition to this, adopters of rice technology had participated more in off-farm activities, access to credit, contact with extension agents, rice related training and field days than the non-adopters.

The econometric regression binary logit model result indicated that level of education, perception on rice yield, access to credit service, participation in off-farm activities, participation on field day and participation in training influencing adoption positively and significantly. Furthermore, market distance influences rice technology adoption negatively and significantly. However, the remaining seven explanatory variables: sex of the household, age, farm size, labor availability, livestock holding, farm income and contact with extension agents were not significant. This implies they do not determine the farmers' sustained adoption decision behavior of rice technology in the study area.

This suggested that, in line with the formal education, focus should be given for adult and continues the training program to bring a change in the adoption of new technologies. Strengthening of rural cooperatives which can deliver inputs and credit on reliable market price should be enhanced. The quantity and quality of rice produced at the farm level affected marketable supply, household income, and its contribution to food security and self-sufficiency. Thus, all stakeholders, especially the agricultural extension service, need to carry out more aggressive promotion of the improved rice technologies through appropriate mechanisms in input delivery, processing facilities and creating market opportunities. Proper attention is also needed for the development of income generating activities through constructing rural infrastructures in order to increase the participation of farmers in off-farm income activities. Moreover, frequent training of the rice farmers in the study area should be given attention, so that the farmers can obtain optimum yield from the adoption of improved rice varieties and attract other farmers. The participation of farmer in field days has to be strengthened so as to improve farmers' access to practical information and extension advice to adopt and expand the new innovation in a sustainable way.

Authors' contributions

HH carried out the study, worked out almost all of the technical details, and performed the numerical calculations for the suggested experiment. EN and JY analyzed the data and wrote the manuscript. EN contributed to the design and implementation of the research, to the analysis of the results and to the writing of the manuscript. JY devised the proposal, the main conceptual ideas and proof outline. All authors read and approved the final manuscript.

Author details

[1] Socioeconomic and Research Extension Directorate, Mekelle Agricultural Research Center, P.O. box 492, Mekelle, Ethiopia. [2] Department of Rural Development and Agricultural Extension, Haramaya University, Tselemti, Ethiopia.

Acknowledgements

The authors would like to thank Tigray Agricultural Research Institute for financing the research work. Similarly, we are grateful to the respondent farmers who have spent their precious time to respond attentively to questionnaire. Moreover, the agricultural office of the district had direct and indirect contribution for the successfulness of the survey work.

Competing interests

The authors declare that they have no competing interests.

Funding

No funding was received for this publication.

References

1. ATA (Agricultural Transformation Agency). Transforming agriculture in Ethiopia. 2014; Annual Report 2013/2014, Addis Ababa, Ethiopia.
2. Gujarati DN. Basic econometrics. 4th ed. New York: McGraw-Hill; 2004.
3. Hegde S, Hegde V. Assessment of global rice production and export opportunity for economic development in Ethiopia. Int J Sci Res. 2013;2(6):2319–7064.
4. Langyintuo A, Mungoma C. The effect of household wealth on the adoption of improved maize varieties in Zambia. Food Policy. 2008;33(6):550–9.
5. Leake G, Adam B. Factors determining allocation of land for improved wheat variety by smallholder farmers of northern Ethiopia. J Dev Agric Econ. 2015;7(3):105–12.
6. Minten B, Barret BC. Agricultural technology, productivity, and poverty in Madagascar. World Dev. 2008;36(35):797–822.
7. MoARD (Ministry of Agriculture and Rural Development). National Rice Research and Development Strategy. Ministry of Agriculture and Rural Development, the Federal Democratic Republic of Ethiopia. Addis Ababa, Ethiopia, 2010.
8. MoFED (Ministry of Finance and Economic Development). National income accounts, Addis Ababa, Ethiopia. www.mofeddatabase.com. Accessed 4 Feb 2010.
9. Mustapha SB, Makinta AA, Zongoma BA, Iwan AS. Socioeconomic factors affecting adoption of soya-bean production technologies in Takum local government area of Taraba State, Nigeria. Asian J Agric Rural Dev. 2012;2(2):4–6.
10. Ogutu WN, Obare GA. Crop choice and adoption of sustainable agricultural intensification practices in Kenya. Adoption pathways project discussion paper 10. Egerton University, Kenya; 2015.

11. Olalekan AW, Simeon BA. Discontinued use of improved maize varieties in Osun state, Nigeria. J Dev Agric Econ. 2015;7(3):85–91.
12. Onyeneke RU. Determinants of Adoption of Improved Technologies in Rice Production in Imo State, Nigeria. Afr J Agricultural Res. 2017;12(11):888–96.
13. Osanyinlusi OI, Adenegan KO. The determinants of rice farmers' productivity in Ekiti State, Nigeria. Greener J Agric Sci. 2016;6(2):049–58.
14. Rahman MR, Bulbul SH. Adoption of water saving irrigation techniques for sustainable rice production in Bangladesh. Environ Ecol Res. 2015;3(1):1–8.
15. Solomon A, Bekele S, Franklin S, Mekbib G. Agricultural technology adoption, seed access constraints and commercialization in Ethiopia. J Dev Agric Econ. 2011;3(9):436–47.
16. TBoPF (Tigray Bureau of Planning and finance). Atlas of Tigray. Mekelle, Ethiopia; 2014.
17. Tiamiyu SA, Usman A, Ugalahi UB. Adoption of on-farm and post-harvest rice quality enhancing technologies in Nigeria. Tropicultura. 2014;32(2):67–72.
18. Tsado JH, Ojo MA, Ajayi OJ. Impact of training the trainers' programme on rice farmers' income and welfare in North Central, Nigeria. J Adv Agric Technol. 2014;1(2):157–60.
19. Tura M, Aredo D, Tsegaye W, Rovere RL, Tesfahun G, Mwangi W, Mwabu G. Adoption and continued use of improved maize seeds: case study of Central Ethiopia. Afr J Agric Res. 2010;5(17):2350–8.
20. Umeh GN, Chukwu VA. Determinants of adoption of improved rice production technologies in Ebonyi state of Nigeria. Int J Food Agric Vet Sci. 2013;3(3):126–33.
21. Wen-chi H, Ghimire R, Shrestha R. Factors affecting adoption of improved rice varieties among rural farm households in central Nepal. Rice Sci. 2015;22(1):35–43.
22. Yemane A. Determinants of adoption of upland rice varieties in Fogera district, Ethiopia. J Agric Ext Rural Dev. 2014;8(12):332–8.

Assessment of production potential and post-harvest losses of fruits and vegetables in northern region of Ethiopia

Hagos Abraha Rahiel[1*], Abraha Kahsay Zenebe[2], Gebreslassie Woldegiorgis Leake[2] and Beyene Weldegerima Gebremedhin[3]

Abstract

Background: Horticultural crops are sources of vitamins, minerals and dietary fiber, but their cultivation is not widely practiced in developing countries, like Ethiopia due to small-scale farming systems and poor pre- and post-harvest handling techniques. In Ethiopia, particularly in northern region, the production of horticultural crops usually practiced in very few pocket areas, such as at river and lakesides. Thus, the production of fruits and vegetables is just at the beginning stage and getting momentum by governmental and non-governmental organizations. To assess the production potential and post-harvest losses of fruits and vegetables, a survey research was conducted in Tigray Regional State, northern Ethiopia. From the study area, *Atsbiwenberta* district was selected with its four purposely selected *Kebeles* (Kebelle is the lowest administrative division of Ethiopia next to districts in each administrative region) (*Ruwafeleg, Felegewoni, Golgolnaele* and *Hayelom*) in which 120 respondents (30 households from each *Kebele*) were participated. Data were collected from both primary and secondary data and analyzed using simple descriptive statistics like frequency, mean and percentage.

Results: From this survey research it was found that the region has the potential to produce both temperate and subtropical fruits and vegetable crops. However, their production potential was limited by different constraints starting from cultivation to consumption. Focus group discussion reveals that farmers obtained high production of potato and apple with average yield of 300–400 and 25–130 qt/ha, respectively. Conversely, post-harvest loss was significantly affected in the study area due to lack of awareness, market access, inadequate water supply and poor post-harvest handling practices. As a result, the loss of potato and other vegetable crops was ranged from 30 to 50 and 0.25 to 5 qt/ha, respectively.

Conclusions: Therefore, designing further research projects is recommended on production and post-harvest handling of fruits and vegetables. In addition, all stakeholders should be designed market linkage and involvement of female farmers in production of horticultural crops.

Keywords: Fruits and vegetables, Production potential, Post-harvest losses, Tigray, Ethiopia

Background

The production and marketing of horticultural crops constitutes a major source of cash income for the households and an opportunity to increase smallholder farmers' participation in the market. More than 85% of Ethiopians in rural areas rely on agricultural production for their sustainable livelihood [1]; thus, the Ministry of Agriculture and Rural Development (MOARD) focuses on market-led agricultural development and the government pledges support to market integration and agro-enterprise development. However, horticultural crop production in northern Ethiopia faces many challenges even though farmers have willingness to increase the

*Correspondence: rahel4ever@gmail.com
[1] Department of Dryland Crop and Horticultural Sciences, Mekelle University College of Dryland Agriculture and Natural Resources, Arid-Main Campus, P.O. Box 231, Mekelle, Tigray, Ethiopia
Full list of author information is available at the end of the article

production and productivity of the crops. Because they are needed to produce potential fruits and vegetables for marketing and consumption, i.e., they have a comparative advantage to generate income as compared to cereals, and specially vegetables require shorter time for production, yield more and generate higher income and the market outlet. Though there is enough labor available to increase horticultural production, there are other factors of production and marketing functions constitute serious challenges to promote horticultural crops in small-scale farming communities. Hence, farmers rely only on traditional production skills and methods [2] to produce horticultural crops in their small sort of lands.

Ethiopia's Ministry of Agriculture is striving to minimize post-harvest losses, which is causing up to 20–40% losses in sub-Saharan Africa [3] and 20–30% production loss even as the country's grain output continues to increase. The post-harvest losses of perishable (vegetable and fruits) food crops amounted to be about 30% [4] due to the presence of high moisture content (65–95%), insect infestation and damage during post-harvest handling techniques (packaging, storage and transportation). However, use of appropriate packaging materials, proper storage facilities and transportation can help to minimize these losses. In addition to this, modern food processing techniques and post-harvest technologies are the main tools to reduce perishable food losses and maintain the quality of products [5]. On the other hand, generating efficient, low-cost and indigenous technology minimizes post-harvest loss of fruits and vegetables as the largest groups of people in Ethiopia who suffer from food and nutrition insecurity are the rural poor who have insufficient land and lack of resources to provide sufficient income generation through production of fruits and vegetables with integrated post-harvest technology [6, 7].

According to Panhwar [8], it was reported that farmers sometimes do not even get the two-way transportation cost (cost to transport), so they would rather leave their produce near the market area than bearing the transportation cost required for taking the produce back. However, improving post-harvest management practices will be reduced post-harvest losses, and hence, production of value-added products with effective and efficient research programs should be strengthened and promoted in developing countries [5, 9].

In the country level, a number of research works have been done on horticultural crops, but little or no more research was done on the production potential and post-harvest losses of horticultural crops in northern part of Ethiopia, particularly in Tigray Region. *Atsebiwonberta* district is one of the districts of Tigray that has been introduced few intensive interventions and successfully applied in the value chains of fruit and vegetable crops. However, individual farmers have limited skill and knowledge on modern cultivation systems and post-harvest handling practices to increase production and post-harvest losses of fruits and vegetables. Therefore, the objective of this research was to assess the production potential and post-harvest losses of fruits and vegetables at *Atsbiwemberta* district farmer's field and to identify the technological gaps in the existing production and post-harvest handling techniques.

Methods
Description of the study area
Atsbiwenberta district is located in eastern zone of Tigray Regional State, northern of Ethiopia. This study was carried out in the highland and lowland areas of the district (Fig. 1). The district is located at 13°52′N and 39°44′E, about 65 km far from Mekelle, Capital city of Tigray Regional State, and shares a border with Afar Regional State to the east, *Wukro* to the west, *Enderta* to the south and *Saesie Tsaedaemba* district to the north. In addition, the district has an altitude between 2300 and 3200 m above sea level with mild temperatures and a mean annual rainfall ranging from 400 to 600 mm [10].

Sources of data
In this survey research, both primary and secondary data were collected. The primary data were obtained from farmer's field visit, group discussion, through interviewing farmers, development agents and Bureau of Agriculture and Rural Development (BoARD). Whereas the secondary data were collected from relevant documents of regional bureaus, zonal agricultural offices, districts agricultural and rural development offices, available literature reviews and internet web pages were also searched to consolidate the research findings.

Sampling techniques and data collection
Prior to the main sampling attempted, there was an informal survey discussion with the agricultural extension staff especially, horticulture experts and development agents. The elders and those who have better experience in cultivation of horticultural crops were participated in the discussion which helped to identify potential areas where horticultural crop production widely practiced. At this stage, the objective and scope of the study was explained briefly to the selected informal discussion group. Based on the understanding and agreement with these officials, community elder and leaders, the real survey was identified.

Four representative *Kebeles (Ruwafeleg, Felegewoni, Golgolnaele* and *Hayelom)* of the district were selected by purposive sampling technique where horticultural crop production is dominantly practiced. A total of 120

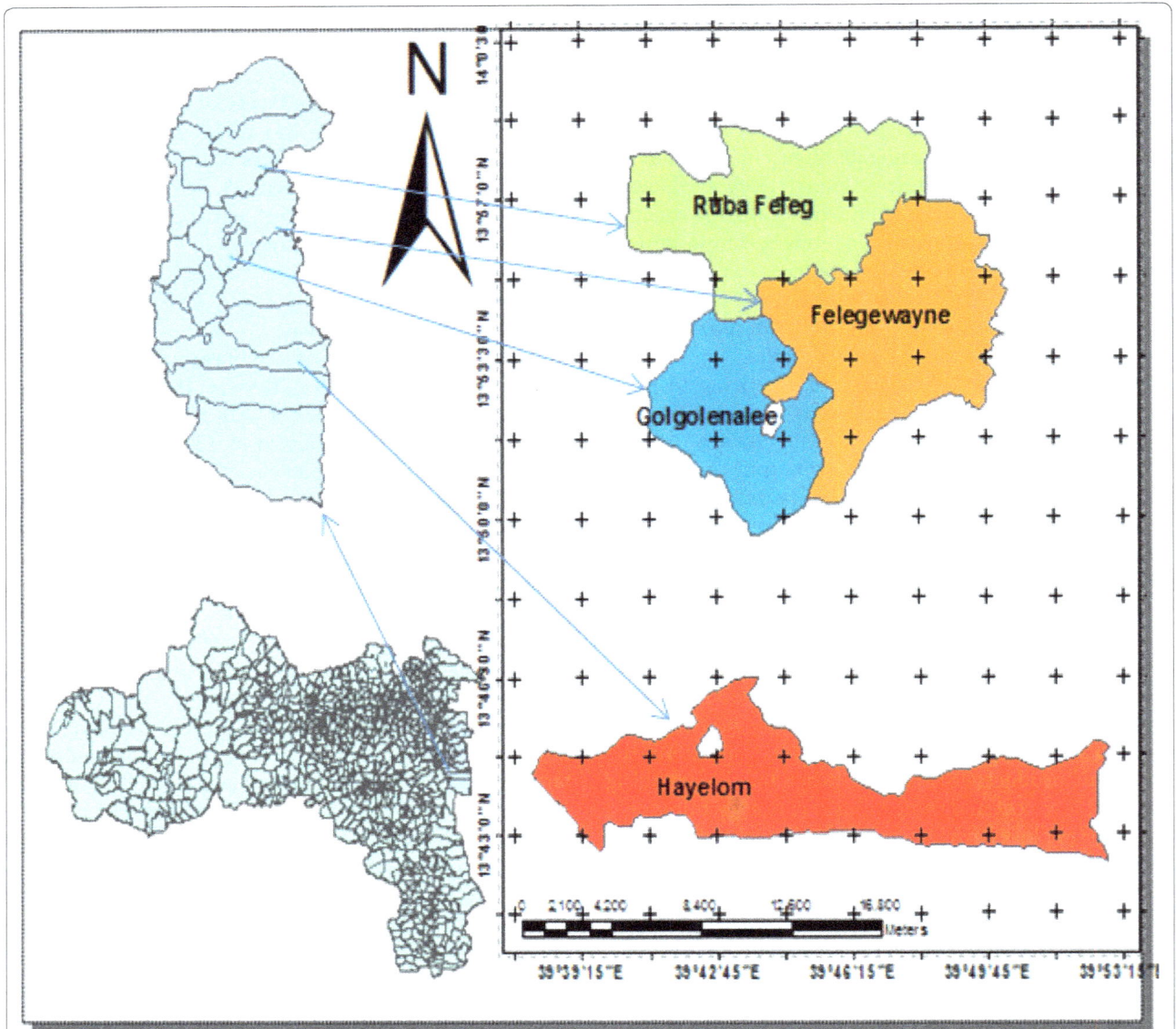

Fig. 1 Map of the study area

households in the district (30 households from each *Kebele*) were selected based on their involvement in fruits, vegetables and root and tuber crops production to address the objective. The questionnaire was mainly focused on vegetables and fruits production, pre-harvest cautions, post-harvest losses at different stages of handling and production constraints of the crops. The data were collected through semi-structured questionnaires (cross-sectional survey), focus group discussion (FGD) and field observation on farmers' cultivation site via formal and informal survey. The questionnaires were pretested and then translated into the local language "Tigrigna" in order to help understand the questions easily.

Data analysis

The collected data were analyzed using simple descriptive statistics like frequency, mean and percentage using SPSS (version 20.0) package software, and data were coded for analysis. Descriptive statistics was used to describe quantitative factors. Frequencies and percentages were used for describing qualitative characteristics. The data were analyzed using one-way analysis of variance (one-way ANOVA) at significance difference $p < 0.05$ level.

Results and discussion

Socio-demographic characteristics of fruit and vegetable producers

Among 120 sample respondents, 92 were males and the remaining 28 were females. Result showed that males involved much more in fruit and vegetable production than female farmers. With regard to age groups of fruit and vegetable producers, 47 respondents were ranged from 31 to 40 years old followed by 36 respondents grouped to ≥ 51 years old and lastly about 30 respondents were ranged their age group from 41 to 50 years old. The level of education was also varying among the respondents in which the majority of them (67) were illiterates, and the rest 41 and 12 of the respondents attended grade 1–6 and grade 7–12, respectively. Most of the respondents were married (101) followed by divorce (14), widowed (3) and unmarried (1). Majority of the respondents (79.4%) have access to market in nearest area to sell their produce, followed by 12.4% of the respondents who have intermediate access to market. However, 8.2% of them travel more than 31 km far away from their home to the market area as shown in Table 1.

Source of income generation

The income generation of sample respondents was mainly focused on cereal crops (36.7%) followed by horticultural crops, particularly fruits and vegetables (28.3%), and 16% of them were generated their income from both cereals and horticultural crops. On the other hand, respondents that have both fruit and vegetable production with grain and pulses production were obtained 16%. In the study area, apple and potato were observed as a significant income generating crops compared to other crops. That is farmers earned an average income of 65.22 USD from seedlings and 565.22 USD from selling apple fruits per cycle. In addition to this, individual potato producers also earned an average income of 869.57 USD per production cycle.

Farming system of individual households

The farming system in both highland and mid-lowland of the study areas is mixed farming, and farmers are own very small farmland; thus, they keep rearing animals in their homestead and provide feed by cut and carry system. In addition, farmers produced different crop varieties and horticultural crops during rainfed season and/ or by irrigation in order to sustain their family food supply and to cover various household expenses. This study showed that farmers are cultivated mostly potato, local cabbage, apple, tomato, onion, garlic, Swiss chard, chili and lettuce together with cereals and pulses within the year life time. According to the respondents' preference, fruit and vegetable crops are mostly produced by irrigation. Hence, about 51% of respondents cultivate mostly vegetables followed by 32% of them cultivate mostly fruits and rarely vegetables. In addition to this, 13% of the respondents are cultivated mostly vegetables and rare fruits, whereas 4% of the respondents are produced mostly fruits as compared to vegetables (Table 2).

Cultural practices and production constraints of fruits and vegetables

Due to arid and semiarid agroecology with small-scale farming system and erratic rainfall of the region, most of

Table 1 Socio-demographic characteristics of fruit and vegetable producers. Source: Own survey (November 2017)

Socio-demographic characteristics of fruits and vegetables	Frequency	%
Gender		
Male	92	76.7
Female	28	23.3
Age group		
≤ 30	7	5.9
31–40	47	39.1
41–50	30	25
≥ 51	36	29
Level of education		
No formal education	67	55.8
1–6 grade	41	34.2
7–12 grade	12	10
Marital status		
Married	101	84.2
Unmarried	1	0.8
Divorce	14	11.7
Widow	3	2.5
Nearest distance to the market (km)		
≤ 10	95	79.4
11–30	15	12.4
≥ 31	10	8.3

Table 2 Frequency and percentage of fruit and vegetable crops production of the respondents in the study area. Source: Own survey (November 2017)

Types of horticultural crops	Frequency	%
Mostly fruits	5	4
Mostly vegetables	61	51
Mostly vegetables and rare fruits	16	13
Mostly fruits and rare vegetables	38	32

Table 3 Production potential, price cost of the produce and post-harvest losses of fruits and vegetables in *Atsebiwonberta* district with respective to each *Keble*'s. Source: Own survey (November 2017)

Types of fruits	Types of vegeta-bles	R/feleg Yld/ha (Qt)	R/feleg Lss/y (Qt)	F/woni Yld/ha (Qt)	F/woni Lss/y (Qt)	Golgolnaele Yld/ha (Qt)	Golgolnaele Lss/y (Qt)	Hayelom Yld/ha (Qt)	Hayelom Lss/y (Qt)	Current price cost (Birr/kg)	Causes of losses
No of producers		35		15		16		49			
Apple		7	2	–	–	130	–	4.5	0.2	25–40	Due to lack of well-defined storage rooms, lack of awareness in the producers and communities, lack of market demand
Banana		–	–	–	–	–	–	2.25	–	12–14	
Guava		4	0.5	–	–	–	–	1–10.8	0.9	10	
Orange		–	–	–	–	–	–	9	0.2	18–25	
No of producers		136		70		14		493			
Potato	300–400	5–35	250–400	12	400		–	–	5–8		Due to their nature of perishability, lack of PHM, mechanical damage, lack of quality seed supply, improper irrigation, market problem and diseases infection due to mechanical and physical damage of the produce, poor transportation techniques
Tomato	280	0.125	3.5–7	1	200		350	12.38	10–18		
Cabbage	120	0.25	25	5	300		300–400	2	7–8		
Swiss chard	130	0.5	2	0.25	150		220	0.25	3–5		
Lettuce	130	0.5	4	0.25	150		120	1	2–4		
Spinach	110	0.25	3	0.3	1–3		140	–	3–4		
Chili	140–180	0.5	–	–	1–2		240	0.125	8–15		
Onion	250	–	4–5	–	250		480	–	10–12		
Carrot	0.5–2.0	–	0.5	–	200		–	–	5–6		
Sugar beet	–	–	–	–	180		–	–	5–6		
Garlic	230–240	–	1	–	200		520	–	32–35		

No number, Qt quintal, Yld/ha yield per hectare, Lss/y loss per year, (–) 'denotes not reported to be present'

Assessment of production potential and post-harvest losses of fruits and vegetables in northern...

187

the arable land of the study area is fragmented and prone to drought. Thus, there were a major production constraint on fruit and vegetable crops production and losses as it is presented in Table 3. The result reveals that there was shortage of improved cultivars of fruits and vegetable seeds supply in the study area, since there was no governmental or non-governmental organization responsible for the multiplication and distribution of horticultural crop seeds specially fruits and vegetables. Consequently, farmers are restricted to use local horticultural crop seeds with lower productivity and prone to most of the

diseases and insects (Fig. 2). Nevertheless, nowadays, for instance, projects like International Center for Potato (CIP), the shortage of quality potato seed supply is solved by half except the problem of post-harvest losses and market access of the crop.

Cultural practices are the basic management practices help to prolong shelf life of perishable horticultural crops. For example, as it was stated by Kader and Rolle [11] and Kamrul [12] pre-harvest techniques such as pruning and thinning increase the fruit size and decrease TSS and acidity of fruits. From the study, it was found that

Fig. 2 **a** Poor cultural practices, **b** weeds invaded cabbage, **c** and **d** late blight (*Phytophthora infestance*) in potato and tomato leaf, **e** head splitting due to improper water irrigation, **f** tip burn due to nutrient deficiency (Ca) in cabbage. Source: From farmer's backyards field (November 2017)

irregular irrigation before and after harvesting decreases the shelf life and sensory quality of fruits and vegetables which is similar with findings of Shimilis [6] and Mulatu [7]. Likewise, other serious problems were also identified from respondents' field due to the presence of nutrient deficiency, diseases and pests on the critical time besides to their poor cultural practices (Fig. 2). This was affected on potato tuber, cabbage heading, tomato fruit size and shape, and immature fruit drop in apple and citrus fruits.

Out of the sample respondents, 44.17% of them were cultivating fruits and vegetables manually and 36.67% of them were used integrated pest management (IPM) practices to control pets. However, 19.17% of the sample respondents are used different pesticides. This indicates that some horticultural crop producers are not practiced well in organic farming system rather they cultivated on inorganic farming which is not necessary for environmental protection and soil microorganisms. Meanwhile, the time of cropping cycle of fruits and vegetables differs with cultivars, type of edible parts and market access. About 93.3% of the interviewed respondents cultivated and harvested two times in a year with the help of irrigation in the winter season.

Even though a lot of challenges and constraints were affected the cultivation and production potential of the crops in the study area, it has the ability to produce apple (130 qt/ha), potato (300 qt/ha), tomato (240 qt/ha), onion (200 qt/ha) and garlic (230–240 qt/ha). However, there was an annual loss per individual crop (Table 3) of the produce due to market problem and poor post-harvest management practices. For instance, the loss of potato was ranged from 5 to 35 quintal per annum. In addition, lack of disease and pests control methods, post-harvest management practices and lack of demand in the market were the main constraints of potato production in the region, specifically in the study area. Moreover, very few of the respondents said (data not shown) that there was no a serious problem during their cultivation of produces, except wilting in leafy vegetables, because they were produced low yield and loss was not much more happened.

From the study, it was also observed that there were poor cultural practices in the farmers' backyard site (Fig. 2). Meanwhile, respondents have their own preferences to know the critical physiological maturity and ripening period of fruit and vegetable that is the crucial art of the producers [11]. So that most respondents were decided to harvest fruits by looking and take sampling judgments such as color, size and shape, touch–texture, hardness or softness, smell–odor/aroma and taste–sweetness, sourness and bitterness. On the other hand, vegetable growers were also harvested over a wide range of maturity stages when they attained physiological maturity and but still tender depending upon the part of the plant used as food as it was cited by Rural Agricultural Development Authority (RADA) [13, 14] which was also used by the fruit and vegetable growers in the study area. In addition, 99% of the respondents are identified and known their fruit and vegetable maturity harvest by visual, smelling, testing, feeling, eating or nutritional quality and harvest their produce by using either hand or supporting equipment's for home consumption as well as for marketing which was also applicable in other horticultural crop growers as cited by Arah et al. [15].

Generally, farmers have their own preferences on how to, when and what type of crops to be cultivated for better yield enhancement by using local cultivars in their small farmlands. In contrast, there are knowledge gaps and challenges along with production, harvesting, storage, transportation and marketing of the produces. As it was obtained from the research, the cultivation and production potential of fruits and vegetables were constrained by different factors such as lack of improved cultivars/seeds, lack of market access, lack of post-harvest managements, lack of equipment's and materials, shortage of insect and pest control, lack of irrigable land (i.e., some producers use temporarily rent farmlands which is not enhanced to grow permanent fruit crops) and problem of late sowing due to erratic rainfall. Indeed, there were also other production constraints in some fruits and vegetables such as fruit ball worm in guava, rust disease and pests in lettuce and immature fruit drop in apple (i.e., decreased the quality and shelf life of the crops, and hence, their prices become low).

Pre-harvest cautions practiced to fruits and vegetables

Pre-harvest factors or activities could affect the post-harvest shelf life and qualities of perishable horticultural crops [16]. Hence, pre-harvest cautions are very important for longevity and quality maintenance of fruits and vegetables [17]. But, results showed that 66 of the respondents were not using pre-harvest cautions to protect their horticultural crops after harvest, whereas 54 of the respondents were using different traditional cultural practices, starting from selection of appropriate season to harvesting time. Moreover, most of the producers were also vulnerable to exhaust their cost of production due to different challenges such as lack of appropriate and well-advanced cultural practices, lack of market demand and market link accesses and lack of post-harvest storage conditions (except few farmers were organized in association) and they were benefited from seed production and multiplication of potato and some cereal crops as shown in Fig. 3b.

On the other side, producers minimize their cost of production through prior production of nursery

Assessment of production potential and post-harvest losses of fruits and vegetables in northern...

189

Fig. 3 Potato storage room prepared from local available materials in *Tabia Felegewoni*; HH—households. **a** (an individual HH potato storage room). **b** (Local seed businesses (LSBs) in Shewit Association). Photograph Source: Own survey (November 2017)

seedlings and selection of appropriate harvesting time on some fruit crops. For example, apple producers were obtained an average income of 652 USD per annum from both seedlings and fruit sell. In deeded, they were protected fruit loss by covering local plastic materials, but there was the problem of fungal formation in the external fruit part. Another pre-harvest caution practiced in the study area was removing of diseased standing plants from the plant population (especially for late and early blight of potato and tomato, bacterial wilt of potato), harvesting of fruits before full ripening (i.e., tomato) was also the most practiced activities.

In the study area, respondents were used pre- and PHM practices supported and established by the CIP project specially for potato and some cereal crops (mainly barley and wheat), that is, in *Shewit*, *Kaly Chema* and *Habes*, and it was focused on production and multiplication of local quality seed in association (both in the field and out of the field) to control pre- and post-harvest losses and to access quality seed supply to the region, specifically to the district: for example, preparation of local aerated storage rooms (Fig. 3), separation and grading of potato tuber, harvesting of vegetables early morning or late evening were practiced for marketing, seed multiplication and consumption purposes.

According to Kitinoja and Kader [18], there are some fruits that had low relative perishability like apple which has a potential storage life of 8–16 weeks, but the research showed that there were losses due to lack of knowledge on post-harvest management, rodents and market access. In the contrary, some respondents in the district said that they did not take any pre-harvest cautions, but they cultivate as they want by using traditional cultural techniques. It is clear that there is lack of awareness, skill and knowledge in the producers' perception.

Method of harvesting and improper harvesting techniques practiced by the respondents

There are several harvesting methods on fruit and vegetable depending on the technology advance, but in the study area, respondents harvested by using hand (74.1%), supporting equipment (22.5%) and by using both hand and supporting equipment (3.3%). This study reveals harvesting using hand had very high (58.3%) market value as compared to using supporting equipment's in the study areas. Respondents also collected the produce, stored and/or covered in temporarily local packaging materials to protect from wilting and contamination before submitted to the traders and end users immediately after harvesting. Besides to this, harvesting at the critical ripening period, collect the yield and pack using local packaging materials (basket, *zembil*) was their day-to-day activity of the potential producers in the study area. Preharvest before ripening to control over ripening in some vegetables also practiced in tomato to protect from perishability and to increase its shelf life.

Harvesting practices should cause a little mechanical damage to produce as possible. Gentle digging, picking and handling will help reduce crop losses as cited by Kader [19] and chilling leafy vegetables by using cold water at harvest will help to maintain quality and prevent wilting [11]. In addition to this, use of suitable harvesting containers like buckets not baskets, since they do not collapse and squeeze the produce harvesting containers like buckets not baskets, since they do not collapse and squeeze the produce, but all of the above harvesting techniques as well as use of appropriate harvesting materials was not still properly applied in the study area (Table 4). However, when the attitude of the respondents was surveyed, it was found that inefficient that is improper handling systems contributing 48.4% of the existing postharvest losses for fruits and vegetables.

Table 4 Improper handling of fruits and vegetables during transportation and marketing in the study area. Source: Own survey (November 2017)

No.	Interview statements	Response		
		Always	Sometimes	Never
1	Have you ever received muddy and dusty fruits from the producer?	30 (24.2%)	63 (50.8%)	27 (21.8%)
2	Have you ever received muddy and dusty vegetables from the producer?	48 (38.7%)	58 (46.8%)	14 (11.3%)
3	Have you ever received physically injured fruits from the producer?	50 (40.5%)	46 (37.1%)	23 (18.5%)
4	Have you ever received physically injured vegetables from the producer?	23 (18.5%)	39 (31.5%)	58 (46.8%)
5	Have you ever received both spoiled and healthy fruits in the same container from the producer?	60 (48.4%)	43 (34.7%)	17 (13.7%)
6	Have you ever received both spoiled and healthy vegetables in the same container from the producer?	48 (38.7%)	51 (41.1%)	21 (16.9%)
7	Have you ever received fruits and vegetables in packed form?	35 (28.2%)	58 (46.8%)	27 (21.8%)
	Overall average	42 (34%)	51 (41.23%)	27 (21.54%)

Numbers out of the parenthesis are frequencies and numbers inside the parenthesis are percentages

Fig. 4 The top **a** and **b**: mixing of spoiled fruits with healthy ones; **c–f**: different packaging materials used for leafy vegetables in *Haykimeshal* market area (*Kebele Hayelom*). Source: Own survey (November 2017)

As shown in Table 4, improper handling of fruit and vegetable during harvesting, packaging and transportation was a serious problem to the producers as well as to the traders, that is, 34% of the respondents were responded that there is a muddy and dust, physically injured and mixing of spoiled with healthy fruits (Fig. 4a, b) and vegetables during packaging (Fig. 4c–f), whereas 41.23% of the respondents were responded that there was the problem of transportation from the field to the market nearby area.

In *Atsbiwonberta* district, post-harvest handling was beginning to potato and some cereal crops, but it is not practically distributed to other perishable horticultural crops which has been affected the shelf life and quality of the produce. For example, according to the district annual report about 36% of vegetables decay due to soft-rot bacteria, but the losses can be minimized by proper pre- and post-harvest treatments (pre-harvest application of Maleic hydrazide reduces sprouting of onions and potatoes during storage, and N-benzyladenine, i.e., prolongs shelf life of leafy vegetables) [11] but not still applicable in the study area.

Causes of post-harvest losses and management practices of fruits and vegetables

As shown in Table 5, 29.8% of the producers were lost their produce due to marketing problems (lack of demand and sanitation) which is followed by injury and infection 18.5% of the yield loss during harvesting, transportation and marketing which is similar to findings of Kitinoja and Kader [8, 20] in developing countries ranges about 20–40% losses, but improper harvesting was the least (15%) loss of fruits and vegetables.

Fruit and vegetable producers as well as traders in *Atsbiwonberta* district were maintained their produce shelf life and quality before marketing by using local storage rooms, sacks and nets, polyethylene bags, carts, carton, local storage materials like baskets, but they were stored them in an unclean area. One of the harvesting methods commonly practiced in the study area was harvesting manually, in which the commodity protects from mechanical damaging and excess moisture loss, but it is tedious to the workers. In addition, leafy vegetables were maintained in an exposed area, but they were traveled it early in the morning; moreover, there was no permanent cool area for selling to protect from excess moisture loss in some marketing areas of the district.

Even though insignificant respondents had established market linkage with whole sellers before they harvest their produce, there was loss of fruits and vegetables without consumption due to shortage of storage rooms, diseases and pests attack and local market access. For example, 30–50 quintal of potato was lost in 2014, which is still the problem of potato producer perceptions, because predominantly there was lack of local market access. On the other hand, there was low yield of vegetable production due to shortage of water supply during the growing period. Thus, as a remedy, it is important to improve post-harvest management practices with aim to reduce post-harvest losses, production of value-added products, effective and efficient research programs to strengthen and promote fruits and vegetables as it was discussed in EARO [5] and Azizah et al. [9] in the study area.

Mode of packaging materials and transportation from the field to the nearby market

During transportation and marketing, fruits and vegetables desired packaging materials and transporters; as a result, producers sell their produces in their nearby market. Leafy vegetables such as cabbage, lettuce, spinach and Swiss chard were packed by local accessory materials (basket, plastic materials and nets), and fruits, root and tuber crops were also packed by using local sacks, wooden boxes, polyethylene bags and other packaging materials.

The survey result showed that 46.6% of the respondents used packaging accessory materials, while they transported from the field to the marketing area, whereas the remaining 26.7% of the respondents were not using packaging materials. On the other hand, there was no much lost in some fruits such as apple and fresh maize corn due to its high market demand and resistant to perishability.

Currently, the production of fruit and vegetable is very low due to improper management practices before and after harvest. But there were few model farmers trained on how to produce and manage their crops and applied their experience for better cultivation and prefer to sell immediately especially for vegetable crops unless they used it for feeding animals and preparation of compost. As shown in Table 6, about 47 of the respondents were transported their produce to the market by equine followed by human back and equine (41), whereas very few farmers were used vehicles (20) as mode of transportation if their production is high.

Focus Group Discussion (FGD)

From the FGD, participants were identified the main constraints on horticultural crop production potential and post-harvest management practices in the study area. The discussion was focused on cultural practices, water availability, harvesting techniques, post-harvest losses and management, storage rooms and market accessibilities. As they were told that there was a high yield production of potato in 2012–2014, but it was loss due to lack

Table 5 Major causes of post-harvest loss of fruits and vegetables in the district. Source: Own survey (November, 2017)

No.	Causes of PHL	Frequency	Response (%)
1	Infection	20	16.1
2	Injury and infection	23	18.5
3	Nature of the produce	21	16.9
4	Improper harvesting	19	15.3
5	Marketing problem	37	29.8

PHL post-harvest loss, % percentage

Table 6 Type of packaging materials and transporter used during transportation of the produce from the field to the marketing area Source: Own survey (November 2017)

No.	Type of packaging materials	Frequency	%
1	Packaging accessories	56	46.6
2	Hand package	32	26.7
3	Without package	32	26.7
No.	Mode of transportation	Frequency	%
1	Equine	47	37.9
2	Human (labor)	12	9.7
3	Vehicles	20	16.1
4	Human and Equine	41	33.1

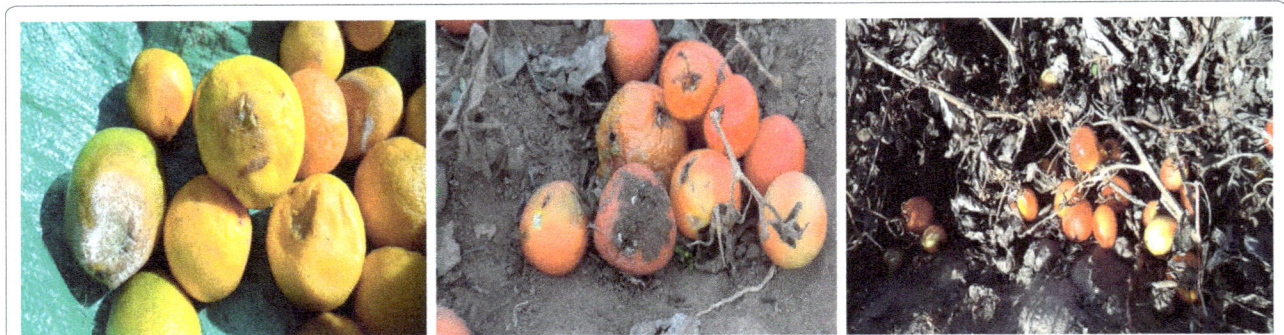

Fig. 5 Improper PHM of fruit left side (orange) and vegetable (tomato) right side brought loss of the produce. Source: Own field survey (November 2017)

of available storage houses, market access and linkage. This indicates that there was a high post-harvest loss and many producers lost their produce near around their field as shown in Fig. 5. Therefore, it is needed to practice and prepare PHM techniques in their locality with clear practical support for easy identification of the basic pre- and post-harvest management practices. This was also used to sustainable the production system of fruits and vegetables. In addition, participants said that the main reasons for loss of horticultural crops were water scarcity, lack of improved resistant varieties to shelf life and diseases and pests as compared to cereal crops. But they concluded that the production of horticultural crops has threefold economic benefits to individual households as compared to cereal crops.

From the FGD, it can be concluded that fruits and vegetables have the potential to change health and economic status of poor societies than cereal crops because they can be grow in marginal lands with low production cost in a sustainable way of production system. Indeed, they have a great role in socioeconomic aspects of the community so as to promote gender mainstreaming in production and expansion of horticultural crops that is better when the agriculture sector supports to them on

technical knowledge, water harvesting techniques and market linkage facilities. Besides to this, it is needed to create job opportunities (i.e., entrepreneurial skills), and focusing on sustainable income instead of short-term incomes was also a reason for promoting horticultural crops in region.

Conclusions and recommendations

Fruits and vegetables are the main sources of vitamins, minerals and dietary fiber in human diet. In Ethiopia, production of horticultural crops is in its infant stage particularly in Tigray region, which has the potential to produce fruits and vegetables. However, due to different social, agroecological, and economical constraints, it is not well adapted and distributed the cultivation of fruits and vegetables. Previously in the region, the production of fruits and vegetables was cultivated in some very small pocket areas near to rivers and lakesides for local consumption in rare societies. But recently, the production of horticultural crops becomes a small-scale farmer's business sector. It is just at the beginning stage and getting momentum from governmental and non-governmental organizations, especially in irrigated areas.

Pre- and PHM practices are essential for fruits and vegetables to supply a quality yield to the market. However, there was a great deal on post-harvest loss of potato (reaches up to 50 quintal/production cycle), in the study area mainly due to lack of market access and linkage and lack of awareness, and it which was mostly happed on small group of producers in association with local seed production and multiplication of potato. Indeed, this was regressive that other individual stakeholders not to produce fruits and vegetables for long-term return.

Generally, it could be concluding that horticulture crop producers in northern part of Ethiopia, particularly in Tigray region, are used different cultural practices adopted from innovative technologies and their own tradition cultural practice, and hence they had been increasing the potential for production of fruits and vegetables. However, there are many production constraints during cultivation, harvesting, transportation, marketing and consumption of fruits and vegetables, and hence post-harvest loss was listed as a main challenge to perishable crop producers in the study area. Therefore, designing further research on innovative technologies for access supply and consumption of horticultural crops should be required because they are more cash crops than other crops are used. In addition, capacity building on post-harvest handling techniques of fruits and vegetables should be promoted. Likewise, all stakeholders (both governmental and NGOs) should work in collaboration to facilitate a sustainable production of fruits and vegetables in small-scale farmers with a long-term market access. Besides to this, involvement and participation of female headed households on production of horticultural crops should be practiced to create awareness and sustain the livelihood of the community.

Authors' contributions
Conception and design of the research proposal, collection of data, data import to excel, analysis and interpretation of data by using SPSS software version 20.0. Authors also work on writing the manuscript article, revising it critically for intellectual content and final approval of the version to be published. Finally, before submission for publication, all authors are read and approved the final manuscript.

Author details
[1] Department of Dryland Crop and Horticultural Sciences, Mekelle University College of Dryland Agriculture and Natural Resources, Arid-Main Campus, P.O. Box 231, Mekelle, Tigray, Ethiopia. [2] Department of Natural Resource, Economics and Management, Mekelle University College of Dryland Agriculture and Natural Resources, Arid-Main Campus, P.O. Box 231, Mekelle, Tigray, Ethiopia. [3] Department of Animal, Rangeland and Wildlife Sciences, Mekelle University College of Dryland Agriculture and Natural Resources, Arid-Main Campus, P.O. Box 231, Mekelle, Tigray, Ethiopia.

Acknowledgements
We would like to give our sincere appreciation to Mekelle University, Institute of Environment, Gender and Development Studies (IEGDS) for its financial support. Our sincere gratitude also goes to *Atsbiwonberta* district administrator and Development Agents (DAs) immeasurable collaborated us on selecting the sample respondents and giving on secondary data sources. Finally, but not the last our great sincere also goes to appreciate farmers/respondents for their contribution on integrity and honesty in answering the questionnaire and taken focus group discussions.

Competing interests
The authors declare that they have no competing interests.

Ethics approval and consent to participate
The study was undertaken after the approval of the university research and publication team, ethical review committee of Mekelle University, Institute of Environment, Gender and Development Studies (IEGDS). Official letter was written from the institute with the detailed explanation of the purpose and importance of the study to Tigray Bureau of Agriculture and distributed it to the selected study areas. The purpose of the study was explained to each study area, and participant informed verbal consent was obtained from all study subjects before conducting the actual interview and discussions. For this purpose, a consent form was attached to each questionnaire which explained about the purpose and importance of the study, confidentiality and the respondent's full right to answer the questions or not before the beginning or at any time of the study. Each interview was conducted after informed verbal consent was secured. All the collected data are kept in safe custody by the responsibility of the primary investigator.

Funding
This study was financially supported by Mekelle University, Institute of Environment, Gender and Development Studies (IEGDS).

References
1. MOARD (Ministry of Agriculture and Rural Development) Vegetables and Fruits Production and Marketing Plan (Amharic Version), Ministry of Agriculture and Rural Development, Addis Ababa, Ethiopia; 2005.
2. Emana B, Gebremedhin H. Constraints and opportunities of horticulture production and marketing in eastern Ethiopia, Drylands Coordination Group report no. 46; 2007.
3. FAO (Food Association Organization). Declaration of the world summit on food security. World Summit on Food Security, 16–18 November, Rome; 2009.
4. Fekadu M, Dandena G. Review of the status of vegetable crops production and marketing in Ethiopia. Uganda J Agric Sci. 2006;12(2):26–30.
5. EARO (Ethiopian Agricultural Research Organization). Food Science and Post-Harvest Technology Research Strategy, Nazareth, Ethiopia; 2000.
6. Shimilis A. Post-harvest sector challenges and opportunities in Ethiopia, Food Technologist, Ethiopian Agricultural Research Organization Addis Ababa, Ethiopia; 2001.
7. Mulatu W. Solar drying of fruits and windows of opportunities in Ethiopia. Afr J Food Sci. 2010;4(13):790–802.
8. Panhwar F. Post-harvest technology of fruits and vegetables, 2006. www.eco-web.com/editorial/060529.html. Accessed 25 May 2016.
9. Azizah O, Nazamid S, Rosli S, Jamilah B, Noor DZ, Masturina Y. Post harvest handling practices on selected local fruits and vegetables at different levels of the distribution chain. J Agribus Mark. 2009;2:39–53.
10. TLZR (Tigray Livelihood Zone Reports pdf) Atsbi Wonberta Highland Livelihood Zone, 2006. http://www.dppc.gov.et/Livelihoods/Tigray/Downloadable/Tigray%20Woreda%20Reports/Atsbi%20Wenberta.pdf.
11. TLP (Tigray Livelihood Profile). Draft Atsbi Wonberta Highland Livelihood Zone, 2006. www.heawebsite.org/download/file/fid/359. Accessed 27 Mar 2018.
12. Kamrul H.A Guide to postharvest handling of fruits and vegetables, Department of Horticulture Bangladesh Agricultural University Mymensingh 2202, Bangladesh; 2010, p. 4.
13. RADA (Rural Agricultural Development Authority). RADA DIARIES: Post-harvest management of fresh produce; 2015. www.rada.gov.jm. Accessed 18 Apr 2015.
14. Wills R, McGlasson B, Graham D, Joyce D. Post-harvest: an introduction to the physiology and handling of fruit, vegetables and ornamentals. Walling ford: CAB International; 1998. p. 262.
15. Arah IK, Harrison A, Ernest KK, Hayford O. Preharvest and postharvest factors affecting the quality and shelf life of harvested tomatoes: a mini review; 2015.

16. Kader AA. Influence of pre-harvest and postharvest environment on nutritional composition of fruits and vegetables. In: Quebedeaux B, Bliss FA, editors. Horticulture and human health—contributions of fruits and vegetables. Englewood Cliffs: Prentice-Hall; 1988. p. 18–32.

17. Kitinoja L, Kader AA. Chapter 3: packinghouse operations. In: Small-scale postharvest handling practices: a manual for horticultural crops, 4th ed. Post-harvest Horticulture Series No. 8E, University of California Postharvest Research & Information Center, UC Davis, 2002, p. 35–58.

18. Kitinoja L, Kader AA. Small-scale post-harvest handling practices: a manual for horticultural crops. 4th ed. Davis: University of California Davis, California, USA; 2003.

19. Kader AA. Assessment of Post-Harvest Practices for Fruits and Vegetables in Jordan, University of California at Davis; 2006.

20. Kitinoja L, Kader AA. Measuring postharvest losses of fresh fruits and vegetables in developing countries. Postharvest education foundation paper 15-02; 2015.

Effect of inter- and intra-row spacing on yield and yield components of mung bean (*Vigna radiata* L.) under rain-fed condition at Metema District, northwestern Ethiopia

Asaye Birhanu[1], Tilahun Tadesse[2] and Daniel Tadesse[3]* ⓘ

Abstract

Background: The study was conducted in 2017 main cropping season at two locations in North Gondar Zone, Ethiopia, to determine the optimum inter- and intra-row spacing of mung bean for maximum yield and yield components. The experiment was laid in a randomized complete block design with three replications in a factorial arrangement of four inter-row (20, 30, 40 and 50 cm) and three intra-row (5, 10 and 15 cm) spacing using mung bean variety Rasa (N-26).

Results: Significant interaction effect of inter- and intra-row spacing was observed for days to maturity, number of branches per plant, number of pods per plant, grain yield, harvest index, days to flowering, plant height and above-ground dry biomass yield. The highest grain yield (1882.67 kg ha^{-1}) was obtained at interaction of 40 × 10 cm spacing, while the lowest (1367.8 kg ha^{-1}) was obtained from 20 × 5 cm spacing. However, the result of economic analysis showed that the maximum net benefit was obtained at spacing of 40 × 15 cm.

Conclusions: Based on agronomic performance and economic analysis, use of 40 × 15 cm is promising for mung bean production in Metema District and similar agroecologies.

Keywords: Economic analysis, Intra-row spacing, Plant spacing, Mung bean, Row spacing

Background

Mung bean (*Vigna radiata* L.) is one of the most important pulse crops, grown from tropical to subtropical areas around the world [1]. It is an important wide-spreading, herbaceous and annual legume pulse crop cultivated mostly by traditional farmers [2]. At present, mung bean cultivation spreads widely in Africa, South America, Australia and in many Asian countries [3]. The world annual production area of mung bean is about 5.5 million hectare with a increase in rate of 2.5% per annum [4]. Its requirement in growing in areas where the rainy season is short, and its wide adaptability together with

its digestibility makes mung bean cultivated all over the world [3]. Mung bean is utilized in several ways; seeds, sprouts and young pods are all consumed and provide a rich source of amino acids, vitamins and minerals [5]. The grain contains 24.2% protein, 1.3% fat and 60.4% carbohydrate [6]. It has low calories and is rich in fiber and easily digestible crop without causing flatulence as happens with many other legumes [7]. The crop is characterized by fast growth under warm conditions, low water requirement and excellent soil fertility enhancement via nitrogen fixation [8]. Currently, mung bean is produced in different parts of Ethiopia. It is mainly cultivated in North Shewa, Harerge, Illubabor, Gamo Gofa, Tigray and Gondar. Farmers in some moisture stress areas of Ethiopia effectively use and produce it to supplement their protein needs [9]. Ethiopia's green mung bean export has grown slightly from 822 tons in (2001) to 26,743 tons in

*Correspondence: kaleabfather@gmail.com
[3] Department of Plant Sciences, University of Gondar, P.O. Box 196, Gondar, Ethiopia
Full list of author information is available at the end of the article

2014 to fulfill the demand of India, Indonesia, Belgium and the UAE [10, 11]. In Ethiopia, mung bean covers about 27,086 ha of land and produces 241,589.90 tons in main cropping season per annum with average productivity of 0.9 ton ha^{-1} [12].

Even though there are high potential uses and export demand, the productivity of mung bean is low in Ethiopia (0.9 ton ha^{-1}) as compared to the world average productivity of 1.2 ton ha^{-1}, which is due to abiotic and biotic factors such as lack of improved variety, disease, insect and agronomic practices like in optimum row spacing and plant population per unit area. Proper method of sowing is among the important biotic factors that determine the proper plant population, which improves the performance and productivity of plants in the field. Plant population plays an important role as it is one of the most important yield contributing characters [13]. Combinations of researches on inter- and intra-row spacing is lacking in boosting productivity of mung bean in northwestern Ethiopia. Therefore, the present study was conducted to investigate the optimum inter- and intra-row spacing of mung bean for maximum yield and yield components and to evaluate the economic feasibility of different inter- and intra-row spacing of mung bean.

Materials and methods

The field experiment was conducted at Kumer and Kokit areas of Metema District from July to September 2017 in the main cropping season. Metema District is located in North Gondar Zone of Amhara Regional State, Ethiopia. Metema has latitude, longitude and elevation of 12°47″38′N, 36°23″41′E and 760 m above sea level, respectively. It has average annual rainfall of 1030 mm. It has maximum and minimum temperatures of 40.0 and 15.0 °C, respectively [14]. The soil type is sandy loam and reddish brown color. The mung bean variety *Rasa* (N-26) was used for the experiment.

Factorial combinations of four inter-rows (20, 30, 40 and 50 cm) and three intra-rows (5, 10 and 15 cm) spacing were laid out in a randomized complete block design with three replications. Each experimental plot had 2.5 and 4 m length and width, respectively. Spacing between plots and replications were 1 and 1.5 m, respectively. Planting was done when there was adequate soil moisture in the field. DAP fertilizer was applied at a rate of 100 kg ha^{-1}, and all other management practices were performed as per the general recommendations for green bean [15].

Data were collected on major agronomic and phenological characters. Analysis of variance (ANOVA) was done using SAS software [16]. Homogeneity of variances was tested using F test as described by Gomez and Gomez [17]. According to the homogeneity test, all parameters were homogenous except the number of branches per plant and number of pods per plant. As a result of it, the two parameters were analyzed in separate locations. Least significant differences (LSD) test at 5% level of probability was used. The partial budget analysis as described by CIMMYT [18] was done to determine the economic feasibility. The net benefit was calculated as the difference between the gross field benefit (Ethiopian Birr ha^{-1}) and the total costs (Ethiopian Birr ha^{-1}) that varied.

Results and discussion

Days to flowering

The combined main effect of inter- and intra-row spacing was highly significant ($P < 0.01$), while their interaction had no significant effect on days to 50% flowering (Table 1). The longest days to 50% flowering (40.83 cm) were recorded at inter-row spacing of 50 cm, while the shortest days to 50% flowering (36.05) were recorded at 20 cm inter-row spacing (Table 2). The longest days to 50% flowering (39.86) were recorded at wider (15 cm)

Table 1 The combined mean square values of ANOVA for phenology, growth and yield components of mung bean inter- and intra-rows spacing at Metema, 2017

SOV	DoF	DF	DM	PH	PL	SPP	DBY (kg ha^{-1})	AGY (kg ha^{-1})	TSW	HI
REP	2	9.39*	0.79^{ns}	219.50**	0.28^{ns}	0.41^{ns}	108,938.01^{ns}	20,723.51*	0.18^{ns}	0.01^{ns}
PS	2	27.18**	10.13**	337.89**	0.074^{ns}	0.32^{ns}	3,140,693.43**	158,340.93**	7.15^{ns}	0.10**
RS	3	91.13**	40.53**	191.65**	0.40^{ns}	0.36^{ns}	11,827,511.24**	175,899.87**	29.47^{ns}	0.22**
Site	1	66.13**	21.13**	302.99**	0.25^{ns}	0.002^{ns}	375,700.01*	747.56^{ns}	1.75**	0.01*
RS*PS	6	0.79^{ns}	4.31**	44.82^{ns}	0.36^{ns}	0.27^{ns}	20,817.43^{ns}	62,836.63**	1.58^{ns}	0.01*
Error	52	2.14	1.08	23.52	0.52	0.38	59,438.14	5636.839	3.71	0.002

NS nonsignificant, *SOV* source of variation, *CV* coefficient of variation, *Rep* replication, *PS* plant spacing, *RS* row spacing, *DF* days to flowering, *DM* days to maturity, *PH* plant height, *PL* pod length, *SPP* seed per pod, *DBY* dry biomass yield, *AGY* adjusted grain yield, *TSW* thousand seed pod, *HI* harvest index, *LSD* least significant difference at 5%

*Significant difference at $P < 0.05$; **highly significant differences at $P < 0.01$

Table 2 The combined main effects of inter- and intra-row spacing on days to 50% flowering and plant height, and aboveground dry biomass yield mung bean at Metema, 2017

	Days to flowering	Plant height	Dry biomass yield (kg ha^{-1})
Intra-rows (PS)			
5	37.75c	80.43a	4009.92a
10	38.92b	76.63b	3548.21b
15	39.86a	72.92c	3296.67c
LSD (5%)	0.85	2.81	141.23
Inter-rows (RS)			
20	36.05c	79.83a	4503.11a
30	38b	77.16a	3962.39b
40	40.5a	77.5a	3400.44c
50	40.83a	72.09b	2607.11d
LSD (5%)	0.98	3.25	163.07
CV (%)	3.76	6.33	6.74

Means in the same column and the same letters are not significantly different at 5% level of probability

NS nonsignificant, LSD least significant difference at 5% level of significant, CV coefficient of variation in percent

Table 3 The combined interaction effect of inter- and inter-row spacing on days to 90% maturity of mung bean at Metema, 2017

Inter-rows (RS)	Intra-rows (PS)		
	5	10	15
20	61.17e	61.17e	60.6e
30	60.5e	62.5dc	61.17e
40	61.5de	63.17bc	64.17ba
50	63.33bc	64.17ba	65.17a
LSD (5%)	1.28	CV (%)	1.8

Means in the same column and the same letters are not significantly different at 5% level of probability

NS nonsignificant, LSD least significant difference at 5% level of significant, CV coefficient of variation in percent

intra-row spacing, while the shortest (37.75) were recorded at narrow (5 cm) intra-row spacing (Table 2). The longest days to flowering with a wider inter- and intra-row spacing might be due to the fact that more nutritional area available in the wider row spacing might have caused the crop to flower later than the narrower spacing. Furthermore, this result might be because wider spacing had a better light interception as compared to the narrow row spacing, resulting in more number of days to flowering of mung bean. This result is in line with Samih [19] who reported that when beans are planted at the lower planting densities, the plants required more number of days for flowering.

Days to 90% maturity

The combined main effect of inter- and intra-row spacing and their interaction was highly significant ($P < 0.01$) on days to 90% maturity (Table 1). The longest days to 90% maturity (65.17) were recorded at interaction of 50 cm inter- and 15 cm intra-row spacing, while the shortest days to 90% maturity (60.5 cm) was recorded at 20 cm inter- and 5 cm intra-row spacing (Table 3). The prolonged days to 90% maturity with the lowest population density or wider inter-row and intra-row spacing might be due to less competition of light interception, high availability of growth resources that promote luxurious growth enhanced the lateral growth and prolonged maturity. This result was not in line with that of Oad et al. [20]

who stated that the interaction of wider inter- and intra-row spacing hastened maturity of safflower.

Plant height (cm)

The combined main effect of inter- and intra-row spacing was highly significant ($P < 0.01$), while their interaction had no significant effect on plant height (Table 1). The maximum plant height (79.83 cm) was recorded at inter-row spacing of 20, 30 and 40 cm (Table 2).

With regard to the effects of intra-row spacing, the maximum plant height (80.43 cm) was recorded at narrow intra-row spacing of 5 cm (Table 2). The present result is in line with Shamsi and Kobraee [21] who found taller plants were recorded at narrow spacing, while the shortest plants at wider spacing on mung bean.

Number of branches per plant

Results from the analysis of variance showed that both the main effect and their interaction effect of inter- and intra-row spacing were highly significant ($P < 0.01$) on the number of branches per plant at Metema site two (Table 4), whereas, at site one, the main effect was significant ($P < 0.01$), while their interaction had no significant effect on the number of branches per plant (Table 4). The highest number of branches per plant (10.4) was obtained from 50×15 cm inter- and intra-row spacing followed by 9.67 at 40×15 cm (Table 5), while the lowest number of branches per plant (3.13) was found at 20×5 cm at Metema site two (Table 5). This result was in line with El Naim et al. [22] who reported that the number of branches per plant was reduced with the increase in plant density. Moreover, our finding is in line with Mehmet [23] who stated that as spacing gets wider, there will be more interception of sunlight for photosynthesis, which results in the production of more nutrients for partitioning toward the development of more branches.

Table 4 Mean square values of ANOVA for the number of branches per plant and pods per plant of mung bean as affected by inter- and intra-row spacing at Metema, Locations one and two, 2017

SOV	DF	Location one		Location two	
		NBPP	NPPP	NBPP	NPPP
REP	2	5.04*	22.58^{ns}	2.42*	4.12^{ns}
PS	2	25.88**	206.92**	53.83**	332.01**
RS	3	16.08**	196.11**	19.76**	131.13**
RS*PS	6	0.63^{ns}	5.50^{ns}	3.55**	17.13**
Error	22	1.18	11.57	0.45	2.89

NS nonsignificant at $P < 0.05$, Rep replication, PS plant spacing, RS row spacing, SOV source of variation, DoF degree of freedom, NBPP number of branches per plant, NPPP number of pods per plant

*Significant at $P < 0.05$; **highly significant differences at $P < 0.01$

Table 5 Interaction effect of inter- and intra-row spacing on number of branches per plant of mung bean at Metema Location two, 2017

Inter-rows (RS)	Intra-rows (PS)			Mean
	5	10	15	
20	3.13e	4.33dc	5.27c	4.24
30	3.87dc	4.8dc	6.6b	5.09
40	3.87de	7.53b	9.67a	7.02
50	4.13de	7.33b	10.4a	7.29
Mean	3.75	6	7.98	
LSD (5%)	1.13	CV (%)	11.29	

Means in the same column and the same letters are not significantly different at 5% level of probability

NS nonsignificant, LSD least significant difference at 5% level of significant, CV coefficient of variation in percent, PS plant spacing, RS row spacing

Number of pods per plant

Results from the analysis of variance indicated that both the main effect and their interaction effect of inter- and intra- row spacing were highly significant ($P < 0.01$) on the number of pods per plant (Table 4) at Metema site two. Similarly, the main effect of inter- and intra-row spacing at site two was significant ($P < 0.01$), while their interaction had no significant effect on the number of pods per plant (Table 4). The highest number of pods per plant (26.73) was obtained from 50×15 cm inter- and intra-row spacing, while the lowest number of pods per plant (7.53) was found at 20×5 cm at Metema site two (Table 6). This result was in line with Malek et al. [24] who reported that the number of pods per plant of lentil was significantly influenced by plant density.

Pod length (cm), number of seeds per pod and thousand seed weight (gm)

In the current study, the combined analysis showed that both main effect and their interactions effect of inter- and intra-row spacing were not significant ($P < 0.05$) on pod length, number of seeds per pod and thousand seed weight (Table 1). This result was in line with Ihsanullah et al. [25] who stated that no significant effect of different row spacing and plant densities on pod length of mung bean. This result was in line with the finding of Ihsanullah et al. [25] who reported no significant effect of row spacing on number of seeds per pod of mung bean. The present result was in line with Lemlem [26] who obtained no significant effect of plant density on hundred seeds weight of soya bean.

Aboveground dry biomass yield (kg ha^{-1})

The combined main effect of inter- and intra-row spacing was highly significant ($P < 0.01$), while their interaction had no significant effect on aboveground dry biomass yield (Table 1). The highest aboveground dry biomass yield (4503.11 kg ha^{-1}) was recorded at 20 cm inter-row spacing which was significantly decreased by wider spacing of 30, 40, and 50 cm spacing, while the lowest value for dry biomass yield (2607.11 kg ha^{-1}) was obtained at 50 cm inter-row spacing (Table 2). Regarding the effect of intra-row spacing, the aboveground dry biomass increased with a decrease in intra-row spacing where the highest aboveground dry biomass yield (4009.92 kg ha^{-1}) was recorded at narrow intra-row spacing of 5 cm, which had a significant difference with 10 cm, while the lowest aboveground dry biomass yield (3296.67 kg ha^{-1}) was recorded at wider intra-row spacing of 15 cm (Table 8). This result was in agreement with Solomon [27] who reported that dry biomass per hectare was significantly increased with increased plant density on haricot bean.

Adjusted grain yield (kg ha^{-1})

The combined analysis showed main effect and their interaction effect of inter- and intra-row spacing which were highly significantly ($P < 0.01$) affected on grain yield of mung bean (Table 1). The highest adjusted grain yield of 1882.67 kg ha^{-1} was obtained at interaction of 40×10 cm, while the lowest adjusted grain yields of 1367.83 and 1401.5 kg ha^{-1} were obtained at interaction of 20×5 cm and 50×15 cm inter- and intra-row spacing, respectively (Table 7).

At very higher population (20×5 cm), the adverse effect on the yield was noticed which might be due to intense interplant competition and floral abortion.

Table 6 Interaction effect of intra- and inter-row spacing on number of pods per plant of mung bean crop at Metema location two, 2017

Inter-rows	Intra-rows		
	5	10	15
20	7.53e	11.33dc	13c
30	9.53de	13.13c	19b
40	9.73de	19.87b	21.53b
50	11.73dc	19.73b	26.73a
LSD (5%)	2.88		
CV (%)	11.16		

Means in the same column and the same letters are not significantly different at 5% level of probability

NS nonsignificant, LSD least significant difference at 5% level of significant, CV coefficient of variation in percent, PS plant spacing, RS row spacing

Table 7 The combined interaction effect of intra- and inter-row spacing on adjusted grain yield (kg ha⁻¹) of mung bean at Metema, 2017

Inter-rows (RS)	Intra-rows (PS)		
	5	10	15
20	1367.83f	1492.67ed	1609.33cb
30	1429.17ef	1551.33cd	1597.17cb
40	1536.17cd	1882.67a	1652.17b
50	1491.83ed	1533.83cd	1401.5f
LSD (5%)	87.731		
CV (%)	4.88		

Means in the same column and the same letters are not significantly different at 5% level of probability

NS nonsignificant, LSD least significant difference at 5% level of significant, CV coefficient of variation in percent, PS plant spacing, RS row spacing

Table 8 The combined interaction effect of intra- and inter-row spacing on harvest index of mung bean at Metema, 2017

Inter-rows (RS)	Intra-rows (PS)		
	5	10	15
20	0.28g	0.38fg	0.39fe
30	0.33g	0.40e	0.44de
40	0.41e	0.58ab	0.54bc
50	0.49cd	0.62a	0.62a
LSD (5%)	0.06	CV (%)	10.57

Means in the same column and the same letters are not significantly different at 5% level of probability

NS nonsignificant, LSD least significant difference at 5% level of significant, CV coefficient of variation in percent

Besides, at spacing of 20 cm × 5 cm, the grain yield ha⁻¹ was significantly higher as compared to the interaction of wider inter- and intra-row spacing (50 × 15 cm), which showed that the main determinant of yield was the plant population which along with other yield attributes contributed toward a significant increase in grain yield (Table 7). This result was in line with Abuzar et al. [28] who observed that the lowest grain yield was recorded at the highest population. In Bangladesh, as well as in other countries, various experiments with improper spacing reduced the yield of mung bean up to 20–40% [29].

Harvest index

The combined main effect of intra- and inter-row spacing had a significant ($P < 0.01$) effect on harvest index (Table 1). Moreover, the interaction effect of intra- and inter-row spacing effect on harvest index was significant ($P < 0.05$) (Table 1). The highest harvest index (62%) was obtained at interaction of (50 × 10) cm followed by (50 × 15) cm spacing, while the lowest (28% and 33%) harvest index was obtained at interaction of (20 × 5) and (30 × 5) cm, respectively (Table 8). Similar result was reported by Khan et al. [30] who recorded maximum harvest index (41.66%) at the highest row spacing (45 cm) of chickpea than 15 cm row spacing (32.6%).

Economic analysis

Results of the economic analysis of the present experiment showed that the maximum net benefit of Ethiopian birr (ETB) 22,185.7 ha⁻¹ with an acceptable marginal rate of return (MRR) (> 100%) was obtained from planting of mung bean with 40 × 15 cm spacing (Table 9). Farmers may have the opportunity to decrease seed and labor cost and to increase mung bean yield and ultimately improve their livelihoods through adopting the appropriate management practices. The present finding demonstrated that optimum economic plant densities of mung bean are often less than densities that result in maximum yield, because of higher labor and seed costs at higher densities. This result is in line with [31] who reported that changes in seeding rates contributed to significant yield changes but not to changes in profitability.

Conclusions

Generating reliable information on agronomic management practices such as appropriate row and plant spacing is quite important to come up with profitable and sustainable mung bean production and productivity. In view of this, an experiment was conducted to determine the effect of intra- and inter-row spacing on the yield and yield components of mung bean.

Table 9 Economic analysis for inter- and intra-row spacing of mung bean at Metema, 2017

	Average yield (kg ha^{-1})	Adjusted yield 10% (kg ha^{-1})	GB (ETB/ha)	TVC (ETB/ha)		DA	
50X15	1367.83	1231.05	22,158.85	3599.305	19,105	–	
40X15	1492.67	1343.40	24,181.25	4579.456	22,185.7	–	314.308
50X10	1609.33	1448.40	26,071.15	5621.72	19,226.33	D	
30X15	1429.17	1286.25	23,152.55	5956.173	19,917.98	D	
40X10	1551.33	1396.20	25,131.55	7144.15	23,355.1	–	45.596
20X15	1597.17	1437.45	25,874.15	9155.312	16,915.83	D	
30X10	1536.17	1382.55	24,885.95	9294.595	15,836.95	D	
50X5	1882.67	1694.40	30,499.25	11,243.44	12,924.21	D	
20X10	1652.17	1486.95	26,765.15	14,288.3	9892.954	D	
40X5	1491.83	1342.65	24,167.65	14,288.3	10,597.65	D	
30X5	1533.83	1380.45	24,848.05	18,587.39	4565.164	D	
20X5	1401.5	1261.35	22,704.30	28,576.6	− 6417.75	D	

GB gross benefit, *ETB* Ethiopian birr, *TVC* total variable cost, *NB* net benefit, *DA* dominance analysis, *MRR* marginal rate of return

This study provides evidence that inter- and intra-row spacing has influence on the phenology, growth, yield and yield components of mung bean. However, for maximum net benefit, we recommend spacing of 40×15 cm as the optimum spacing for the cultivation of mung bean variety Rasa (N-26) under the rain-fed in North Gondar Zone, Metema Districts, and similar agroecologies.

Abbreviations
ANOVA: analysis of variance; ETB: Ethiopian birr; MRR: marginal rate of return.

Authors' contributions
AB initiated the research, wrote the research proposal, conducted the research, did data entry and analysis and wrote the manuscript. TT and DT were involved in analysis, methodology, writing, reviewing and editing of research proposal and manuscript. All authors read and approved the final manuscript.

Author details
[1] Gondar Agricultural Research Center, P.O. Box 1337, Gondar, Ethiopia. [2] Fogera National Rice Agricultural Research Center, P.O. Box 1937, Woreta, Ethiopia. [3] Department of Plant Sciences, University of Gondar, P.O. Box 196, Gondar, Ethiopia.

Acknowledgements
We would like to acknowledge Gondar Agricultural Research Center for providing seed and equipment for measuring various traits. University of Gondar is also acknowledged for accepting AB as a Master of Science student in Agronomy discipline.

Competing interests
The authors declare that they have no competing interests.

Funding
No funding was received toward this study.

References
1. Kumar SG, Gomathinayajam P, Rathnaswmy R. Effect of row spacing on a dry matter accumulation of black gram. Madras Agric J. 1997;84(3):160–2.
2. Ali MZ, Khan MA, Rahaman AK, Ahmed M, Ahsan AF. Study on seed quality and performance of some mung bean varieties in Bangladesh. Int J Exp Agric. 2010;1(2):10–5.
3. Gebre Wedajo. Adaptation study of improved mung bean (*Vigna radiata* L) varieties at Alduba, south Omo Ethiopia. Res J Agric Environ Manag. 2015;4(8):339–42.
4. Tomooka N, Vaughan DA, Kaga A. Mung bean (*Vigna radiata* L. Wilczek). In: Singh RJ, Jauhar PP, editors. Chromosome engineering and crop improvement: grain legumes genetic resources. Boca Raton: CRC Press; 2005.
5. Somta P, Srinives P. Genome research in mung bean (*Vigna radiata* L. Wilczek) and black gram (*Vigna mungo* L. Hepper). Sci Asia. 2007;33(1):69–74.
6. Hussain M, Mehmood Z, Khan MB, Farooq S, Lee DJ, Farooq M. Narrow row spacing ensures higher productivity of low tillering wheat cultivars. Int J Agric Biol. 2012;14:413–8.
7. Minh NP. Different factors affecting to mung bean (*Phaseolus aureus*) tofu production. Int J Multidiscip Res Dev. 2014;1(4):105–10.
8. Yagoob H, Yagoob M. The effects of water deficit stress on protein yield of mung bean genotypes. Peak J Agric Sci. 2014;2(3):30–5.
9. Asfaw A, Fekadue Gurum F, Alemayehu S, Rezene Y. Analysis of multi-environment grain yield trials in mung bean (*Vigna radiate* L. Wilczek) based on GGE bi plot in Southern Ethiopia. J Agric Sci Technol. 2012;14:389–98.
10. MoARD (Ministry of Agriculture and Rural Development). Crop variety register. Animal and Plant Health Regulatory Directorate, Issue No. 12. Addis Ababa, Ethiopia; 2008. p. 96–103.
11. MoARD (Ministry of Agriculture and Rural Development). Crop variety register. Animal and Plant Health Regulatory Directorate, Issue No. 12. Addis Ababa, Ethiopia; 2015. p. 96–103.
12. CSA (Central Statistical Authority). Agricultural sample survey 2008/2009: report on area and production of crops (private peasant holdings 'Meher' season). Addis Ababa Ethiopia, the FDRE statistical bulletin, Vol. IV; 2016.
13. Rafiei M. Influence of tillage and plant density on mung bean. Am Eur J Sustain Agric. 2009;3:877–80.
14. NMSA (National Metrological Services Agency) Bahir Dar Branch. 2017.
15. Dessalegn L. Snap bean production and research status (*Phaseolus vulgaris* L.) in Ethiopia. Nazret: Melkassa Agricultural Research Center; 2003.
16. SAS (Statistical Analysis System). SAS Version 9.1 © 2001–2002. Cary, NC: SAS Institute, Inc.; 2002.

17. Gomez KA, Gomez AA. Statistical procedures for agricultural research. 2nd ed. New York: Wiley; 1984.

18. CIMMYT (International Maize and Wheat Improvement Center). From agronomic data to farmer recommendations: an economics training manual. Texcoco: CIMMYT (International Maize and Wheat Improvement Center); 1988 **(completely revised edition)**.

19. Samih A. Effect of plant density on flowering date, yield and quality attribute of bush beans (*Phaseolus Vulgaris* L.) under center pivot irrigation system. Am J Agric Biol Sci. 2008;3(4):666–8.

20. Oad FC, Samo MA, Qayylan SM, Oad NL. Inter and intra row spacing effect on the growth, seed yield and oil continent of safflower. Asian Plant Sci J. 2002;1:18–9.

21. Shamsi K, Kobraee S. Effect of plant density on the growth, yield and yield components of three soybean varieties under climatic conditions of Kermanshah, Iran. Anim Plant Sci J. 2009;2(2):96–9.

22. El-Noemani AA, El-Zeiny HA, El-Gindy AM, Sahhar EA, El-Shawadfy MA. Performance of some bean (*Phaseolus Vulgaris* L.) varieties under different irrigation systems and regimes. Aust J Basic Appl Sci. 2010;4:6185–96.

23. Mehmet OZ. Nitrogen rate and plant population effects on yield and yield components in soybean. Afr Biotechnol J. 2008;7(24):4464–70.

24. Malek MFN, Majnonhoseini N, Alizade H. A survey on the effects of weed control treatments and plant density on lentil growth and yield. Eur J Clin Pharmacol. 2013;6:135–48.

25. Ihsanullah F, Taj H, Akbar H, Basir A, Noor U. Effect of row spacing on the agronomic traits and yield of mung bean (*Vigna radiata* L. Wilczek). Asian J Plant Sci. 2002;1(4):328–9.

26. Lemlem HG. Effect of N fertilizer and plant density on yield, seed quality, and oil content of soybean (*Glycine max* L. Merr) at Hawassa, southern Ethiopia. M.Sc. thesis presented to Haramaya University, Ethiopia; 2011.

27. Abate S. Effects of irrigation frequency and plant population density on growth, yield components and yield of haricot bean (*Phaseolus vulgaris* L.) in Dire Dawa area. M.Sc. thesis presented to Haramaya University, Ethiopia; 2003.

28. Abuzar MR, Sadozai GU, Baloch MS, Baloch AA, Shah IH, Javaid T, Hussain N. Effect of plant population densities on yield of maize. J Anim Plant Sci. 2011;21(4):692–5.

29. Mondal MMA. A study of source-sink relation in mung bean. Ph.D. dissertation. Department of Crop Botany, Bangladesh Agriculture University, Mymensingh; 2007. p. 82–84.

30. Khan EA, Aslam M, Ahmad HK, Ayaz M, Hussain A. Effect of row spacing and seeding rates on growth yield and yield components of chickpea. Sarhad J Agric. 2010;26(2):201–11.

31. De Bruin JL, Pedersen P. Effect of row spacing and seeding rate on soybean yield in the Upper Midwest. Agron J. 2008;100:696–703.

32. Hashemijazi SM, Danesh A, Gani B. Effects of plant arrangement on yield and yield components of Soya bean. In: Proceedings of the 1st Iranian Pulse Symposium, November 20–21, 2005, IEEE Xplore; 2005. p. 208–209.

Effect of split application of different *N* rates on productivity and nitrogen use efficiency of bread wheat (*Triticum aestivum* L.)

Fresew Belete[1,2]* **(iD)**, Nigussie Dechassa[1], Adamu Molla[3] and Tamado Tana[1]

Abstract

Background: Bread wheat is an important staple and cash crop grown by smallholder farmers in the central highlands of Ethiopia. However, the productivity of the crop is constrained by low soil fertility and poor nitrogen fertilizer management in the area. For example, there is limited information on optimum rates and timing of nitrogen fertilizer application in the area. Therefore, a field experiment was conducted for two consecutive years (2014 and 2015) under rain-fed condition to determine the effect of *N* fertilizer rate and timing of application on grain yield and nitrogen use efficiency of bread wheat. Factorial combinations of three *N* levels and five application times plus one control were laid out in a randomized complete block design with four replications.

Results: The optimum grain yield (6060.04 kg ha^{-1}) was recorded when 240 kg N ha^{-1} was applied ¼ at sowing, ½ at tillering and ¼ at booting, and it showed no significant additional response to *N* fertilizer above this rate. Higher *N* level (360 kg N ha^{-1}) always increased *N* content in the grain and nitrogen uptake by wheat crop. The best recovery of nitrogen (59.74%) by wheat was found when 120 kg of nitrogen was applied (¼ at sowing, ½ at tillering and ¼ at booting). The nitrogen use efficiency traits decreased with increased *N* rate (120–360 kg N ha^{-1}) indicating poor *N* utilization. The split application of nitrogen (¼ at sowing, ½ at tillering and ¼ at booting) produced the highest nitrogen use efficiency traits.

Conclusion: The application of 240 kg. N ha^{-1} in three split doses (T_5) was required to obtain optimum wheat yield. In addition, increasing the rate of nitrogen beyond 120 kg N ha^{-1} decreased nitrogen use efficiency traits.

Keywords: Efficiency, Recovery, Split, Uptake, Yield

Background

Bread wheat (*Triticum aestivum L.*) is one of the most important cereal crops in the world in terms of area coverage and production. It is a major source of nutrition for humans and livestock, estimated to contribute as much as 60 million tonnes of protein per year [1]. The total worldwide production of wheat in 2012 was around 671 million tonnes on an area of 215 million ha [2]. In Ethiopia, wheat is grown approximately by 4.8 million farmers on 1.6 million hectares representing 13.33% of total crop area [3]. Data aggregated at a worldwide level over several decades have shown a strong link between agriculture production and fertilizer use [4]. Of the nutrients, nitrogen (N) is frequently regarded as the single most important mineral nutrient limiting crop production in many agricultural crops worldwide, and it is needed in large amount, as it constitutes 1–4% of the plant dry matter [5]. However, the average yield of wheat in Ethiopia is very low; it is about 2.5 ton/ha as compared to the world's average of about 3.4 ton/ha [2]. The low mean national yield of wheat is mainly the result of depleted soil fertility, especially nitrogen (*N*) deficiency, which is often encountered in cool wet areas or in soils that are frequently water logged such as the highland *Vertisols*. Therefore, greater usage of chemical fertilizer has been advocated as

*Correspondence: beletefresew@gmail.com
[1] School of Plant Sciences, College of Agriculture and Environmental Sciences, Haramaya University, P.O. Box 138, Dire Dawa, Ethiopia
Full list of author information is available at the end of the article

a primary means of increasing wheat grain yield in Ethiopia [6].

Although N is the key element in increasing productivity and the increase of agricultural food production worldwide over the past four decades, a small fraction of this fertilizer is taken up by the plant [7], being 33% for wheat [8]. Poor N recovery is a function of N flows to competing pathways such as gaseous N losses, leaching and biological immobilization and in-efficiencies in crop N uptake and utilization [9, 10]. However, adoption of appropriate N fertilizer management practices is reported to increase N recovery up to 70–80% [11]. Split application of N is one of the methods to improve N use by the crop while reducing the nutrient loss through leaching, denitrification, runoff and volatilization [12]. Some research findings indicated that late season N application as dry fertilizer material was effective in attaining higher N recovery and use efficiency [13]. In addition, determining the right N fertilizer rates and timing of application is decisive factor in obtaining higher yields [14].

In many parts of the world, limited research has been done on the effect of split application of N for wheat and its association with grain yield and NUE [15], which is also true in Ethiopia where information on the subject is meager. Besides, matching crop N demand with available soil N has been challenging for wheat producers in Enewari due to the susceptibility of Vertisols to water logging, which leads to denitrification, leaching and runoff losses during heavy rainfall [12]. According to Molla [16], this forced farmers of Enewari to apply as large as 256 kg N ha^{-1} (some even apply more) which is by far higher than the blanket recommendation (87 kg N ha^{-1}). However, the optimum rate of nitrogen fertilizer for wheat production in the study area and its

time of application are not yet investigated. This study was, therefore, conducted to determine the effect of N levels and time of application on yield and nitrogen use efficiency of bread wheat.

Methods

The study site

The study was conducted for two consecutive years during 2014 and 2015 main cropping seasons in the district of Moretina Jiru at the Enewari experimental field station in the central highlands of Ethiopia. Enewari is located at 9° 52′ N latitude, 39° 10′ E longitude at an altitude of 2680 meters above sea level. This area is typical of the rain-fed wheat-growing regions of Ethiopia with average annual rainfall of 1153.69 mm. The dominant soil type of the area is Vertisols which are known for their high water logging and drainage problems. Figure 1 shows monthly total rainfall and monthly mean temperatures at the experimental site over the 2-year study period.

Prior to planting, the surface (0–20 cm) soil samples from ten spots across the experimental field were collected, composited and analyzed for determining selected soil physicochemical properties at Debre Berhan Agricultural Research Center following the procedure outlined by [17]. Values for the selected physicochemical properties are presented in Table 1.

Description of the study materials

Fertilizer sources were urea (46% N) for nitrogen fertilizer and triple superphosphate (46% P_2O_5) for phosphorus fertilizer. A wheat variety called Menze (HAR-3008) was used as a test crop which was developed and released by DBARC (Debre Berhan Agriculture Research Center) in 2007. It has been widely promoted

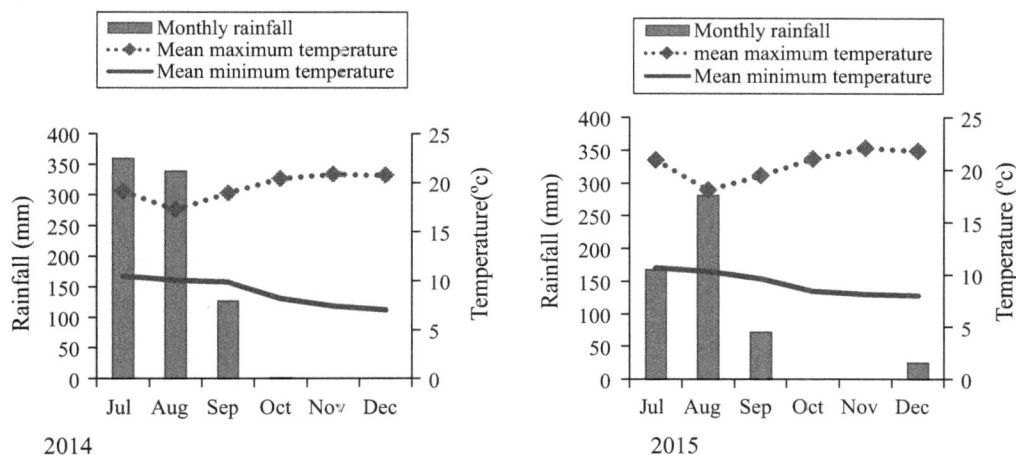

Fig. 1 Monthly total rainfall and average maximum and minimum temperatures in 2014 and 2015 growing seasons at Enewari, central highlands of Ethiopia

Table 1 Soil physicochemical properties at the depth of 0–20 cm during the years of 2014 and 2015 before sowing of bread wheat

Parameter	Value		Rating	References
	Year 2014	Year 2015		
Sand	15	12	–	–
Silt	18	17	–	–
Clay	67	71	–	–
Texture class	Clay	Clay	–	–
pH	7.02	7	Neutral	Tekalign Tadesse [42]
Organic carbon (%)	1.08	1.15	Low	Tekalign Tadesse [42]
Total N (%)	0.08	0.06	Low	Tekalign Tadesse [42]
Av. P (ppm)	6.54	7.82	Low	Olsen et al. [43]
CEC [cmol(+)/kg soil]	48.75	45.25	Very high	Metson [44]
Exchangeable K	0.23	0.2	Low	Metson [44]

for its resistance to yellow rust and with a yield potential of 1900–3300 kg ha^{-1} high yielder in Moretina Jiru district, Enewari area. The variety is medium in maturity (154 days), with a medium stature of 64 cm [18].

Treatments and experimental design

The treatments consisted of complete factorial combinations of three N fertilizer rates and five split N applications, plus one unfertilized control. The three N-fertilization levels were 120, 240 and 360 kg N ha^{-1}. The five N split application timings were adjusted according to Zadoks decimal growth stage for wheat [19] at the time when the moisture is available for nutrient dissolution and absorption. These application timings were: $T_1 = N$ applied ½ at sowing and ½ at tillering (Zadok scale 21–22); $T_2 =$ all N applied at tillering (Zadok scale 21–22); $T_3 = N$ applied ½ at tillering (Zadok scale 21–22) and 1/2 at booting (Zadok scale 41–45); $T_4 = N$ applied 1/3 at sowing, 1/3 at tillering (Zadok scale 21–22) and 1/3 at booting (Zadok scale 41–45); and $T_5 = N$ applied 1/4 at sowing, ½ at tillering (Zadok scale 21–22) and 1/4 at booting (Zadok scale 41–45).

These treatments were laid out in a Randomized Complete Block Design (RCBD) with four replications. The gross plot size of each treatment was 2 m × 3 m (6 m^2) accommodating eight rows spaced 20 cm apart. The plot size for planting was 1.6 m × 3.0 m (4.8 m^2), and four central rows were used for data collection and measurement. The distance between the plots and replications was kept at 0.5 m and 1 m apart, respectively.

Crop management

Wheat seed was sown by drilling in rows at the recommended rate of 150 kg ha^{-1} on July 24 in both years.

Each year, all the wheat plots were supplied with triple superphosphate (TSP) at a recommended rate of 138 kg P$_2$O$_5$ ha^{-1} [20]. Similarly, the N was applied in the form of urea (as per the treatment) at planting and the later stage splits were applied by side dressing at the specified Zadoks growth stages. Plots were kept free of weeds by hand weeding. No insecticide or fungicide was applied since there was no outbreak of any insect or disease incidence. Harvesting was done manually using hand sickle in late December.

Data collection and measurements

In both years, gain yield (kg ha^{-1}) was determined from the harvested net plot area of 2.4 m^2 and was adjusted to 12.5% moisture content. At crop maturity, a subsample from each net plot was harvested at ground level, oven-dried at 70 °C until constant weight was reached for dry weight determination and partitioned into straw and grain. The dried samples were milled and the grain and straw N content of the plant samples was determined using the micro-Kjeldahl method as stated by American Association of Cereal Chemists (AACC) [21]. Total grain N uptake (GNUP) in kg ha^{-1} was calculated by multiplying grain yields by N content percentage. Total nitrogen uptake (TNUP) was calculated as the sum of the respective GNUP and SNUP values.

Nitrogen use efficiency traits

The following N-efficiency parameters were calculated for each treatment following Fageria [22]:

1. Agronomic efficiency (AE, kg kg^{-1}) $= \frac{G_f - G_u}{N_a}$, where G_f is the grain yield of the fertilized plot (kg), G_u is the grain yield in the unfertilized plot (kg) and N_a is the quantity of N applied.
2. Agro-physiological efficiency (APE, kg kg^{-1}) $= \frac{G_f - G_u}{N_f - N_u}$, where G_f is the grain yield of the fertilized plot (kg), G_u is the grain yield in the unfertilized plot (kg), N_f is the N accumulation in the fertilized plot (kg) and N_u is the N accumulation in the unfertilized plot (kg).
3. Apparent recovery efficiency (ARE, %) $= \frac{N_f - N_u}{N_a} * 100$, where N_f is the N accumulation by straw and grain in the fertilized plot (kg), N_u is the N accumulation by the straw and grains in the unfertilized plot (kg) and N_a is the quantity of N applied (kg).
4. The nitrogen harvest index (NHI) was determined as the ratio of nitrogen uptake by grain and nitrogen uptake by grain plus straw as described by [22].

Effect of split application of different N rates on productivity and nitrogen use efficiency...

205

Data analysis

After verifying the homogeneity of error variances, combined analysis of variance was done using the procedure of SAS [23], and to facilitate factorial analysis, the control was excluded. Mean comparisons were done by Duncan's multiple range tests Gomez and Gomez [24] at the 5% level.

Results

Grain yield

Grain yield was significantly affected by the main effects of year, N rate, time of application as well as the interaction effect of N rate × time of application. What is more, the interaction effect of year × N rate, year × time of application and year × N rate × time of application

did not affect this parameter (Table 2). The split application of the different N fertilizer rates significantly ($P < 0.01$) affected grain yield. The highest grain yield was obtained in response to the application of 360 kg N ha^{-1} in three splits of ¼ at sowing, ½ at tillering and ¼ at booting, which was in statistical parity with the grain yield obtained in response to the application of 240 kg N ha^{-1} ¼ at sowing, ½ at tillering and ¼ at booting (Table 3).

Nitrogen uptake
Grain N uptake

Grain N uptake (GNUP) was significantly influenced by the rate and timing of N application. The interaction effect of N rate × time of application and year × time of application also revealed a significant effect on nitrogen

Table 2 Mean squares of analysis of variance for year, N fertilizer rate and time of N application, and their interaction

Source	DF	GY	GNUP	TNUP	AE	RE	APE	NHI
Y	1	3,386,971**	789.77**	6813.22**	506.38**	4492.67**	270.07ns	839.08**
Rep (Y)	6	61,018	30.39	120.53	18.85	35.95	51.62	18.28
N	2	12,053,826**	22,875.82**	43,818**	2090.92**	2817.12**	3064.21**	120.10**
T	4	7,906,065**	4969.21**	10,914**	127.93**	1622.94**	287.29**	108.04**
N × T	8	1,127,802**	720.28**	1632.7**	6.12ns	68.09*	11.19ns	26.01*
Y × N	2	233836ns	211.25ns	552.37*	43.59**	142.47*	0.26ns	31.07*
Y × T	4	311205ns	254.56*	59.89ns	4.32ns	24.27ns	167.89**	43.21**
Y × N × T	8	231,387ns	70.27ns	115.94ns	4.89ns	32.98ns	6.73ns	11.28ns
Error	84	130,052	94.9	124.13	3.07	27.57	11.52	9.13

Y year, *Rep* replication, *N* N rate, *T* timing of N application, *Df* degree of freedom, *GY* grain yield, *GNUP* grain nitrogen uptake, *TNUP* total nitrogen uptake, *AE* agronomic efficiency, *RE* recovery efficiency, *APE* agro-physiological efficiency, *NHI* nitrogen harvest index

*Significant at the 0.05 probability level; **Significant at $P < 0.01$ probability level

Table 3 Grain yield (GY), grain nitrogen uptake (GNUP) and total nitrogen uptake (TNUP) as influenced by the interaction effect of N fertilizer rate and time of N application

N timing	N rate (kg ha^{-1})			N rate (kg ha^{-1})			N rate (kg ha^{-1})		
	120	240	360	120	240	360	120	240	360
	GY (kg ha^{-1})			GNUP (kg ha^{-1})			TNUP (kg ha^{-1})		
T_1	4436.40efg	4948.16def	5468.85cd	78.04fg	106.19de	127.5bc	96.27fg	131.45de	161.02bc
T_2	4076.89fg	4356.21fg	4406.88efg	71.09g	91.56ef	100.24de	85.18g	114.12ef	124.20e
T_3	4307.19fg	5050.70de	4688.13efg	68.08g	91.22ef	94.79def	83.94g	116.11ef	121.24e
T_4	4362.17fg	5538.53bcd	6189.36ab	70.96g	110.15cd	134.17ab	92.97g	144.62cd	175.63b
T_5	4756.73ef	6060.04abc	6436.00a	82.49fg	128.96b	148.55a	100.05fg	170.04b	201.47a
Treated mean	5005.48a			100.27			127.76a		
Control mean	1307.96b			24.89			28.36b		
		NR × NT	Treated versus control		NR × NT	Treated versus control		NR × NT	Treated versus control
CV (%)		7.2	3.84		9.72	2.32		8.72	1.51

Means followed by the same letters for the same parameter are not significantly different at $P \leq 0.05$

CV Coefficient of variation, *NR* nitrogen rate, *NT* time of nitrogen application

T_1 = N application of ½ at sowing and ½ at tillering; T_2 = N application at tillering; T_3 = N application of ½ at tillering and ½ booting; T_4 = N application 1/3 at sowing, 1/3 at tillering and 1/3 at booting; and T_5 = N application ¼ at sowing, ½ at tillering and ¼ at booting

uptake by the grain. However, the effect of year, year $\times N$ rate and year $\times N$ rate $\times$ time of application on grain nitrogen uptake was nonsignificant (Table 2). Nitrogen uptake by the grain tended to increase in response to the level of N as it rises from 120 to 360 kg ha^{-1} in both growing years. The maximum grain N uptake value (148.55 kg ha^{-1}) was obtained when 360 kg N ha^{-1} was applied in three splits (¼ at sowing, ½ at tillering and ¼ at booting) while the lowest value (68.0 kg ha^{-1}) was recorded when 120 kg N ha^{-1} was applied equally at tillering and booting (T_3) (Table 3). With regard to the interaction effect of year $\times$ time of N application, split application of N three times at sowing, tillering and booting (T_5) produced the highest N uptake value in both growing years while the lowest grain N uptake (78.61 kg ha^{-1}) was recorded when N was applied equally at tillering and booting (T_3) in the year 2014 (Table 4).

Total N uptake

The analysis of variance indicated that year, N rate, time of N application had highly significant effect on total

Table 4 Interaction effect of the year $\times$ N rate and year $\times$ N timing on GNUP and TNUP of bread wheat

N timing	Year	
	2014	2015
	GNUP (kg ha^{-1})	
T_1	101.66c–e	106.15b–d
T_2	82.1fg	93.15def
T_3	78.61g	90.78efg
T_4	106.66bc	103.52c–e
T_5	119.45ab	120.55a
Treated mean	100.27a	
Control mean	24.89b	
CV (%)	9.72	
N rate (kg ha^{-1})	**Year**	
	2014	2015
	TNUP (kg ha^{-1})	
120	88.29d	95.07d
240	124.33c	145.46b
360	148.07b	165.36a
Treated mean	127.76a	
Control mean	28.36b	
CV (%)	8.72	

Means followed by the same letters for the same parameter are not significantly different at $P \leq 0.05$

CV Coefficient of variation, NR nitrogen rate, NT time of nitrogen application

$T_1 = N$ application of ½ at sowing and ½ at tillering; $T_2 = N$ application at tillering; $T_3 = N$ application of ½ at tillering and ½ booting; $T_4 = N$ application 1/3 at sowing, 1/3 at tillering and 1/3 at booting; and $T_5 = N$ application ¼ at sowing, ½ at tillering and ¼ at booting

nitrogen uptake of wheat. Likewise, the interaction of N rate $\times$ time of N application, year $\times N$ rate also revealed a significant effect on total nitrogen uptake. But, the interaction effect of year $\times N$ rate $\times$ time of application (Table 3) was not significant. The highest total N uptake value (201.47 kg ha^{-1}) was attained when 360 kg N ha^{-1} was applied three times at sowing, tillering and booting (T_5) while the lowest (83.94 kg ha^{-1}) was recorded when 120 kg N ha^{-1} was applied equally at tillering and booting (T_3) (Table 3). The year $\times N$ rate interaction shows that wheat N uptake had the highest value (165.4 kg N ha^{-1}) in the year 2015 at the highest N rate while the lowest value (88.3 kg N ha^{-1}) was recorded in 2014 at a rate of 120 kg N ha^{-1} which was statistically similar to that of 2015 under the same N rate (Table 4).

Nitrogen use efficiency traits
Agronomic efficiency

Nitrogen agronomic efficiency (AE) represents the ability of the plant to increase yield in response to N applied [25]. AE varied significantly according to year, N rates and timing of application, as well as by the interaction of year $\times N$ rate. The interaction between year $\times$ time of application, N rate $\times$ time of application and year $\times N$ rate $\times$ time of application did not show a significant effect on this parameter (Table 2). In 2015, the year with the highest grain yield, the value recorded for AE was significantly higher than 2014 under all N rates. The application of 120 kg N ha^{-1} produced the highest AE value (28.8 kg ha^{-1}) in 2015. The lowest (10.47 kg kg^{-1}) value was recorded when 360 kg N ha^{-1} was applied in 2014 (Table 5).

Nitrogen agro-physiological efficiency

Nitrogen agro-physiological efficiency (APE) represents the ability of a plant to transform N acquired from fertilizer into economic yield (grain) [26]. APE was also influenced by the main effects of N rate and time of application and by the interaction of year $\times$ time of N application. However, the effect of year, the interaction of N rate $\times$ time of application and year $\times N$ rate $\times$ time of N application had no significant effect on this index (Table 2). As to the interaction of year $\times$ time of N application, the highest APE (49.75 kg kg^{-1}) was obtained when N was applied in equal split at sowing and tillering in the year 2014 while the lowest value (35.4 kg kg^{-1}) was recorded in response to the application of nitrogen only once at tillering (T_2) in 2015 (Table 6).

Nitrogen apparent recovery efficiency

Nitrogen apparent recovery efficiency (RE) depends on the congruence between plant N demand and the quantity of N released from applied N [27]. RE varied

Table 5 Interaction effect of year × N rate on agronomic efficiency (AE), apparent recovery efficiency (RE) and nitrogen harvest index (NHI) of bread wheat

N rate (kg ha^{-1})	Year		Year		Year	
	2014	2015	2014	2015	2014	2015
	AE (kg kg^{-1})		RE (%)		NHI (%)	
120	22.58b	28.75a	44.69c	59.85a	84.13a	78.07cd
240	14.1d	18.25c	37.36d	50.92b	82.13ab	75.60d
360	10.47e	12.47d	31.5e	39.48d	79.32bc	76.05d
CV (%)	9.86		11.94		3.82	

Means followed by the same letters for the same parameter are not significantly different at $P \leq 0.05$

CV Coefficient of variation

Table 6 Interaction effect of year × N timing on agro-physiological efficiency (APE) and nitrogen harvest index (NHI) of bread wheat

	Year		Year	
	2014	2015	2014	2015
N timing	APE (kg kg^{-1})		NHI (%)	
T_1	37.21d	38.43cd	82.8a	78.05bc
T_2	45.23ab	35.4d	83.35a	80.06ab
T_3	49.75a	42.2bc	81.43ab	78.34bc
T_4	39.62cd	39.35cd	81.43ab	71.72d
T_5	36.61d	38.02cd	80.31ab	74.69cd
CV (%)	8.45		3.82	

Means followed by the same letters for the same parameters are not significantly different at $P \leq 0.05$

CV Coefficient of variation

$T_1 = N$ application of ½ at sowing and ½ at tillering; $T_2 = N$ application at tillering; $T_3 = N$ application of ½ at tillering and ½ at booting; $T_4 = N$ application 1/3 at sowing, 1/3 at tillering and 1/3 at booting; and $T_5 = N$ application ¼ at sowing, ½ at tillering and ¼ at booting

significantly according to year and treatment (N application timing and N fertilizer rates) as well as by the interaction of N rate × time of N application ($P < 0.05$) and year × N rate ($P < 0.05$). But, the interaction effect of year × time of N application and year × N rate × time of N application did not affect recovery efficiency of nitrogen (Table 2). With regard to the interaction effect of year × N rate, the highest RE (59.85%) was recorded in the year 2015 with the application of 120 kg N ha^{-1}. The lowest RE (31.5%) was recorded with the application of 360 kg N ha^{-1} in 2014 (Table 5). As to the interaction of N rate × application timing, the highest RE (59.7%) was obtained from the application of 120 kg N ha^{-1} three times in split (¼ at sowing, ½ at tillering and ¼ at booting). However, the lowest value (25.64%) was obtained from the application of 360 kg N ha^{-1} two times equally at tillering and booting (T_3) which is statistically similar to the recovery efficiency recorded when the highest level of nitrogen was applied only once at tillering (Table 7).

Table 7 Apparent nitrogen recovery efficiency (RE) and nitrogen harvest index (NHI) as influenced by the interaction effect of N fertilizer rate and time of N application

N timing	N rate (kg ha^{-1})			N rate (kg ha^{-1})		
	120	240	360	120	240	360
	RE %			NHI (%)		
T_1	56.1ab	42.71d–f	36.68ef	80.99a–c	81.08a–c	79.21a–d
T_2	46.85b–d	35.48fg	26.46gh	83.75a	80.55a–c	80.81a–c
T_3	45.81c–e	35.53fg	25.64h	81.45a–c	80.28a–c	78.22b–d
T_4	53.34a–c	48.19b–d	40.74d–f	76.66cd	76.68cd	76.39cd
T_5	59.24a	58.78a	47.92b–d	82.65ab	76.04cd	73.82d
CV (%)	11.94			3.82		

Means followed by the same letters for the same parameter are not significantly different at $P \leq 0.05$

CV Coefficient of variance

$T_1 = N$ application ½ at sowing and ½ at tillering; $T_2 = N$ application at tillering; $T_3 = N$ application ½ at tillering and ½ at booting; $T_4 = N$ application 1/3 at sowing, 1/3 at tillering and 1/3 at booting; and $T_5 = N$ application ¼ at sowing, ½ at tillering and ¼ at booting

Nitrogen harvest index

Nitrogen harvest index is a measure of N partitioning in the crop, which provides an indication of how efficiently the plant utilized the acquired N for grain production [26]. Nitrogen harvest index (NHI) was significantly influenced by year, N rate and timing of N application. A generally significant effect of two-way interactions was also observed. However, the interaction effect of year $\times N$ rate $\times$ time of application was not significant (Table 2). With regard to the interaction effect of year $\times N$ rate, the highest value of NHI (84.13%) during the first growing year (2014) and the lowest value (75.6%) of NHI in the second growing year (2015) for wheat were recorded with the application of 120 kg N ha^{-1} and 360 kg N ha^{-1}, respectively (Table 5). In general, application of nitrogen beyond 120 kg N ha^{-1} did not significantly affect NHI in the second growing season while the application of 360 kg N ha^{-1} significantly produced lower NHI as compared to the application of 120 kg N ha^{-1}. With regard to the year $\times$ time of N application, the highest NHI (83.35%) was recorded when the whole nitrogen was applied only once at tillering (T_2) which was statistically similar to the rest timing treatments in 2014, while the lowest NHI (71.72%) was recorded with the split application of nitrogen three times (T_4) in 2015 (Table 6). As to the interaction effect of N rate $\times$ time of application, the highest value (83.75%) was recorded due to the application of 120 kg N ha^{-1} only once at tillering (T_2), while the lowest nitrogen harvest index (73.82%) was produced from the application of 360 kg N ha^{-1} three times in split (¼ at sowing, ½ at tillering and ¼ at booting) (Table 7).

Discussion

Variations in climatic conditions registered during the cropping periods (Fig. 1) induced large variations in grain yield and the efficiency of N use by wheat. This agrees with Lopez-Bellido [28], who found a relationship between nitrogen fertilizer, wheat yield and seasonal trend, where there is a decline in yield during the wet years while little or no effect of N fertilizer during the dry years.

Grain yield

In the current experiment, increase in the N rate up to 240 kg N ha^{-1} and splitting it three times (T_5) had a positive effect on grain yield of wheat and were not significantly different with the application of 360 kg N ha^{-1} with the same timing averaged over years. In general, the highest grain yield obtained in this experiment exceeds the yield obtained in response to the application of 120 kg N ha^{-1} all at once at tillering by 57.8% (Table 3). Compared to the grain yield obtained from the control

plot, the grain yields obtained from the aforementioned most productive treatments were superior by 392.1% and 372%, respectively (Table 3). The optimum wheat grain yield was obtained in response to applying 240 kg N ha^{-1} applied in three splits ¼ at sowing, ½ at tillering stage of growth and ¼ at booting. This optimum yield exceeds the national average wheat yield of the country by about 152.5%, which is about 2.4 ton ha^{-1} [3]. It also exceeds the world's average yield of 3.4 t ha^{-1} by about 78% [29]. This indicates that evaluated N rates positively affected grain yield. This dramatic yield increase with N fertilizer application is the reason why farmers in the study area use higher rates of nitrogen (256 kg ha^{-1}) than the blanket recommendation (87 kg ha^{-1}). The increased grain yield due to the increased application of nitrogen might be attributed to the high concentration of N in the leaves which increased and prolonged the photosynthesis ability of the plant which leads to an increase in grain yield. In agreement with the present result, Abedi [30] reported that different N rates (120, 240 and 360 kg ha^{-1}) had a significant effect on grain yield increment in wheat (46% at $N=120$, 72% at $N=240$, and 78% at $N=360$) compared to control.

The result of the current experiment also revealed that the application of N three splits yielded more grain than application of nitrogen only once at tillering or just in two splits. The increase in grain yield due to the trice split application of nitrogen might be the better matching of N availability with the crop needs during the growing period. Similarly, higher grain yield of wheat was reported when N was applied in three splits (at planting, tillering and post-anthesis) compared with two splits (at planting and tillering) and one-time application (at planting) Brian [31]. Contrary to the current result, Chen and Neil [32] reported that split application of N did not affect wheat grain yield significantly. Similarly, there was a report where applications of all N rates at planting and twice split application timing showed the same significance effect on grain yield (each 5.4 t ha^{-1}) with 8% higher yield over trice split N timing [33]. The lowest value of grain yield in this experiment was recorded with the full application of N only at tillering, where the applied N was likely susceptible to leaching, denitrification and runoff loss as the amount of rainfall was higher during this period.

Nitrogen uptake

In this study, the highest GNUP was 118% higher than the lowest value which was obtained with the application of 120 kg N ha^{-1} equally at tillering and booting (Table 3). The overall higher grain N uptake due to the split application of the highest N rate (360 kg N ha^{-1}) at sowing, tillering and booting (T_5) can be explained by the more efficient N mobilization to the grain at grain filling stage. This is in conformity with Jan [34] who reported higher efficiency of N partitioning to the grain when N was supplied in splits (at planting, tillering and stem elongation). Similarly, Fageria and Baligar [35] reported that split applications of nitrogen fertilizer cause high amount of nitrogen content to be taken by the grain rather than by straw of wheat. The present experiment also revealed nitrogen fertilizer rates significantly increased wheat N uptake. Uptake values were similar in both growing years at lower N rate, whereas significant differences were recorded between all fertilizer rates, rising as fertilizer rates increased. This might be because application of extra N through increased levels increased the concentration of N in the soil and led to greater absorption of nutrients, which ultimately resulted in vigorous growth of bread wheat in terms of higher dry matter accumulation and enhanced the total uptake of nitrogen. The result also revealed that the split application of the highest dose (360 kg N ha^{-1}) and applying it three times (T_5) increased wheat N uptake. The increased N uptake of wheat due to the split application of nitrogen (T_5) could be ascribed to the continuous supply of N which may have increased the synchrony between plant N demand and supply from the soil coupled with the reduction of N losses via denitrification, leaching or runoff [4]. This proposition is consistent with the report of N uptake by wheat crop which is significantly enhanced when application of the highest dose of N fertilizer was done and synchronized with the time of high demand of the plant for uptake of the nutrient [36].

Nitrogen use efficiency traits

The present study demonstrated that a significant variation existed in the nitrogen use efficiency traits for year, rate and timing of N applications. In 2015, the year with the highest grain yield had the highest AE and RE of wheat under the rate of 120 kg ha^{-1} which were notably higher than the year 2014 under the same N rate. The increase in AE and RE in the second growing season might be due to the absence of water logging which reduces the availability of nitrogen which is the problem of the first growing season. In general, AE diminished as the N fertilizer rates increased in both growing seasons, with significant differences among all the levels of nitrogen. This result is in agreement with the finding of Roberts [37] who reported that increase in N fertilizer rates resulted in a decline in agronomic efficiency. Higher AE could be obtained if the yield increment per unit N applied is high because of reduced losses and increased uptake of N [25]. Nitrogen agronomic efficiency value ranging from 10 to 30 is common, and values higher than 30 indicate efficiently managed systems [26]. Consistent with this suggestion, in this study N application resulted

in AE between 10.47 and 28.75 kg kg^{-1} in both the growing seasons, showing the importance of appropriate management system in wheat production.

The highest APE recorded in this study during the first wet growing season (2014) as a result of splitting nitrogen equally at tillering and booting implies that there was a higher loss of nitrogen in treatments where N was applied during sowing time. However, in the second growing year (2015), time of application had less impact on APE since the loss of nitrogen was minimized as a result of reduction in waterlogging problem due to a lower amount of rainfall. In addition, the higher APE due to the split application of nitrogen in two splits (at tillering and booting) in both growing seasons might be attributed to adequacy of available nitrogen during grain development stage that might have increased the assimilation and redistribution of N from the vegetative plant component to wheat grain. In contrast to the present finding, lower nitrogen utilization efficiency was reported in the early N applications at planting and tillering compared with additional split application at anthesis [34].

The current experiment also revealed that the highest value of 59.8% for recovery efficiency (RE) was obtained with the triplicate application of 120 kg N ha^{-1} (T_5) and it is 131% higher than the lowest value of 25.64% which was obtained with the application of the highest dose in two equal splits at tillering and booting. In line with the present result, Haile [36] reported 13.7% rise in recovery efficiency of nitrogen as a result of N application three times (¼ at sowing, ½ at mid-tillering and ¼ at anthesis) at lower N rate. The application of N three times in split (T_5) produced higher RE for all the N rates tested in the current experiment. This implies if N is applied in several small doses during the period of rapid crop growth, rather than as a single large dose at the beginning of rapid crop growth, then losses are minimized and crop recovery is maximized. Furthermore, the highest RE in the second growing season might be due to lower rainfall which improved the availability of nitrogen than the first growing season; thus, the crop has used the applied nitrogen more efficiently. The highest RE was recorded at a rate of 120 kg N ha^{-1} in both growing seasons. In line with the current experiment, increase in apparent nitrogen recovery efficiency was reported at the rate of 150 kg N ha^{-1} for wheat and barley [38]. In contrast, lower NUE (27.10%) with the highest nitrogen rate of 120 kg N ha^{-1} and the highest value (39.27%) at the lowest N rate of 30 kg N ha^{-1} were reported on bread wheat [36]. However, nitrogen recovery may vary with the location, soil type, crop variety and the environmental conditions prevailing during the crop growth [39]. In conformity with this result, studies from Ethiopia reported highest apparent nitrogen recovery efficiency of 65.8% Selamyihun [40] and 39.27% [25] on wheat in Ethiopia. However, the common apparent recovery N-efficiency values ranging between 30 and 50%, and 50 and 80% indicate well-managed system [27].

In the current experiment, application of nitrogen beyond 120 kg N ha^{-1} did not significantly affect NHI in the second growing season while the application of 360 kg N ha^{-1} significantly produced lower NHI as compared to the application of 120 kg N ha^{-1}. The first growing season produced the highest NHI than the second growing season under all the levels of N. This showed a strong influence of rainfall, in the variable response of NHI to time of application. The higher NHI during the first growing season might be due to the production of a lower aboveground biomass yield due to water logging, which resulted from higher rainfall. The lower NHI in the second growing as compared to the first growing season might be attributed to the increase in aboveground biomass yield. In general, the highest NHI value was recorded when nitrogen was applied only at tillering during both growing seasons. This might be due to the lowest aboveground biomass and grain yield produced by this treatment. Similarly, a higher nitrogen harvest index for wheat was obtained with treatments which produced the least aboveground biomass and grain yield [41].

Conclusion

The results of this study have demonstrated that application of a large quantity of nitrogen (a minimum of 240 kg N ha^{-1}) in three split doses (T_5) was required to obtain optimum wheat yield, which is about 2.5-fold higher than the national average yield of the crop in Ethiopia. The soil requires application of as much as 240 kg N ha^{-1} to produce about 6 tons of wheat per hectare which implies that the soil is productive unless the nitrogen uptake efficiency of the crop possibly is reduced as a result of its characteristic waterlogging condition. The importance of splitting nitrogen in three split doses (1/4th at sowing, ½ at tillering and the other 1/4th at booting) was also evidenced in the optimum yields and improving nitrogen recovery. Nitrogen fertilizer led to a general decrease of nitrogen use efficiency traits in both growing years. Higher N level increased N content in the grain and nitrogen uptake by wheat crop. In view of the current result, the significant interaction with year indicates that the efficiency of broad bed and furrows to drain excess soil moisture is lower in years which receive a higher amount of rainfall. Therefore, it should be supported by developing wheat varieties tolerant or resistant to such shocks.

Abbreviations

AE: agronomic efficiency; APE: agro-physiological efficiency; CV: coefficient of variation; GNUP: grain nitrogen uptake; GY: grain yield; MoARD: Ministry of Agriculture and Rural Development; N: nitrogen; NHI: nitrogen harvest index; NUE: nitrogen use efficiency; RE: recovery efficiency; TNUP: total nitrogen uptake.

Authors' contributions

FB conceived the study and design, collected the data, performed the analysis on all samples, interpreted the data, wrote the manuscript and acted as corresponding author. ND, AM and TT assisted in analysis and interpretation of data and drafting of the manuscript. All authors read and approved the final manuscript.

Author details

[1] School of Plant Sciences, College of Agriculture and Environmental Sciences, Haramaya University, P.O. Box 138, Dire Dawa, Ethiopia. [2] Department of Plant Sciences, College of Agriculture and Natural Resource Sciences, Debre Berhan University, P.O. Box 445, Debre Berhan, Ethiopia. [3] Chickpea and Malt Barley-Faba Bean Projects ICARDA, Addis Ababa, Ethiopia.

Acknowledgements

The authors would like to thank Debre Berhan Agricultural Research Center for providing the land and allowing us to use their facilities and Debre Berhan University for allowing the corresponding author a leave of absence to conduct PHD dissertation research from which this article was emanated.

Competing interests

The authors declare that they have no competing interests.

Funding

All authors dedicated their additional working hours to develop this paper with no specific grant from any funding agency.

References

1. Shewry PR. Wheat. J Exp Bot. 2009;60(6):1537–53.
2. FAOSTAT. online [Internet]. 2014 [cited 2015 Mar 6]. Available from: http://faostat.fao.org/site/291/default.aspx.
3. CSA (Central Statistic Agency). AGricultural Sample Survey Report on Area and Production of Major Crops for the period 2015/2016 cropping season volume I. Statistical bulletin 584, Addis Ababa, Ethiopia. vol. 584, Statistical Bulletin. 2016.
4. Tilman D, Cassman K, Matson P, Naylor R, Polasky S. Agricultural sustainability and intensive production practices. Nature. 2002;418:671–7.
5. Good A, Shrawat K, Muench G. Can less yield more? Is reducing nutrient input into the environment compatible with maintaining crop production? Trends Plant Sci. 2004;9(12):597–605.
6. Teklu E, Hailemariam T. Agronomic and Economic Efficiency of Manure and Urea Fertilizers Use on Vertisols in Ethiopian Highlands. Agric Sci China. 2009;8(3):352–60. https://doi.org/10.1016/S1671-2927(08)60219-9.
7. Carranca C. Nitrogen use efficiency by annual and perennial crops. In: Lichtfouse E, editor. Farming for food and water security. Berlin: Springer; 2012. p. 57–82.

8. Raun WR, Johnson GV. Improving nitrogen use efficiency for cereal production. Agron J. 1999;91:357–63.

9. Moll RH, Kamprath EJ, Jackson WA. Analysis and interpretation of factors which contribute to efficiency of nitrogen utilization. Agron J. 1982;74(3):562.

10. Huggins D, Pan W. Key indicators for assessing nitrogen use efficiency in cereal-based agroecosystems. J Crop Prod. 2003;8(1–2):157–85.

11. Legg JO, Meisinger JJ. Soil nitrogen budgets. In: Stevenson FJ, editor. Nitrogen in agricultural soils. Madison: American Society of Agronomy; 1982. p. 503–66.

12. Gehl RJ, Schmidt JP, Maddux LD, Gordon WB. Corn yield response to nitrogen rate and timing in sandy irrigated soils. Agron J. 2005;97(4):1230–8.

13. Anthony G, Woodard B, Hoard J. Foliar N application timing influence on grain yield and protein concentration of hard red winter and spring wheat. Argon J. 2003;95:335–8.

14. López-Bellido JRR, López-Bellido L. Efficiency of nitrogen in wheat under Mediterranean conditions: effect of tillage, crop rotation and N fertilization. F Crop Res. 2001;71(1):31–46.

15. Brian NO, Mohamed M, Joel KR. Seeding rate and nitrogen management effects on spring wheat yield and yield components. Am J Agron. 2007;99:1615–21.

16. Molla A. On-farm participatory evaluation of bread wheat productivity under different NP levels, precursor crops, and Vertisols types in the highlands of central Ethiopia. In: Achievements of integrated crop, soil and water management research activities on wheat. EAAP/EIAR. Addis Abeba, Ethiopia; 2015.

17. Page AL, Miller RH, Keeney. DR. Methods of Soil Analysis. Part 2. Chemical and Microbiological Properties. 2nd edn. Soil Science Society of America; Madison, Wisconsin, U.S.A. 1982.

18. MoARD (Ministry of Agriculture and Rural Development. Crop variety register. Animal and plant health regulatory directorate Issue No.12. June, 2009. Addis Ababa, Ethiopia. 2009.

19. Zadoks JC, Chang TT, Konzak CF. A decimal code for the growth stages of cereals. Weed Res. 1974;14:415–21.

20. Molla A. Farmers' knowledge helps develop site specific fertilizer rate recommendations, central highlands of Ethiopia. World Appl Sci J. 2013;22(4):555–63.

21. American Association Cereal Chemists (AACC). Approved Methods of the American Association Cereal Chemists. American Association of Cereal Chemists, Inc., St. Paul, Minnesota; 2000.

22. Fageria NK. Nitrogen management in crop production. Boca Raton: CRC Press; 2014.

23. SAS Institute Inc. SAS® 9.3 Companion for Windows. Cary, NC: SAS Institute Inc. Cary, NC, USA; 2011.

24. Gomez KA, Gomez AA. Statistical procedures for agricultural research. New York: Wiley; 1984.

25. Craswell ET, Godwin DC. The efficiency of nitrogen fertilizers applied to cereals in different climates. Adv Plant Nutr. 1984;1:1–55.

26. Dobermann AR. Nitrogen Use Efficiency – State of the Art. Univ Nebraska. 2005;17.

27. Fageria NK, Baligar VC, Jones CA. Growth and mineral nutrition of field crops. 3rd ed. New York: Taylor, Francis Group; 2011. p. 530.

28. López-Bellido L, López-Bellido RJ, Redondo R. Nitrogen efficiency in wheat under rainfed Mediterranean conditions as affected by split nitrogen application. F Crop Res. 2005;94:86–97.

29. MAFAP (Monitoring African Food and Agriculture Policies). Improving incentives to expand wheat production in Ethiopia. Policies Brief, #9. 2013.

30. Abedi T, Alemzadeh A, Kazemeini SA. Wheat yield and grain protein response to nitrogen amount and timing. Aust J Crop Sci. 2011;5(3):330–6.

31. Brian NO, Mohamed M, Joel KR, Otteson BN, Mergoum M, Ransom JK, et al. Seeding rate and nitrogen management effects on spring wheat yield and yield components. Am J Agron. 2007;99(6):1615–21.

32. Chen C, Neill K. Response of spring wheat yield and protein to row spacing, plant density, and nitrogen application in Central Montana. In: Fertilizer Fact: No.37. Montana State University, Agricultural Experiment Station and Extension Service, USA. 2006;.

33. Chibsa T, Gebrekidan H, Kibret K, Tolessa D. Effect of rate and time of nitrogen fertilizer application on durum wheat (*Triticum turgidum Var* L. Durum) grown on Vertisols of Bale highlands, southeastern Ethiopia. Am J Res Commun. 2016;5(1):39–56.

34. Jan MT, Khan JM, Khan A, Arif M, Shafi M, Nullah N. Wheat nitrogen indices response to nitrogen source and application time. Pak J Bot. 2010;42(6):4267–79.

35. Fageria NK, Baligar VC. Enhancing nitrogen use efficiency in crop plants. Adv Agron. 2005;88:97–185.

36. Haile D, Nigussie D, Ayana A. Nitrogen use efficiency of bread wheat: effects of nitrogen rate and time of application. J Soil Sci Plant Nutr. 2012;12(123):389–409.

37. Roberts TL. Improving nutrient use efficiency. Turk J Agric. 2008;32:177–82.

38. Delogu G, Cattivelli L, Pecchioni N, De Falcis D, Maggiore T, Stanca AM. Uptake and agronomic efficiency of nitrogen in winter barley and winter wheat. Eur J Agron. 1998;9(1):11–20.

39. Sinebo W, Gretzmacher R, Ede bauer A. Genotypic variation for nitrogen use efficiency in Ethiopian barley. F Crop Res. 2004;85:43–60.

40. Selamyihun K, Tanner DG, Tekalign M. Residual effects of nitrogen fertilizer on the yield and N composition of succeeding cereal crops and on soil chemical properties of an Ethiopian highland Vertisol. Can J Soil Sci. 2000;80:63–9.

41. López-Bellido L, López-Bellido RJRJ, Redondo R. Nitrogen efficiency in wheat under rainfed Mediterranean conditions as affected by split nitrogen application. F Crop Res. 2005;94(1):86–97.

42. Tadesse T. Soil, plant, water, fertilizer, animal manure and compost analysis manual: Plant Science Division Working Document No. 13. Addis Ababa, Ethiopia: International Livestock Center for Africa. 1991.

43. Olsen SR, Cole CV, Watandbe FS, Dean LA. Estimation of available Phosphorus in soil by extraction with Sodium Bicarbonate. J Chem Inf Model. 1954;53(9):1689–99.

44. Metson AJ. Methods of chemical analysis for soil survey samples. Soil Bureau Bulletin No. 12, New Zealand Department of Scientific and Industrial Research. Government Printer: Wellington, New Zealand; 1961. p. 168–75.

Survey of mushroom consumption and the possible use of gamma irradiation for sterilization of compost for its cultivation in Southern Ghana

Nii Korley Kortei[1*], George Tawia Odamtten[2], Mary Obodai[3], Michael Wiafe-Kwagyan[2] and Juanita Prempeh[4]

Abstract

Background: Mushroom cultivation is increasingly becoming a serious agribusiness in Ghana, especially at the time when entrepreneurship is being encouraged to reduce the pressure of employment in the government sector and also due to its nutritional and medicinal attributes.

Methods: A survey was carried out using the rapid appraisal method to review the existing methods of sterilization, use of gamma radiation in substrate sterilization and food preservation, preference of mushrooms in Ghana by consumers and nutritional and medicinal attributes of the mushroom.

Results: The survey demonstrated the popularity of drum (moist heat) technique of sterilization in Ghana. Majority (64%) of the respondents were dissatisfied with the method of sterilization of compost and spawn substrate, while 36% indicated the method was alright by them. Majority (82%) of the respondents had never heard of sterilization of substrates for cultivation and its subsequent preservation of food or mushroom by gamma irradiation technique. All consumers (100%) desired to see their favorite mushroom produced all year round, and this constituted a significant ($p < 0.05$) viewpoint. Furthermore, a significant ($p < 0.05$) majority (90%) of the respondents were all for promotion of the consumption of mushroom, while a small percentage (10%) were noncommittal.

Conclusion: Information and knowledge on the gamma irradiation technique for substrate production and consumption patterns of *P. ostreatus* mushrooms were not widely disseminated as anticipated.

Keywords: Moist heat, Gamma radiation, Compost, Mushroom, Survey, Ghana, Consumption

Background

Mushroom cultivation has become a profitable commercial agribusiness in many developing and developed countries such as the USA, Great Britain, China, Asia, Japan and Europe. Many indigenous and commercial cultivation methods have been employed over the years to domesticate many species of edible mushrooms for human consumption [1]. Fungi in the Basidiomycota are large and diverse which include forms commonly known as mushrooms, boletes, earthstars, stinkhorns, bird nest fungi, jelly fungi, bracket or shelf fungi and rusts and smuts [2]. The mushrooms are sought eagerly for human consumption. Consequently, the medical community widely recognizes the health-stimulating properties of mushrooms [3].

Mushroom cultivation as an agribusiness has gained some modest success in Ghana because many of our forest reserves which support rich wild growth of the mushrooms are depleting fast in rich biodiversity of indigenous mushroom species. It has now become necessary to adopt modern mushroom cultivation strategies to sustain supplies of the species for human consumption in Ghana such as those obtained in the developed economies.

*Correspondence: nkkortei@uhas.edu.gh
[1] Department of Nutrition and Dietetics, School of Allied Health Sciences, University of Health and Allied Sciences, PMB 31, Ho, Ghana
Full list of author information is available at the end of the article

The evolution of technical knowledge for commercial cultivation of mushrooms was a product of human need to extend the period of availability (seasonality) and reduce the inherent risk of mushroom hunting, thus meeting the growing demands for diverse culinary mushrooms [4]. Some modern pretreatments of substrates for mushroom cultivation include the use of chemical amendments, steam and gamma irradiation.

Gamma radiations come from the spontaneous disintegration of radioactive nuclides (Cobalt 60 or Cesium 137) as their energy source. The high-energy gamma rays have high penetrating power and pass through any matter without leaving any radioactivity within the product. The deep penetrating power of gamma rays could serve as a decontamination agent of food items [5]. Currently, over 34 countries are using gamma irradiation for this purpose [5]. The gamma rays can also be used as a hydrolytic agent due to its unique ability to enhance depolymerization of lignocellulose linkages without the need to aid exogenous toxic chemicals [6–9].

Recent studies in Ghana have demonstrated the ability of gamma radiation to depolymerize lignocellulosic mushroom compost into inexpensive protein per unit area of mushroom of the genus *Pleurotus* [10–14] and produced quantities of the cherished mushroom.

The use of this gamma irradiation technology to facilitate the cultivation of mushrooms on agricultural lignocellulose in Ghana promises to be faster and a more reliable technique to reduce the humdrum tasks associated within the existing conventional production chain [10–14]. The objective of this paper was to assess the knowledge of the respondents to a questionnaire on the benefits of mushroom consumption and the possibility of the use of gamma radiation sterilization of substrates and spawns for mushroom cultivation.

Materials and methods
Sample area for questionnaire administration
The study areas chosen for the collection of data were Greater Accra, Central and Eastern regions where majority of the producers and consumers of mushrooms are located. Accra is the capital city of Ghana, and it is in close proximity with the Eastern and Central regions. Majority of the studies were conducted in Accra because the national irradiation facility is located at the Ghana Atomic Energy Commission (GAEC) at Kwabenya which was used for this study. Data from mushroom cultivators were collected from mushroom farms in the above mentioned regions of Ghana. Questionnaires were administered to obtain information on method of sterilization, aspect of production which needs much attention, general perception about

production, popularity of gamma irradiation technique, etc. A total of 150 volunteers were interviewed.

Consumers were interviewed from the following locations: Ofankor, Adenta-Frafraha, Cape Coast Polytechnic, Somanya, Nkawkaw, CSIR-FRI, Shiashie, Graduate School of Nuclear and Allied Sciences, Kwabenya, University of Ghana, Abelemkpe and Dome market, etc. (Greater Accra, Eastern and Central regions of Ghana) (Fig. 1). Questionnaires were administered to obtain information on the benefits of consumption of mushrooms, methods of preservation, qualities looked out for in mushrooms localities where mushrooms were obtained, etc. A total of 120 volunteers were interviewed. The cultivators of mushroom were selected from the members of the Mushroom Growers' Association of Ghana found in the selected regions.

Study design
Convenient cross-sectional sampling method was used.

Statistical analysis
Graphical presentations of data were done with Excel for Microsoft windows (version 10). Descriptive statistics were employed with SPSS 16 (Chicago, USA). Parameters investigated were subjected to analyses of variance (one-way ANOVA) at significant difference ($p < 0.05$).

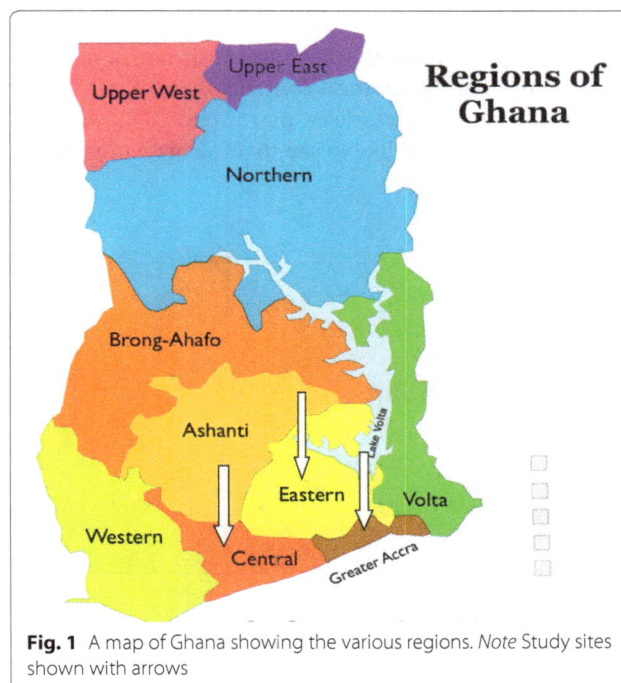

Fig. 1 A map of Ghana showing the various regions. *Note* Study sites shown with arrows

Results

Producers

Majority of the respondents (70%) use drum pasteurization (moist heat), while 26% treat the substrate with unspecified chemicals. The remaining 4% use other unspecified methods of sterilization (Fig. 2). Majority of the respondents (64%) were not satisfied with the sterilization methods; 36% were in agreement with the sterilization procedure. The differences observed were significant ($p < 0.05$) (Fig. 3).

A significant ($p < 0.05$) majority of 82% have not heard of the sterilization of mushroom compost by gamma irradiation. The remaining 18% were aware of its use in sterilization of compost before spawning (Fig. 4). The survey also showed that 34% of the respondents believe that the sterilization process needs more attention in mushroom cultivation industry, while 50% desired the inoculation process to be perfected to exclude contaminants (Fig. 5). About 8% wanted more attention to be paid to packaging. The rest were noncommittal. Majority (92%) of mushroom cultivators agreed to the need to achieve better sterilization of compost bags and other ancillary methods, while an insignificant ($p > 0.05$) minority of the cultivators disagreed (Fig. 6).

Because fresh mushrooms are perishable, shelf-life extension is vital to the industry. Exactly 54% of the respondents were able to preserve all their produce, while 46% were unable to preserve all their produce (Fig. 7).

Consumers

Majority of consumers interviewed (72%) intimated that they prefer eating oyster mushrooms (*Pleurotus* spp.), while 20% patronized the termite mushroom (*Termitomyces* spp.) (Fig. 8). Domo (*Volvariella volvacea*) or the oil palm mushroom was least patronized (8%). All consumers (100%) would like to see their favorite mushroom

produced all year round, and this constituted a significant ($p < 0.05$) viewpoint (Fig. 9). Furthermore, a significant ($p < 0.05$) majority (90%) of the respondents were all for promotion of the consumption of mushroom (Fig. 10), while a small percentage (10%) were noncommittal.

On the question of the medicinal and nutritional benefits derived from eating mushrooms, 54% stated that mushrooms have medicinal values; 34% agreed mushrooms have nutritional attributes not excepting 10% who assigned other unspecified benefits of eating mushrooms. The remaining 2% said that mushrooms have both medicinal and nutritional values (Fig. 11). It was shown by the survey that majority (64%) of mushroom consumers obtained them directly from the market place. Interestingly, 34% obtained their supply from backyard garden or from the farm (Fig. 12) and the remaining 2% harvested from the wild in the mushroom season.

The aesthetic appearance of mushrooms contributed to the choice of mushroom for consumption; 34% of the consumers considered appearance as important and 30% considered the taste as a determining factor, while

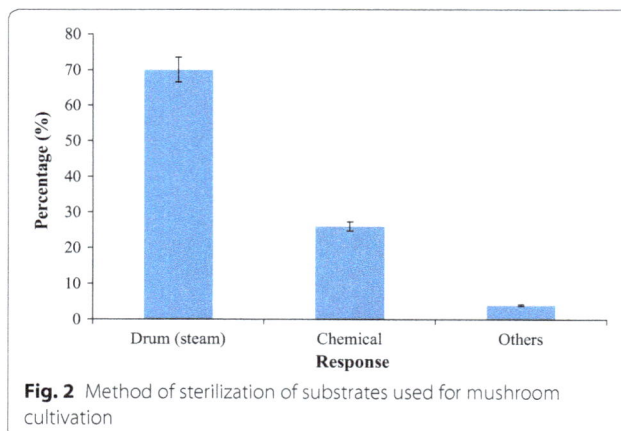

Fig. 3 Mushroom farmer's satisfaction on the reliability of their sterilization methods in mushroom production

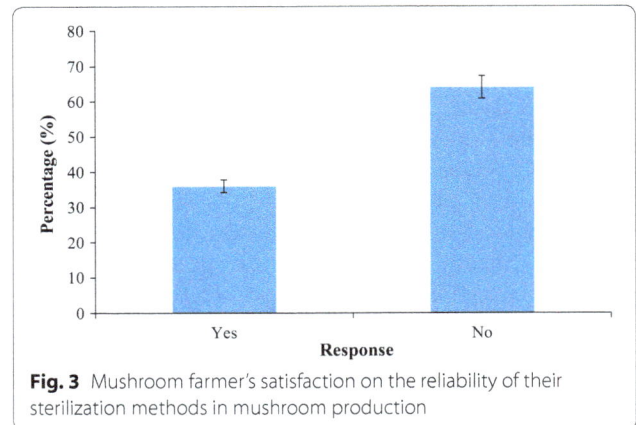

Fig. 2 Method of sterilization of substrates used for mushroom cultivation

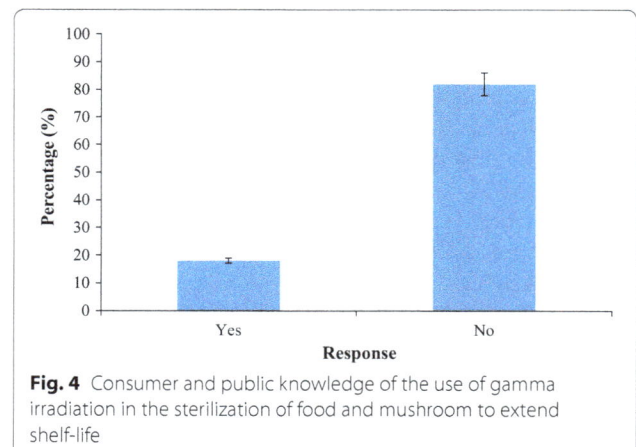

Fig. 4 Consumer and public knowledge of the use of gamma irradiation in the sterilization of food and mushroom to extend shelf-life

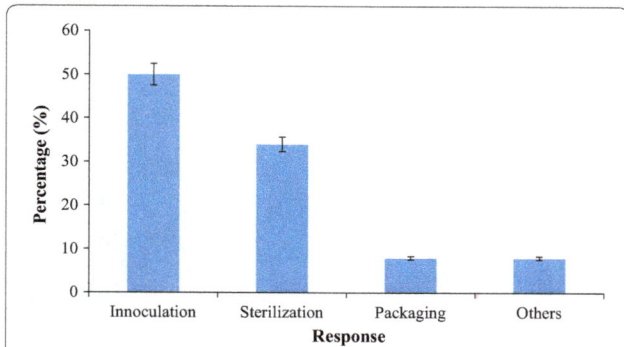

Fig. 5 Aspects of production which needs much attention according to respondents

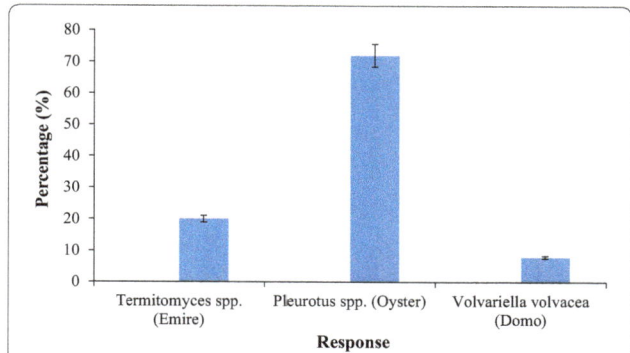

Fig. 8 Types of mushrooms most patronized by consumers

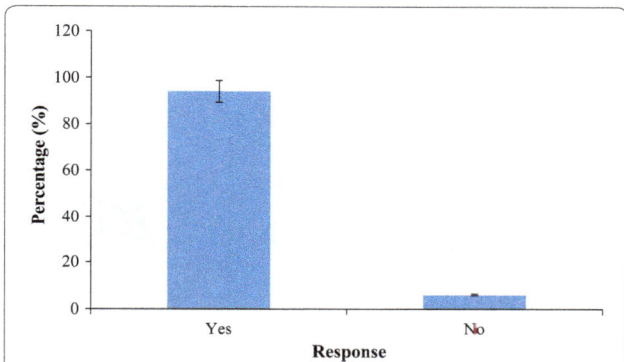

Fig. 6 Response of mushroom cultivators on the need to achieve better sterilization of compost and other ancillary methods used in the production of mushrooms

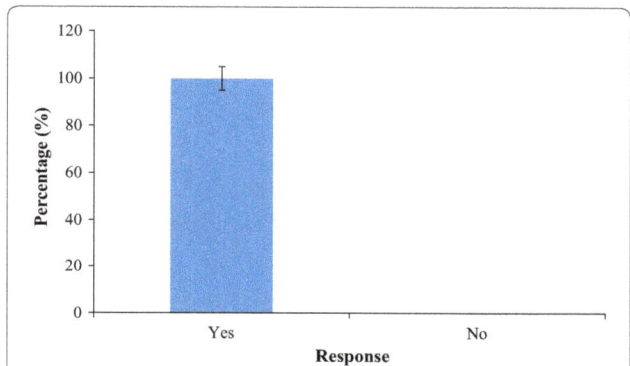

Fig. 9 Consumer preference to see their favorite mushroom produced throughout the year

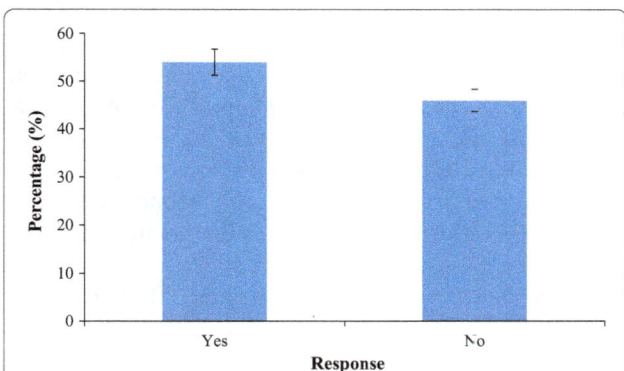

Fig. 7 Percentages of farmers who were able to preserve unsold mushrooms (Yes) and those unable to preserve (No)

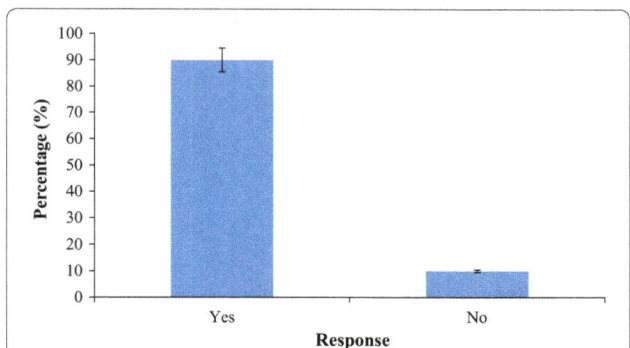

Fig. 10 Respondents view of the promotion of mushroom consumption

12% considered both texture and taste as equally important. Only, 10% used texture as a criterion for choosing mushroom and 6% considered both texture and appearance concurrently before choosing a mushroom for purchase (Fig. 13). Currently in the country, mushrooms are preserved by drying by 55% of the people, while 32% use refrigerators for preservation. The rest preserve mushrooms by smoking after blanching in brine (Fig. 14). Exactly 45% of the respondents stated that mushroom production was laborious, while 10% believed the process was expensive. About 45% found the process and costing normal (Fig. 15).

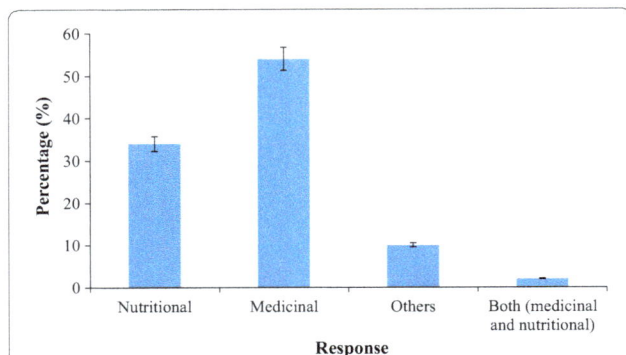
Fig. 11 Respondents view on the benefits of eating mushrooms

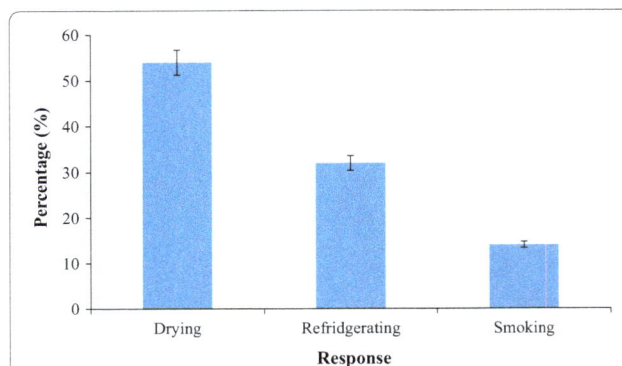
Fig. 14 What are the methods used to preserve mushrooms?

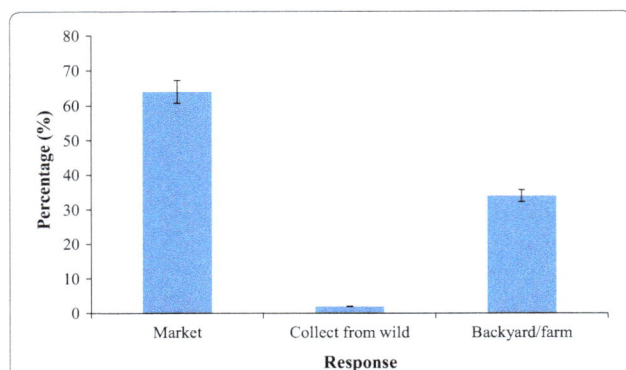
Fig. 12 Source of mushrooms for use by consumers

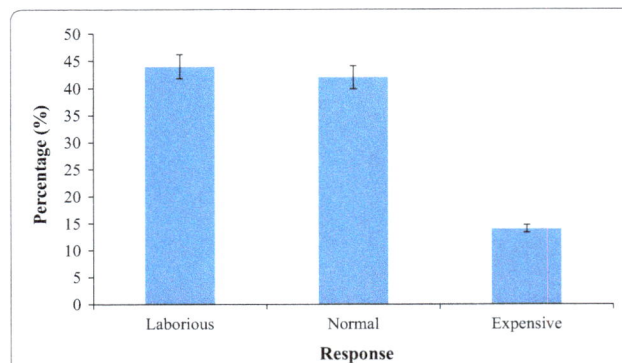
Fig. 15 Public opinion on labor and cost of mushroom production in Ghana

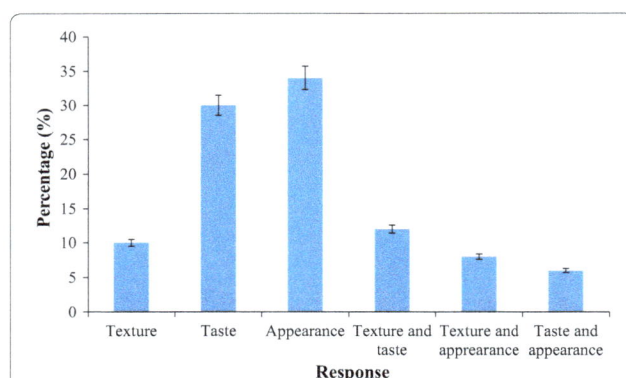
Fig. 13 Choice of mushroom for consumption by consumers based on the listed criteria

Discussion

Results obtained from the structured questionnaire survey (Fig. 2) demonstrated the popularity of drum (moist heat) technique of sterilization both locally and internationally which agrees with findings by several researchers [9, 10, 15, 16]. Majority (60%) of the respondents were

not satisfied with their method of sterilization of compost, while 30% found it alright (Fig. 3).

This also gave an indication of general unsatisfactory results obtained from their sterilization methods and therefore their desire to improve upon the improper or incomplete sterilization resulting in partial or incomplete elimination of contaminants. Residual contaminants from incomplete pasteurization of compost and spawn substrates could result in reduction in economic yield of mushrooms on compost substrates due to antibiosis and competition for nutrients.

Interestingly, majority of the respondents (82%) have not heard about sterilization of mushroom compost using gamma irradiation (Fig. 4) although gamma irradiation sterilizing technique has now gained some credence among mushroom farmers in Ghana. In spite of this, some researchers [6, 8, 9, 11–14] have reported success in the use of gamma radiation to decontaminate and depolymerize different lignocellulosics agrowastes for *P.ostreatus* cultivation, thus achieving growth and good of fruiting bodies yield comparable to the moist heat (drum) technique. Gamma irradiation facility technique allows a greater volume and quantity of sterilized compost bags per unit time, and is less laborious and more effective decontamination of substrate bags. This could probably reassure the majority of cultivators of mushrooms who expressed the need to achieve better sterilization technique for mushroom compost based on their experience with moist heat sterilization (Fig. 6).

Shelf-life extension of mushroom is economically vital to the industry. The most common method for the preservation of mushrooms is drying since it is the most economical and oldest method [17]. Data from this paper show that drying is the most popular preservation method of mushrooms in Ghana (Fig. 14). Several edible mushrooms are consumed by respondents, but the most preferred one was oyster mushrooms (*Pleurotus* spp.) (Fig. 8). Afetsu [18] reported similar results in a survey of mushroom consumption conducted in the Volta region of Ghana which agrees with the results obtained in this work. The popularity of mushroom to Ghanaian was shown in the survey where all respondents wanted an all year round supply (Fig. 9).

Obodai et al. [19] attributed oyster mushroom's emerging popularity among Ghanaians to its comparatively easy method of cultivation. This agrees with findings in this paper. However, results obtained in this paper (Fig. 8) contrast the findings of Apertorgbor et al. [20] who reported *Termitomyces* spp. as the most preferred mushrooms in the Eastern and middle belts of Ghana. Dijk et al. [21] made similar findings in South Cameroun where they very often consumed this mushroom. This difference in preference might be attributed to the availability of particular kinds of mushroom species, taste and/or medicinal attributes. Mushrooms have a myriad of benefits derived from its consumption. Mushrooms have been found to have some medicinal values [22]. The Asantes and Sefwis in Ghana believe that mushrooms lower blood pressure in hypertensive patients. The globular subterranean sclerotium of *P. tuber-regium* is chewed by local people to alleviate heart pains, and the powder is taken in warm water to lower blood pressure in hypertensive patients [23]. The medicinal value of mushrooms for Ghanaian consumers was evident from the questionnaire survey (Fig. 11). Completely colonized composted bags are readily available for sale at numerous mushroom farms and some research institutions such as the Council for Scientific and Industrial Research-Food Research Institute (CSIR-FRI) Ghana. With the current rate of deforestation caused by urbanization, bush fires and mining companies, the collection of wild mushrooms by the rural folk is greatly threatened leaving government-protected areas (forest reserves) as the only remaining areas where non-timber forest products can be collected [24].

The depletion of our forest is a major cause of scarcity of most mushroom species [20]. Rigorous scientific research into bioconversion of lignocellulosic waste by mushroom is making it possible to cultivate different species of mushrooms all year round with specific emphasis on Jun-Cao technology which involves the use of plastic bag and agro, industrial, forest lignocellulosic wastes. This practice is advantageous since it is more efficient and does not require so much space. Small-scale mushroom farms have emerged in Southern Ghana as a result of the introduction of the National Mushroom Development Project aimed at promoting the economic welfare of rural communities [23]. Mushroom production is a demand-driven enterprise and so requires the appropriate technologies to keep up with its supply. Consumers seem to prefer mushrooms on the basis of taste, appearance, texture or combination of these qualities (Fig. 13). Although production of oyster mushroom is laborious, it is also very capital intensive (Fig. 15).

Conclusions

The survey carried out showed among other things, the humdrum tasks of preparing the compost and spawn nutrient and the sterilization method, not excepting the short shelf-life of the harvested fruit bodies. Majority of the farmers were not aware of the possible use of gamma irradiation for the sterilization/pasteurization of the spawn nutrient and the 'wawa' sawdust compost as well as the preservation of both fresh and dry fruitbodies. However, consumers prefer to see their favorite mushroom on the market throughout the year.

Abbreviations
CSIR-FRI: Council for Scientific and Industrial Research-Food Research Institute; GAEC: Ghana Atomic Energy Commission; PMB: Private Mail bag; LG: Legon.

Authors' contributions
NKK, GTO and MO were involved in the conception of the research idea, design of the experiments, data analysis and also drafting of the paper. MW-K and JP participated in the design of the experiments and data collection. GTO and MO provided guidance and supervision of the entire research and critically reviewed the manuscript. NKK, GTO, MO and MW-K read, reviewed and amended manuscript. All authors read and approved the final manuscript.

Author details
[1] Department of Nutrition and Dietetics, School of Allied Health Sciences, University of Health and Allied Sciences, PMB 31, Ho, Ghana. [2] Department of Plant and Environmental Biology, College of Basic and Applied Sciences, University of Ghana, P. O. Box LG 55, Legon, Ghana. [3] Food Microbiology Division, Council for Scientific and Industrial Research-Food Research Institute, P. O. Box M20, Accra, Ghana. [4] Department of Food Science, Royal Agricultural University, Cirencester, Gloucestershire GL7 6JS, UK.

Acknowledgements
We are grateful to all the participants of this study and specifically acknowledge the technical assistance of our technical staffs at the University of Ghana, CSIR-FRI and the University of Health and Allied Sciences.

Competing interests
The authors declare that they have no competing interests.

Funding
Not applicable.

References

1. Stamets P. Growing gourmet and medicinal mushrooms. Olympia: Ten Speed Press and Mycomedia™; 1993. p. 55.
2. Pipenbring M. Introduction to Mycology in the Tropics. St. Paul Minnesota: American Phytopathological Society; 2015. p. 366.
3. Harpen GM. Healing mushrooms. Effective treatments for Todays illnesses. Garden City Park: Square One Publishers; 2007. p. 182.
4. Osarenkhoe OO, John OA, Theophilus DA. Ethnomycological conspectus of West African mushrooms: an awareness document. Adv Microbiol. 2014;4:39–64.
5. IAEA-TECDOC-1530. Use of irradiation to ensure hygienic quality of fresh, pre-cut fruits and vegetables and other minimally processed food of - plant origin. Proceedings of a final research coordination meeting organized by the joint FAO/IAEA Programme of Nuclear Techniques in Food and Agriculture, Pakistan. 2006.
6. Kortei NK, Wiafe-Kwagyan M. Evaluating the effect of gamma irradiation on eight different agro-lignocellulose waste materials for the production of oyster mushrooms (*Pleurotus eous* (Berk.) Sacc. Strain P-31). Croat J Food Technol Biotechnol Nutr. 2014;9(3–4):83–90.
7. Betiku E, Adetunji OA, Ojumu TV, Solomon BO. A comparative study of the hydrolysis of gamma irradiated lignocelluloses. Braz J Chem Eng. 2009;26(2):251–5.
8. Kortei NK, Odamtten GT, Obodai M, Wiafe-Kwagyan M, Dzomeku M. Comparative bioconversion of gamma irradiated and steam sterilized 'wawa' sawdust (*Triplochiton scleroxylon*) by mycelia of oyster mushroom (*Pleurotus ostreatus*) Jacq. Ex.Fr. Kummer. Int Food Res J. 2018;25(3):943–50.
9. Gbedemah C, Obodai M, Sawyer LC. Preliminary investigations into the bioconversion of gamma irradiated agricultural wastes by *Pleurotus* spp. Radiat Phys Chem. 1998;52(6):379–82.
10. Kortei JNK. Growing oyster mushrooms (*Pleurotus ostreatus*) on composted agrowastes: an efficient way of utilizing lignocellulosic materials. Germany: Lambert Academic Publishing; 2011. p. 9.
11. Kortei NK, Odamtten GT, Obodai M, Appiah V, Annan SNY, Acquah SA, Armah JNO. Comparative effect of gamma irradiation and steam sterilized composted 'wawa' (*Scleroxylon triplochiton*) sawdust sawdust on the growth and yield of *Pleurotus ostreatus* (Jacq.Ex.Fr.) Kummer. Innov Roman Food Biotechnol. 2014;14:69–78.
12. Kortei NK, Odamtten GT, Obodai M, Appiah V, Wiafe- Kwagyan M. Evaluating the effects of gamma irradiation and steam sterilization on the survival and growth of sawdust fungi in Ghana. Br Microbiol Res J. 2015;7(4):180–92.
13. Kortei NK, Odamtten GT, Obodai M, Appiah V, Adu-Gyamfi A. Wiafe – Kwagyan M. Comparative occurrence of resident fungi on gamma irradiated and steam sterilized sorghum grains (*Sorghum bicolor L*) for spawn production in Ghana. British. Biotechnol J. 2015;7(1):21–32.
14. Kortei NK, Odamtten GT, Obodai M, Appiah V, Abbey L, Oduro-Yeboah C, Akonor PT. Influence of gamma radiation on some textural properties of fresh and dried oyster mushrooms (Pleurotus ostreatus)(Jacq. Ex. Fr) kummer. Ann Food Sci Technol. 2015;16(1):12–9.
15. Obodai M, Cleland- Okine J, Vowotor KA. Comparative study on the growth and yield of *Pleurotus ostreatus* mushroom on different lignocellulosic by-products. J Ind Microbiol Biotechnol. 2003;30:146–9.
16. Owusu-Boateng G, Dzogbefia VP. Establishing some parameters for the cultivation of oyster mushroom (*Pleurotus ostreatus*) on cocoa husk. J Ghana Sci Assoc. 2005;7(1):1–7.
17. Bala BK, Morshed MA, Rahman MF. Solar drying of mushroom using solar tunnel dryer. In: International solar food processing conference, 2009; 1–11.
18. Afetsu JY. Postharvest losses in oyster mushroom (*P. ostreatus*) produced in the Ho Municipality of the Volta Region of Ghana. Msc. thesis, Kwame Nkrumah University of Science and Technology, Ghana. 2014.
19. Obodai M, Amoa- Awua W, Odamtten GT. Physical, chemical and fungal phenology associated with composting of 'wawa' sawdust (*Triplochiton scleroxylon*) used in the cultivation of oyster mushrooms in Ghana. Int Food Res J. 2010;17:229–37.
20. Apetorgbor MM, Apetorgbor AK, Nutakor E. Utilization and cultivation of edible mushrooms for rural livelihood in Southern Ghana. In: 17th Commonwealth Forestry Conference, Colombo, Srilanka. 2005.
21. van Dijk H, Onguene NA, Kuyper TW. Knowledge and utilization of edible mushrooms by local populations of the rain forest of South Cameroun. Ambio. 2003;32(1):19–23.
22. Singh VK, Patel Y, Naraian R. Medicinal properties of *Pleurotus* species (oyster mushrooms). World J Fungal Plant Biol. 2012;3(1):1–12.
23. Sawyerr LC. Genetic resource aspects of mushroom cultivation on small scale. In: Labarere JE, Menini UG (eds) Proceedings of the 1st international congress for the characterization, conservation, evaluation and utilization of mushroom genetic resources for food and agriculture. Bordeaux (France). 2000. pp. 193–199.
24. Meke G, Lowne J, Ngulabe M. Literature review of indigenous edible fungi from Miombo woodlands in Malawi. In: Boa E, Ngulube M, Meke G, Munthali C, Morris B (eds) Proceedings of forest regional workshop on sustainable use of forest products. Miombo wild edible fungi. 1987. Common mushrooms of Malawi. Fungiflora. Oslo, Norway. 2000. p 108.

Influence of productive resources on bean production in male- and female-headed households in selected bean corridors of Kenya

Scolastica Wambua[1*], Eliud Birachi[2], Ann Gichangi[1], Justus Kavoi[1], Jemimah Njuki[3], Mercy Mutua[2], Michael Ugen[4] and David Karanja[1]

Abstract

Background: Gender-related constraints reflect gender inequalities in access to resources and development opportunities. Access to productive assets is a major issue in the gender empowerment discourse. Despite the significant roles women play in agriculture and food security in many developing countries, they continue to have a poorer command over a range of productive resources, including education, land, information and financial resources compared to their men counterparts. The purpose of the study was to establish the effect of access and control of productive resources on bean production.

Results: Data collected from 412 households in the major bean corridors of Kenya (Homa Bay, Machakos, Bomet and Narok counties) were used to explain the importance of access to productive resources and income use in determining the quantity of beans produced by households. We found that the sex of the respondent was significantly correlated with bean production, with female-headed households producing less beans than the male-headed ones ($p = 0.0.08$). With regard to access and control of productive resources, households with more agricultural incomes and those who put a larger proportion of their land to agriculture produced more beans ($p = 0.008$; $p = 0.000$, respectively). Access and use of fertilized and hired labour was also highly significant. When assessing decision making on the use of income from bean sales was considered, households where the female spouse made decisions produced less beans compared to those that had the male household head being the main decision maker ($p = 0.011$).

Conclusions: We concluded that access and control of productive assets are important in determining the quantity of beans produced at household level. There is a need therefore to come up with interventions which will benefit all the households but are targeted to the needs of the male- and female-headed households.

Keywords: Gender, Access, Control, Productive resources, Beans, Income

Background

In Kenya, common bean (*Phaseolus vulgaris* L.) is the most significant pulse crop with maize being second as a food crop [5]. Beans are a source of cheap dietary protein and thus affordable by most poor households. The crop has multiple uses, the major ones being food and source of income. Access to productive capital such as land, fertilizers, farm equipment, education, technology and financial services is a key element of agricultural productivity [11]. Though agriculture is important to women, they have less access to the resources and services required for agricultural production. It is important to understand the gender differences in access and control of productive resources and how this affects bean production. This paper therefore assessed the effects of access and control of productive resources on bean production for male- and female-headed households. The resources studied here were land, seed, farm equipment and fertilizers. Access to new technology is crucial in maintaining and improving agricultural productivity.

*Correspondence: scolasticawambua@yahoo.com
[1] Kenya Agricultural and Livestock Research Organization (KALRO), P.O Box 57811-00200, Nairobi, Kenya
Full list of author information is available at the end of the article

There gender gaps exist for a wide range of agricultural technologies, including machinery and tools, improved crop varieties and animal breeds, fertilizers, pest and disease control measures and management techniques. "A number of constraints, including the gender gaps described above, lead to gender inequalities in access to and adoption of new technologies, as well as in the use of purchased inputs and existing technologies" [11]. The use of bought inputs depends on the availability of assets such as land, credit, education and labour, all of which happen to be more constrained for female-headed households than for male-headed households. According to Blackden et al., adoption of improved technologies for example crop varieties is positively correlated with education but is also dependent on time constraints. In an activity with long turnaround periods, such as agriculture, working capital is required for purchasing inputs such as chemicals, fertilizers and improved seeds; however, as discussed above, women face more obstacles relative to men in their access to credit. Women's lesser ability to absorb risk may constrain adoption of improved bean varieties and inputs.

Growing populations and declining agricultural productivity are leaving millions without secure sources of food; hence, there is the need to upsurge food production. Advances in food production are constrained by the "invisibility factor" in other words, by women's major but largely unrecognized roles in agriculture. According to Palacios-Lopez et al. [13], average labour contribution to crop production in six Sub-Saharan countries was estimated at 40% instead of the 80% reported by FAO, though differences exist across countries. The female labour share amounts to slightly more than 50% in Uganda, Tanzania and Malawi (56, 52 and 52%, respectively), which is also consistent with the slightly higher share of women in these populations (52, 53 and 51%, respectively) [9]. Failure to recognize this contribution is costly. This results in imprudent policies and programs, forgone agricultural production and associated income flows, higher levels of poverty, and food and nutrition insecurity [20]. In sub-Saharan Africa, it has been estimated that agricultural productivity could increase by up to 20 per cent if women's access to such resources as land, improved seed and fertilizer was equivalent to men's [7]. However, women still face serious constraints in obtaining essential support for most productive resources, such as land, fertilizer, knowledge, infrastructure and market. For households that rely on agriculture for their livelihoods, land is the most critical household asset [8]. It is a basic requirement for farming, and control over it is tantamount to wealth, status and power in many areas. Strengthening women's access to and control over land is an imperative means of raising their status and influence within households and communities. Improving women's access to land and security of tenure has direct effects on farm productivity and can also have important implications for improving household welfare. Allendorf [1] found out that strengthening land ownership by women in Nepal was linked to better health outcomes for children. According to a SNV brief [15], women farmers often have to negotiate or even pay to access productive resources, which are mainly owned by men or controlled by male-dominated authorities.

Methodology
Description of study areas
Study areas were Bomet, Homa Bay, Machakos and Narok Counties, Kenya. These counties form one of the bean corridors in Kenya. A corridor is an area of bean intensification characterized by flows of product from production to consumption in specific bean intensification zones where significant bean activities take place including production, distribution and consumption. The corridor approach is motivated by existence of inefficiencies in production and marketing of bean. This is shown by lack of sufficient tradable volumes of right bean grain to attract major off-takers and reluctance by bean producers and buyers to engage in longer-term contract for supply of bean. The approach is meant to bring about more efficient bean product movement, by reducing the costs of production and improving marketing to enhance incomes of households and other actors participating in the value chain [14]. Bomet County is among the nine counties in the Rift Valley region. It lies between 00 39' and 10 02' South of the Equator and between longitudes 35°' and 32°' East of prime meridian. Agriculture is the backbone of the county with tea farming and dairy production leading in the region. Some of the crops grown in the county include beans, Irish potatoes, millet, cabbages, onions, bananas and pineapples. In Bomet, women form the bulk of the work force in the tea growing areas mostly in weeding and tea picking. According to Kalenjin culture, women do not own land and they also do not make decisions, and if so, they must consult their spouses [12]. Homa Bay County lies between latitude 0°15' South and 0°52' South and between longitudes 34° East and 35° East. The major food crops grown in the area include sorghum, millet, maize, beans, kales, sweet potatoes and cow peas. The vast majority (80%) of the farmers produce maize and beans mainly because they are considered the staple foods of the county. Narok County is situated in the South Rift Valley bordering the Republic of Tanzania to the South, Kisii, Migori, Nyamira and Bomet counties to the West, Nakuru County to the North and Kajiado County to the East. The county lies between latitudes 0°50' and 1°50' South and longitude 35°28' and 36°25'

East. Narok County is occupied by the Maasai community. In Maasai culture, women perform many household chores including constructing huts, fetching water, feeding livestock, gathering firewood, milking, cooking and caring for children and the old. Although they do not own any property, since everything belongs to the man, husbands apportion a number of cows, sheep and goats to them for which they take charge of the products such as milk, butter, meat and skin. Machakos County covers an area of 6208.2 km^2 with most of it being semi-arid. It lies between latitudes 0°45′ South and 1°31′ South and longitudes 36°45′ East and 37°45′ East. Machakos County is mainly occupied by the Kamaba community whose culture is not as strong, and women participate in decision making and even own land. Majority of the women are also literate.

Sampling procedures and sample size

The survey adopted a multi-stage sampling technique from county, sub-county, ward, location, sub-location and then villages. In the four counties of Bomet, Homa Bay, Machakos and Narok, a list of bean-producing sub-counties was obtained and two wards randomly selected, with assistance from both the local administration and Ministry of Agriculture extension staff This formed the sampling frame from which the desired sample size of 440 households was randomly drawn in the second stage at the ward level. The households were proportionately spread across the four counties comprising 70 respondents (74% male-headed, 24% female-headed) from Bomet, 238 respondents (70% male-headed, 30% female-headed) from Homa Bay, 61 (92% male-headed, 8% female-headed) from Machakos and 71 (86% male-headed, 14% female-headed) from Narok. Homa Bay County had more bean farmers than the other three Counties due to its volumes of bean produced.

Data collection method

Data were collected by use of questionnaires on tablets via computer-assisted personal interviewing (CAPI) method. This helped reduce errors by enumerators and also avoided missing data; hence, all questions in the questionnaire were answered. Both husband and wife in the household answered certain sections of the questionnaire, where applicable. Questions answered by both husband and wife included: variety preferences, division of labour, decision making, ownership of assets. The questionnaire collected data on demographic and farm characteristics, household incomes, livestock ownership, farm equipment ownership and decision making.

Data analytical technique

Stata and SPSS packages were used in data analysis. Data were presented by way of means, frequencies, and proportions and cross-tabulations. A multiple regression model was used to determine the effects of access and control of productive resources and household income on quantities of beans produced by households. This was presented as follows;

$$Y_i = a + bx_{1+} bx_2 + \cdots + bx_n + e_i.$$

where Y_i = quantity of beans produced in households, X_1 = sex of respondent, X_2 = literacy level of household head, X_3 = access and use of fertilizer, X_4 = use of pesticides, X_5 = use of hired labour, X_6 = annual household income, and X_7 = proportion of land allocated for bean production.

Results and discussion

This section describes the data used and discusses results from the study.

Demographic characteristics of households

Table 1 presents selected household demographics. Over 70% of the households in all the counties were headed by men. Machakos County had the lowest number of households headed by women (8%) followed by Narok (14%) and Bomet (16%). Homa Bay had the highest female-headed households (30%), and Homa Bay also had the highest number of widows compared to the other counties. The mean age for farmers ranged from 41 years in Narok County to 58 years in Machakos County. Age of

Table 1 Demographic characteristics of respondents

County	Mean age (years)	Sex of H/head (%)		Marital status			Educational level (%)				
		Male	Female	Married	Single	Widowed	Non	Primary	Secondary	Diploma	University
Bomet (N = 70)	46	74	26	89	1	10	10	29	47	14	–
Narok (N = 70)	41	86	14	86	7	7	10	39	37	11	3
Machakos (N = 60)	58	92	8	92	2	6.7	–	27	57	15	2
Homa Bay (N = 228)	50	70	30	77	–	23	6	57	28	7	2

farmers could affect productivity and access to productive resources negatively or positively. In a study by Wiredu et al. [19] on rice cultivation in Ghana, age had positive effect on yield meaning experience in rice cultivation implies accumulated knowledge in rice production. This study showed that the more aged the household head was, the more farm yield was realized. Thamaga-Chitja et al. [17] showed that older household heads had greater access to resources than households headed by younger members, which thus could afford to facilitate production on the farm. Majority of the farmers in all the four counties were literate, and this could have affected access to productive resources. In Vihiga District in Kenya, Waithaka et al. [18] noted the significance of increased education level of the household head on the increased amount of fertilizer used, presumably arising from a better understanding of the usefulness of fertilizers, and it may also imply better crop management. This was echoed by Ariga et al. [2] who found that the level of education has a significant effect on fertilizer use.

Table 2 presents the total land size holding and by sex of household head in the different counties. Narok County had the biggest sizes of land held by households (7 ha), while Homa Bay had the smallest (1.3 ha). Male-headed households owned slightly bigger sizes of land than the female-headed ones. Narok County is known for cash crop farming, especially for wheat and maize. Maize is mostly produced in Transmara sub-counties where data collection was done. In all the counties, all the households grew beans in 2 or 3 different plots either at home or far away. In Narok beans were grown as pure stand and mainly for the market while in the other three counties, beans were intercropped and were grown for both consumption and market. This means that Narok could be targeted for commercial production of beans for the precooked bean industry. Beans take approximately 2 h to boil, but a precooked bean product will take 15 min to boil and be ready for serving. This product will be dried and packed in different sizes and targets career men, women and the youth in urban areas who

would like to consume beans but do not have the time to prepare and cook beans. With growing populations in urban areas, this product will come in handy as nutrition security. Common bean is a major source of protein, and majority of Kenyans are embracing healthy to avoid lifestyle diseases caused by poor eating habits. Farmers in other counties could be encouraged to increase sizes of land under beans by also growing beans under pure stand for commercial purposes. The average land size owned by women seems slightly bigger than that of men, but this is because only 8% of households were headed by females. This could mean higher productivity and improved access to productive resources because they control land use.

Annual income from crops and livestock per household

Households in Bomet and Machakos earned more annual incomes from livestock than all the other counties (USD 1050 and 950), respectively (Fig. 1). It was expected that Narok would have higher incomes from livestock, but this was not the case. This could be explained by the fact that although the residents have big number of livestock, they do not sell frequently due to their cultural belief that the more cattle and sheep you have, the wealthier you are. Bomet County also had a substantial number of dairy cattle, resulting in higher milk sales which contributed to high livestock income. Mua hills in Machakos County

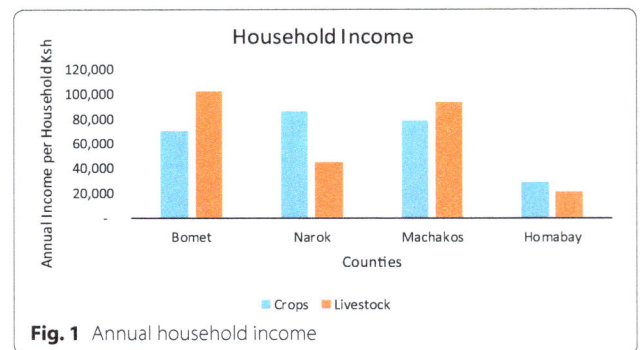

Fig. 1 Annual household income

Table 2 Size of land and proportion under crops

County	Land allocation				
	Size of total land holding (ha)	Proportion of land is under crops (ha)	Plots with crops last season (2014B)	Average land size owned (ha)	
				Men	Women
Bomet	2	0.8	2	1.8	3.0
Narok	6.7	1.6	2	7.0	4.8
Machakos	3.8	1.9	3	4.0	4.2
Homa Bay	1.3	1.1	2	1.5	1.1

receive high rainfall, which hence could have contributed to higher incomes from crop and livestock production. Homa Bay had the lowest income from both crops and livestock (USD 300 and 200, respectively); this could be attributed to the fact most of the households there are female-headed caused by death of husbands due to HIV. Homa Bay County has the highest HIV prevalence rate in Kenya. High household incomes could lead to increased bean production because farmers are able to use the recommended quantities of inputs, e.g. fertilizers, herbicides, labour for weeding, chemicals and proper storage.

Farm input use in households per county

Bomet and Narok had the highest number of households using fertilizers and field pesticides, while Homa Bay had the lowest with 8% of farmers applying fertilizers and none applying field pesticides. In Bomet, 71% of male-headed households and 78% of female-headed households used fertilizers, while in Narok County 71% of male-headed households applied fertilizers on the farms compared to 56% of households headed by females.

In Machakos County, female-headed households did not use fertilizers (0%) and storage pesticides (0%) completely. This could be explained by the fact that the area is semi-arid and sources of income are limited, so women cannot afford fertilizers and pesticides. The other explanation for the low input use even for male-headed households in the county could be that beans are grown for subsistence purposes, so farmers possibly do not see the importance of investing so much in the crop. The scenario was almost the same in Homa Bay County whereby both households did not use fungicides and field pesticides completely. This could be due to the warm weather in both counties as compared to the cold and chilly weather in Bomet and high areas of Narok where this study was carried out. Diseases like halo blight, bean anthracnose and leaf rust are prevalent in cool areas with heavy rainfall [6]. In all counties, herbicide, irrigation and manure use were very low or not there completely. The issue of herbicides use could be explained by the fact that most bean farmers are small holders and use manual weeding in their farms using family labour mostly. Due to the fact that bean is viewed as a food security crop, farmers do not irrigate as compared to horticulture crops which farmers view as high value crops. Low use of fertilizers (32%) and manure (0%) could lead to low productivity, while low use of storage pesticides (14%) could explain the high post-harvest losses farmers incur.

Farm equipment ownership by sex of household head

Table 3 presents ownership of farm equipment by sex of household head. Results show more male-headed households owned weighing scale (9%), knapsack sprayer (43%)

Table 3 Farmer equipment ownership by sex of household

Sex of HH	Farm equipment ownership		
	Weighing scale	Knapsack sprayer	Ox plough
Male (%)	9	43	52
Female (%)	3	24	25
sig	0.098	0.000	0.003

and an ox plough (52%). All these are statistically significant through a Chi-square test. Majority of small holder farmers in these bean-producing areas use ox ploughs to plough and weed their farms. Ownership of an ox plough could have an implication on time of ploughing and planting and eventually quantity harvested. Majority of farmers without ox ploughs tend to plough late because they borrow from their relatives or friends who have to prepare their land first. If rainfall is limited, only the early planters get some harvest. Ownership of a knapsack sprayer means that the farmer sprays his crops against pests which could have a positive effect on yields harvested. For a household, to own a weighing scale could mean that they weigh their produce before selling, hence which cannot be misinformed by the buyers. Cost of equipment could explain the low percentage of ownership of weighing scales. Few female-headed households own farm equipment due to the cost of purchasing, and this could affect negatively their crop productivity. For these households, the implication is that they will plant late, cultivate less land and will not protect their crop from crops and diseases, resulting in low yields.

Effects of access to productive resources on bean production

Table 4 shows a regression analysis which was used to determine variables that influenced bean production. The variables used were sex of respondent, access to improved seed, annual income from crop farming, proportion of land use and number of market information sources. Results show that the sex of the respondent was slightly correlated with bean production, with female-headed households producing less beans than the male-headed households ($p = 0.089$). This could be attributed to the fact that male-headed households tend to have more access and control of key productive resources like land and inputs.

Regarding access and control of productive resources, households who applied fertilizers produced more beans ($p = 0.003$). Utilization of hired labour was highly significant, implying that labour is the key in bean production due to the fact that it is labour intensive and the crop is delicate and needs care, especially

Table 4 Effects of productive resources on bean production

Source	SS	df	MS	Number of Obs.	=	225
Model	11,070,595	10	1,107,059	Prob > F	=	0
Residual	19,939,303	214	93,174.31	R^2	=	0.357
				Adj R^2	=	0.327

	Coef.	SE	t	p > t	[95% CI]	
Sex of respondent (1 = male; 0 = female)	78.85395	46.11763	1.71	0.089	− 12.049	169.7569
If head of household head can read and write (1 = yes; 0 = no)	40.74894	108.4455	0.38	0.707	− 173.009	254.5072
If used improved seed (1 = yes; 0 = no)	− 66.4666	48.31498	− 1.38	0.17	− 161.701	28.76759
If used fertilizer on beans (1 = yes; 0 = no)	167.6574	56.48232	2.97	0.003	56.32443	278.9903
If used pesticides, fungicides, herbicides (1 = yes; 0 = no)	199.804	55.34028	3.61	0	90.72211	308.8858
If used hired labour on beans (1 = yes; 0 = no)	62.65897	92.66948	0.68	0.5	− 120.003	245.3208
Annual income	0.001043	0.000196	5.32	0	0.000656	0.001429
Proportion of land allocated to beans	220.7884	81.88656	2.7	0.008	59.38088	382.1959
If own ox plough (1 = yes; 0 = no)	− 28.9785	42.45408	− 0.68	0.496	− 112.66	54.70318
Number of information sources	− 25.458	17.43061	− 1.46	0.146	− 59.8156	8.899722

during weeding. Another variable which was significant was proportion of land allocated to bean production (0.008). Research findings show that land is the most important household asset for households that depend on agriculture for their livelihoods. Birachi et al. [3] found out that land size influences bean production in Burundi with elasticity of 0.323. This infers an elastic response to bean production; thus, a unit increase in land would increase production by 32%. This concurs with Allendorf [1] that improving women's access to land and security of tenure has direct impacts on farm productivity and can also have far-reaching implications for improving household welfare. Bigger sizes of land under bean production could mean higher yields and eventually higher incomes. Women farmers need more access to land and other productive resources in order to secure livelihoods and food production for their families. However, according to FAO, women have access to only about 20% of all land worldwide, with their allotments generally of smaller size and lower quality [10]. The more land allocated for bean production, the higher the household incomes from crop sales, all other variables being constant. This concurs with a study by Takulder [16] in Bangladesh that found out

that the size of household land had a positive effect on household incomes.

Annual income from cropping was highly significant and influenced bean production in the 4 counties. This could mean that households with higher incomes from sale of crop produce ploughed back some money to bean production. This money could have bought improved certified seed, fertilizers, chemicals and farm equipment. According to a SNV brief [15], women farmers often have to negotiate or even pay to access productive resources, which are mainly owned by men or controlled by male-dominated authorities. In many communities, gender disparities with regard to land and other productive resources are linked to assumptions that men, as heads of households, control and manage land—subliminally reflecting ideas that women are unable to manage productive resources such as land effectively, that productive resources given to women are "lost to another family" in the event of marriage, divorce or (male) death and that men will provide for women's financial security [4].

Table 5 presents the correlation between quantity of beans harvested per season per household and other variables. The results indicate that there was a positive correlation between quantity of beans harvested and land

Table 5 Factors affecting bean production in the selected counties

Variable	Coefficient	SE	t value	p value
Sex of household head	− 88.04712	43.85819	− 2.01	0.046***
Literacy level of household head	− 46.50036	115.8537	− 0.40	0.689
Annual income from crop farming	0.0003993	0.0001754	2.28	0.024***
Proportion of land used for crop	79.41412	12.25358	6.48	0.000***
Number of market information sources	12.26448	18.28571	0.67	0.503

under crops and annual income from crops. This could be explained by the assumption that the more the income, more money is invested in bean production in purchasing inputs and paying labour for weeding, harvesting and storage, all other variables remaining constant. On the other hand, there is a weak negative correlation between the sex of household head and quantity of beans harvested. Male-headed households tended to harvest more quantities than female-headed ones. This could be attributed to the fact that majority of male farmers use farm inputs which help increase yields and productivity.

Conclusions and implications of the study

Demographic results showed that most of the farmers were aged between 41 and 58 years. There is need to involve the youth in bean production, but this can only be possible if the production is commercialized because the youth want to engage in money-making ventures. Majority of the respondents were married; hence, interventions should target both husband and wife for improved bean production. It is important to note that all the respondents were the farmers involved in bean production. As much as the households were headed by men, 68% of the respondents were women who are the major bean producers. This implies that for improved bean production this group needs to be targeted.

It is also clear that access and control of resources is the very key in crop production in this case beans. Some of the resources found statistically significant were land under crops, access and use of fertilizer and annual household incomes from crops. The more land a family allocated to bean production, the higher the quantities harvested. The implication is that farmers both men and women need sensitization on the importance of equitable access and control of productive resources in bean production. Education level though not significant is important because literate farmers tend to apply skills learned better than the illiterate ones. They tend to adopt new technologies faster. Farm input use was low in all counties; hence there is a need for training farmers on good agricultural practices. Fertilizer and storage pesticides use was very minimal, hence leading to low productivity and loss of bean grain, respectively. More male-headed households owned important farm equipment compared to female-headed households. Gender of household head was strongly correlated with bean production with male-headed households producing higher quantities than the female-headed ones. In two counties, female-headed households did not use some key farm inputs like fertilizer, fungicides and storage pesticides. Most female-headed households did not own an ox plough which is the very key in land preparation for bean production. Lack of it could mean late planting, resulting in low quantities harvested. This is a gap that needs to be addressed.

Recommendation

There is a need to come up with interventions which will benefit the men, women and youth to avoid disharmony in the households. Commercialization of bean production and processing should be gender responsive. There is need for capacity building for men, women and youth farmers on good agricultural practices to fill the gap on farm input use, post-harvest practices and agribusiness skills improved productivity. Farmers need to be trained on gender issues, especially on importance equal access and control of resources. Commercialization of the bean crop can be promoted by engaging processors and other value chain actors who will add value to beans and eventually create demand for the bean grain. One way of bringing together the value chain actors is forming a bean stakeholders platform which would comprise farmers, input dealers, transporters, aggregators, traders and service providers. Ownership of farm equipment which affects productivity was very low in female-headed households; hence there is a gap which needs to be filled. This could be done through linking women to credit providers and coming up with affordable collateral so that they can access and afford the services. For policy makers, there is a need to subsidize farm inputs like fertilizers and farm chemicals.

Abbreviations
CAPI: Computer-Assisted Personal Interviewing; KALRO: Kenya Agricultural and Livestock Research Organization; CIAT: International Centre for Tropical Agriculture; IDRC: International Development Research Centre; NARO: National Agricultural Research Organization

Authors' contributions
AG and JK helped in data collection, EB and MM helped in data analysis, and EB, MU, DK and JN reviewed the paper internally. All authors read and approved the final manuscript.

Author details
[1] Kenya Agricultural and Livestock Research Organization (KALRO), P.O Box 57811-00200, Nairobi, Kenya. [2] International Centre for Tropical Agriculture, Nairobi, Kenya. [3] International Development Research Centre (IDRC), Nairobi, Kenya. [4] National Agricultural Research Organization (NARO), Kampala, Uganda.

Acknowledgements
I would like to acknowledge Collins Odhiambo who helped with plagiarism check and the team of enumerators who helped with data collection.

Competing interests
The authors declare that they have no competing interests.

Funding
The research was funded by IDRC and ACIAR. Design, data collection, analysis and interpretation were carried out by the research scientists.

References
1. Allendorf K. Do women's land rights promote empowerment and child health in Nepal? World Dev. 2007;35(11):1975–88.
2. Ariga J, Jayne TS, Kibaara B, Nyoro JK. Trends and patterns in fertilizer use by smallholder farmers in Kenya, 1997–2007. Njoro: Tegemeo Institute of Agricultural Policy and Development, Egerton University; 2009.
3. Birachi EA, Ochieng J, Wozemba D, Ruraduma C, Niyuhire MC, Ochieng D. Factors influencing small holder farmers' bean production and supply to market in Burundi. Afr Crop Sci J. 2011;19(4):335–42.
4. Canadian HIV/AIDS Legal Network. Respect, protect and fulfill: legislating for women's rights in the context of HIV/AIDS, vol. two, family and property issues; 2009.
5. CIAT. Impact of improved bean varieties in Western Kenya, highlights. No. 18; December 2004.
6. CIAT. Bean disease and pest identification and management, hand book for small-scale seed producers. ISSN 2220-3370, Uganda; 2010.
7. Department for International Development. Gender equality at the heart of development: why the role of women is crucial to ending world poverty. London: DFID; 2007.
8. Food Agricultural Organization. Gender mainstreaming in forestry in Africa. Regional Report. Rome; 2007.
9. Food Agricultural Organization. The state of food and agriculture; 2009.
10. Food Agricultural Organization. Gender and land rights: understanding complexities, adjusting policies; Economic and social perspectives. FAO Policy Brief, Rome; 2010.
11. Food Agricultural Organization. Status of food and agriculture, Rome; 2011.
12. Langat HK (2016) Social cultural factors influencing women's participation in food security programs among households in Bomet County, Kenya.
13. Palacios-Lopez A, Christiaensen L, Kilic T. How much of the labour in African agriculture is provided by women? Food Policy. 2017.
14. PABRA. Bean corridors: a novel approach to scale up national and regional trade in Africa. Nairobi: Pan-Africa Bean Research Alliance (PABRA); 2017.
15. SNV Practice Brief. Gender and agriculture; 2012.
16. Takulder D. Assessing determinants of income of rural households in Bangladesh. J Appl Econ Bus Res. 2014;4(2):80–106.
17. Thamaga-Chitja JM, Hendriks SL, Ortmann GF, Green M. Impact of maize storage on rural household food security in Northern Kwazulu-Natal. J Fam Ecol Consum Sci. 2004;32:8–15.
18. Waithaka MM, Thornton PK, Shepherd KD, Ndiwa NN. Factors affecting the use of fertilizers and manure by smallholders: the case of Vihiga, western Kenya. Nitrogen Cycle Agro Ecosyst. 2007;78:211–24.
19. Wiredu AN, Gyasi KO, Marfo KA, Asuming-Brempong S, Haleegoah J, Asuming Boakye A, Nsiah BF. Impact of improved varieties on the yield of rice producing households in Ghana. In: Second Africa rice congress, Bamako, Mali, 22–26 March; 2010.
20. World Bank. World Development Report 2008. Agriculture for development. Washington, DC; 2007.

Permissions

All chapters in this book were first published in A&FS, by BioMed Central; hereby published with permission under the Creative Commons Attribution License or equivalent. Every chapter published in this book has been scrutinized by our experts. Their significance has been extensively debated. The topics covered herein carry significant findings which will fuel the growth of the discipline. They may even be implemented as practical applications or may be referred to as a beginning point for another development.

The contributors of this book come from diverse backgrounds, making this book a truly international effort. This book will bring forth new frontiers with its revolutionizing research information and detailed analysis of the nascent developments around the world.

We would like to thank all the contributing authors for lending their expertise to make the book truly unique. They have played a crucial role in the development of this book. Without their invaluable contributions this book wouldn't have been possible. They have made vital efforts to compile up to date information on the varied aspects of this subject to make this book a valuable addition to the collection of many professionals and students.

This book was conceptualized with the vision of imparting up-to-date information and advanced data in this field. To ensure the same, a matchless editorial board was set up. Every individual on the board went through rigorous rounds of assessment to prove their worth. After which they invested a large part of their time researching and compiling the most relevant data for our readers.

The editorial board has been involved in producing this book since its inception. They have spent rigorous hours researching and exploring the diverse topics which have resulted in the successful publishing of this book. They have passed on their knowledge of decades through this book. To expedite this challenging task, the publisher supported the team at every step. A small team of assistant editors was also appointed to further simplify the editing procedure and attain best results for the readers.

Apart from the editorial board, the designing team has also invested a significant amount of their time in understanding the subject and creating the most relevant covers. They scrutinized every image to scout for the most suitable representation of the subject and create an appropriate cover for the book.

The publishing team has been an ardent support to the editorial, designing and production team. Their endless efforts to recruit the best for this project, has resulted in the accomplishment of this book. They are a veteran in the field of academics and their pool of knowledge is as vast as their experience in printing. Their expertise and guidance has proved useful at every step. Their uncompromising quality standards have made this book an exceptional effort. Their encouragement from time to time has been an inspiration for everyone.

The publisher and the editorial board hope that this book will prove to be a valuable piece of knowledge for researchers, students, practitioners and scholars across the globe.

List of Contributors

Alex Wilhans Antonio Palludeto
Instituto de Economia, Universidade Estadual de Campinas, Rua Pitágoras 353, Campinas, São Paulo CEP 13083-857, Brazil

Tiago Santos Telles
Instituto Agronômico do Paraná, C.P. 10.030, Londrina, Paraná CEP 86057-970, Brazil

Roney Fraga Souza
Faculdade de Economia, Universidade Federal de Mato Grosso, Avenida Fernando Corrêa 2367, Cuiabá, Mato Grosso CEP 78060-900, Brazil

Fábio Rodrigues de Moura
Departamento de Economia, Universidade Federal da Grande Dourados, C.P. 364, Dourados, Mato Grosso do Sul CEP 79.804-970, Brazil

Krishna P. Timsina and Yuga N. Ghimire
Socioeconomics and Agricultural Research Policy Division (SARPOD), Nepal Agricultural Research Council (NARC), GPO 5459, Khumaltar, Lalitpur, Nepal

Devendra Gauchan
Bioversity International Nepal Office, National Agriculture Genetic Resource Centre, Khumaltar, Lalitpur, Nepal

Sanjiv Subedi
Regional Agricultural Research Station (RARS, Nepalgunj), Nepal Agricultural Research Council (NARC), GPO 5459, Khumaltar, Lalitpur, Nepal

Surya P. Adhikari
Regional Agricultural Research Station (RARS, Tarahara), Nepal Agricultural Research Council (NARC), GPO 5459, Khumaltar, Lalitpur, Nepal

Amarjeet Kaur
Food Science and Technology, Punjab Agricultural University, Ludhiana, Punjab 141004, India

Vidisha Tomer, Vikas Kumar and Kritika Gupta
Food Technology and Nutrition, School of Agriculture, Lovely Professional University, Phagwara, Punjab 144411, India

Ashwani Kumar
Food Science and Technology, Punjab Agricultural University, Ludhiana, Punjab 141004, India
Food Technology and Nutrition, School of Agriculture, Lovely Professional University, Phagwara, Punjab 144411, India

Kenneth W. Sibiko
Department of Agricultural Economics and Rural Development, School of Agriculture and Food Security, Maseno University, Private Bag, Maseno, Kenya

Prakashan C. Veettil
International Rice Research Institute, New Delhi, India

Matin Qaim
Department of Agricultural Economics and Rural Development, University of Goettingen, 37073 Göttingen, Germany

Celia A. Harvey and M. Ruth Martinez-Rodríguez
Conservation International, 2011 Crystal Drive Suite 500, Arlington, VA 22202, USA

Barbara Viguera and Francisco Alpizar
Tropical Agriculture and Higher Education Center (CATIE), Apdo 7170, Turrialba, Costa Rica

Milagro Saborio-Rodríguez
Tropical Agriculture and Higher Education Center (CATIE), Apdo 7170, Turrialba, Costa Rica
University of Costa Rica, San Pedro de Montes de Oca 11501, Costa Rica

Adina Chain-Guadarrama
Turrialba, Costa Rica

Raffaele Vignola
Tropical Agriculture and Higher Education Center (CATIE), Apdo 7170, Turrialba, Costa Rica
Wageningen University, Hollandseweg 1, 6706 KN Wageningen, The Netherlands

Terence Epule Epule
Department of Geography, McGill University, 805 Sherbrooke St. W., Burnside Hall 416, Montreal, QC H3A 0B9, Canada

James D. Ford
Department of Geography, McGill University, 805 Sherbrooke St. W., Burnside Hall 416, Montreal, QC H3A 0B9, Canada
Priestley International Centre for Climate, University of Leeds, Leeds, UK

Shuaib Lwasa, Benon Nabaasa and Ambrose Buyinza
Department of Geography, Makerere University, Kampala, Uganda

Gideon Danso-Abbeam and Dennis Sedem Ehiakpor
Department of Agricultural and Resource Economics, University for Development Studies, Tamale, Ghana

Robert Aidoo
Department of Agricultural Economics, Agribusiness and Extension, Kwame Nkrumah University of Science and Technology, Kumasi, Ghana

Honest Machekano, Reyard Mutamiswa and Casper Nyamukondiwa
Department of Biological Sciences and Biotechnology, Botswana International University of Science and Technology (BIUST), Private Bag 16, Palapye, Botswana

Msafiri Yusuph Mkonda
Department of Geography and Environmental Studies, Solomon Mahlangu College of Sciences and Education, Sokoine University of Agriculture, Morogoro 3038, Tanzania

Xinhua He
Centre of Excellence for Soil Biology, College of Resources and Environment, Southwest University, Chongqing 400715, China
School of Plant Biology, University of Western Australia, Crawley 6009, Australia

Nii Korley Kortei
Department of Nutrition and Dietetics, School of Allied Health Sciences, University of Health and Allied Sciences, PMB 31, Ho, Ghana

George Tawia Odamtten and Michael Wiafe-Kwagyan
Department of Plant and Environmental Biology, College of Basic and Applied Sciences, University of Ghana, Legon, Ghana

Mary Obodai and Deborah Louisa Narh Mensah
Food Microbiology Division, Council for Scientific and Industrial Research - Food Research Institute, Accra, Ghana

Caesar Agula and Franklin Nantui Mabe
Department of Agricultural and Resource Economics, University for Development Studies, Tamale, Ghana

Mamudu Abunga Akudugu
Institute for Interdisciplinary Research and Consultancy Services (IIRaCS), University for Development Studies, Tamale, Ghana

Saa Dittoh
Department of Climate Change and Food Security, University for Development Studies, Tamale, Ghana

Bright Masakha Wekesa, Oscar Ingasia Ayuya and Job Kibiwot Lagat
Department of Agricultural Economics and Agribusiness Management, Egerton University, Egerton, Kenya

Zerihun Yohannes Amare
Institute of Disaster Risk Management and Food Security Studies, Bahir Dar University, BahirDar, Ethiopia

Johnson O. Ayoade and Ibidun O. Adelekan
Department of Geography, Faculty of the Social Sciences, University of Ibadan, Ibadan, Nigeria

Menberu Teshome Zeleke
Department of Geography and Environmental Studies, Debre Tabor University, Debre Tabor, Ethiopia

Maria Tesfaye
Department of Plant Science, College of Agriculture, Mekdela Amba University, Tuluawlia, Ethiopia

Derbew Belew
College of Agriculture and Veterinary Medicine, Jimma University, Jimma, Ethiopia

Yigzaw Dessalegn
International Livestock Research Institute, Bahir Dar, Ethiopia

Getachew Shumye
Department of Plant Science, College of Agriculture, Wollo University, Dessie, Ethiopia

Hadush Hagos
Socioeconomic and Research Extension Directorate, Mekelle Agricultural Research Center, Mekelle, Ethiopia

Eric Ndemo and Jemal Yosuf
Department of Rural Development and Agricultural Extension, Haramaya University, Tselemti, Ethiopia

Hagos Abraha Rahiel
Department of Dryland Crop and Horticultural Sciences, Mekelle University College of Dryland Agriculture and Natural Resources, Arid-Main Campus, Mekelle, Tigray, Ethiopia

Abraha Kahsay Zenebe and Gebreslassie Woldegiorgis Leake
Department of Natural Resource, Economics and Management, Mekelle University College of Dryland Agriculture and Natural Resources, Arid-Main Campus, Mekelle, Tigray, Ethiopia

Beyene Weldegerima Gebremedhin
Department of Animal, Rangeland and Wildlife Sciences, Mekelle University College of Dryland Agriculture and Natural Resources, Arid-Main Campus, Mekelle, Tigray, Ethiopia

Asaye Birhanu
Gondar Agricultural Research Center, 1337, Gondar, Ethiopia

Tilahun Tadesse
Fogera National Rice Agricultural Research Center, Woreta, Ethiopia

Daniel Tadesse
Department of Plant Sciences, University of Gondar, Gondar, Ethiopia

Nigussie Dechassa and Tamado Tana
School of Plant Sciences, College of Agriculture and Environmental Sciences, Haramaya University, Dire Dawa, Ethiopia

Fresew Belete
School of Plant Sciences, College of Agriculture and Environmental Sciences, Haramaya University, Dire Dawa, Ethiopia

Department of Plant Sciences, College of Agriculture and Natural Resource Sciences, Debre Berhan University, Debre Berhan, Ethiopia

Adamu Molla
Chickpea and Malt Barley-Faba Bean Projects ICARDA, Addis Ababa, Ethiopia

Nii Korley Kortei
Department of Nutrition and Dietetics, School of Allied Health Sciences, University of Health and Allied Sciences, PMB 31, Ho, Ghana

George Tawia Odamtten and Michael Wiafe-Kwagyan
Department of Plant and Environmental Biology, College of Basic and Applied Sciences, University of Ghana, Legon, Ghana

Mary Obodai
Food Microbiology Division, Council for Scientific and Industrial Research-Food Research Institute, Accra, Ghana

Juanita Prempeh
Department of Food Science, Royal Agricultural University, Cirencester, Gloucestershire GL7 6JS, UK

Scolastica Wambua, Ann Gichangi, Justus Kavoi and David Karanja
Kenya Agricultural and Livestock Research Organization (KALRO), Nairobi, Kenya

Eliud Birachi and Mercy Mutua
International Centre for Tropical Agriculture, Nairobi, Kenya

Jemimah Njuki
International Development Research Centre (IDRC), Nairobi, Kenya

Michael Ugen
National Agricultural Research Organization (NARO), Kampala, Uganda

Index

A

Acdep, 88-96

Adaptation Strategies, 51, 53-54, 61, 66, 68-69, 139, 154-157, 160-164

Adoption, 2, 10, 14-15, 20-22, 49-50, 61, 89-91, 94, 96-97, 119, 129-144, 146-147, 149-150, 152-155, 161-164, 173-175, 177-181, 203, 220

Agricultural Extension, 88-90, 92-96, 132, 135, 137, 163, 180, 183

Agricultural Production Trend, 110

Agro-ecosystem, 129, 139

Agro-vets, 13, 15-20

B

Basal Thermal Tolerance, 101-102, 105

Beans, 34-35, 52, 54, 58-59, 61, 64-65, 112-113, 115, 152, 197, 201, 219-225

Binary Logistic Model, 173

Bioenergy Crops, 1

Biological Efficiency, 121, 123-128

C

Cap Diameter, 121-127

Climate Change, 1, 23, 34, 49-55, 59-61, 63-64, 66-71, 74, 86-87, 89, 108-109, 111-112, 117-120, 135, 138, 140-143, 146-147, 149-155, 158-164

Climate Change Impact, 154, 158, 160, 163

Climate Risk, 37, 39, 49, 70

Climate-smart Agricultural Practices, 140, 143

Climatic Drivers, 71-72, 74, 78, 82

Coffea Arabica, 51

Compost, 4, 76, 121-123, 128, 130, 135, 19-, 211-217

Crop Insurance, 37, 39, 47-49

Crop Management Practices, 146-148, 154

Crop Management Strategies, 154-160, 163

Crop Production, 39, 49, 51-54, 60-61, 66-67, 79, 82, 85-86, 111-112, 114, 117, 119, 129, 146, 148, 183, 191, 202, 211, 220, 225

Crop Yields, 37, 45, 51, 60-61, 63, 66, 71-72, 74-85, 87, 110-112, 115-117, 119-120, 155

Csa Practices, 140-143, 146-150

D

Discrete Choice Experiment, 37, 40

Dry Lands, 23-24

E

Economic Analysis, 78, 80, 195, 199-200

Ecosystem-based Adaptation, 66-67

Ecosystems, 69, 81, 108, 120, 129-130, 133-134, 138-139

F

Farm Management Practices, 66-67, 89, 96, 130, 132, 138

Farm Productivity, 88-90, 92, 94, 96, 220

Farmer Adaptation, 51-54, 63, 66-67

Farmland Prices, 1

Food Security, 21-22, 37, 48-53, 59-60, 66-68, 70-71, 86-88, 110-112, 114, 119-120, 138, 140-144, 146, 150-154, 162-163, 173-174, 180, 193, 219, 223, 226

Fruits and Vegetables, 182-183, 185-194, 218

G

Gamma Radiation, 123, 127, 212-213, 216, 218

Global Change, 11, 98-99, 108, 164

H

Heckman Treatment Effect, 88, 92, 94-96

I

Insect Invasion, 98

Intra-row Spacing, 195-200

L

Land Price, 1-5, 7, 9-10

Land Profitability, 1

Land Use, 1, 3, 10-11, 54, 69, 71-72, 82, 85-87, 139, 223

Land Value, 1-3, 5, 9

Low Fertility, 23, 25

M

Micronutrient Deficiency, 23

Millets, 23-36

Moist Heat, 123, 212, 214, 217

Multinomial Endogenous Switching Regression Analysis, 140

Multivariate Regression Analysis, 13

Mung Bean, 36, 195-201

Mushroom, 121-123, 125, 127-128, 212-218

N

Non-climatic Drivers, 71-72, 74, 78

Nutri-cereals, 23

Nutrition, 23, 26, 34-35, 99, 107, 118, 121-122, 127, 144, 153, 170, 174, 183, 202, 211-212, 217, 220

Nutritional Security, 23, 35, 99

O

Onion, 22, 165-172, 185-186, 188

P

P. Ostreatus, 121, 123-127, 218

Panel Data, 1, 7, 10-12

Plant Spacing, 195-196, 198-199

Planting Date, 157-159, 161-163, 165-168, 170-172

Post-harvest Losses, 182-184, 186-187, 189, 191, 223

Production Potential, 142, 174-175, 183, 186, 188, 191

Productive Resources, 219-225

Productivity, 4, 14, 23, 25, 39, 68, 72, 74, 82, 86-92, 94, 96-97, 111, 133, 140-141, 153, 162, 166, 169, 171, 174, 178, 180, 183, 187, 196, 200, 202, 220, 222-223, 225

R

Recovery, 101, 108, 202-207, 209-210

Regression On Propensity Scores, 88, 92, 94-95

Row Spacing, 195-201

S

Seed Flow, 14-15, 20-21

Seed Quality, 21, 165, 200-201

Seed Yield, 165-172, 201

Smallholder Farmers, 37-38, 47-54, 59-61, 63-64, 66-70, 86, 89, 111, 118, 120, 135, 140-142, 152-153, 164, 166, 180, 202, 226

Solanaceous Plants, 98, 105, 107

Solanum Coccineum, 98, 102

Steam-sterilized Composted Sawdust, 121

Stipe Length, 121-124, 126-128

T

Thermal Tolerance, 98-102, 105-106, 108

Tigray, 96-97, 173-174, 180-183, 192-193, 195

Tomato Leaf Miner, 98-99, 107

Tuta Absoluta, 98-100, 102-103, 106-109

U

Upland Rice, 173, 178-179, 181

W

Weather Index Insurance, 37, 39, 46, 48-49

Y

Yield, 2, 13-15, 24, 31, 41, 44, 49, 61, 66, 72, 76, 80, 82, 87, 91, 94, 101, 112, 119, 123, 125-130, 137, 165-174, 177-180, 183, 189, 191, 195-206, 211, 216, 222, 226

Z

Zea Mays, 31, 51, 54